Ornamental Horticulture

NIPA® GENX ELECTRONIC RESOURCES & SOLUTIONS P. LTD.
New Delhi-110 034

Ornamental Horticulture

Editor
S.S. Sindhu
Head
Division of Floriculture and Landscaping
ICAR –Indian Agricultural Research Institute (IARI)
New Delhi - 110 012

NIPA® GENX ELECTRONIC RESOURCES & SOLUTIONS P. LTD.
New Delhi-110 034

NIPA® GENX ELECTRONIC
RESOURCES & SOLUTIONS P. LTD.

101,103, Vikas Surya Plaza, CU Block
L.S.C. Market, Pitam Pura, New Delhi-110 034
Ph : +91 11 27341616, 27341717, 27341718
E-mail: newindiapublishingagency@gmail.com
web: www.nipabooks.com

For customer assistance, please contact
Phone: + 91-11-27 34 17 17
Fax: + 91-11- 27 34 16 16
E-Mail: feedbacks@nipabooks.com

ISBN: 978-81-19235-96-4

Composed and Designed by NIPA®.

DEDICATED TO

Dr. B.P. PAL
(1906-1989)

Dr. M.S. SWAMINATHAN
(1925 -)

The book Ornamental Horticulture is dedicated to Late Dr Benjamin Peary Pal who initiated applied research in ornamentals especially roses, bougainvilleas and temperate seasonal flowers.

The book is also devoted to Prof. Mankombu Sambasivan Swaminathan whom the Time magazine listed among the most influential 'Time 20' Asian of the 20th century and one among the three most influential Indians, the other two are M K Gandhi and Rabindra Nath Tagore.

Both Dr B P Pal and Prof. M S Swaminathan are Fellows of Royal Society of London and a grateful nation honoured both with Padma Vibhushan, the highest Padma Award. 'Grains alone will not keep man happy, ornamentals add value to living' to quote Prof. M S Swaminathan.

Devotion

The book ORNAMENTAL HORTICULTURE is devoted to the memory of late Benjamin P. Pal who spend 5 decades of his life to roses, bougainvilleas and annual seasonal flowers. He is immortal as long as roses exist. The book is also devoted to great florists of India like Dr. Vishnu Swarup, Dr. M.H. Marigowda and Dr. Manmohan Attavar who spend their whole life to the cause of flowers, gardens and seeds. We remember all our teachers who taught us floriculture and landscape architecture during graduate and post graduate days.

Indo-American
Hybrid Seeds (India) Pvt. Ltd.

Speciality
Hybrid Flower & Vegetable

Branch Office :
POST BOX NO. 7099
7th KM, Banashankari-Kengeri Link Road
Sy. No. 13/4, 14, Channasandra
Rajarajeshwarinagar Post, Bengaluru-560 098
Tel : +91-80-28611499 Fax : 91-80-28612912
Website : www.indamseeds.com
CIN : U02001KA1997PTC023141

Foreword

The Genesis narrates creation of Eden garden as the living habitat for the first man Adam later joined by Eve. The garden was divine in its all senses and meanings. Ornamental values of plant were later understood by men of all cultures, civilizations and continents. Once settled, gardens became an integral part of living providing peace, health and tranquility to mind and body. Formal system of gardening was much cherished by Mughal Kings while informal gardens were preferred by Japanese, Chinese and Thais. Gardens became places for meditation and prayer. The Indian epics Mahabharath and Ramayan expose sacred gardens where divine couples lived in harmony with nature. Public gardens provided the much needed open space for urban settlers. Private gardens were established to meet the special requirements of affluent families to provide privacy and tranquility. There are now specially designed gardens like aquatic gardens, floating gardens, orchid gardens, palm gardens, hanging gardens, vertical gardens and even butterfly gardens, left to dreams, imaginations and situations. Aesthetic, spiritual and healing values of gardens were later understood and garden making therapy was recommended to differentially enabled and depressed people. The features of almost all gardens are arches and pargoals, edges and hedges, lawns, water bodies with lotus, peacons, fish, bird sanctuaries, potted plants, bonsai, animations, climbers and vines *etc.*

The present book 'Ornamental Horticulture' has 20 chapters authored by the best scientists in the field. The first chapter deals with Basic Principles and Elements of Landscape Design. Garden features and types are explained to meet the inquisitiveness of students and new entrants to gardening. Temperate Ornamentals plays a significant role in landscape planning. Green Building Concept - the needs of 21st Century express the readers to measures taken to minimize effects of pollution in areas polluted.

Other chapters are Hanging gardens - A softer touch to carbon landscapes for economic development, lawn development and maintenance, CAD for landscape gardening, LED lighting in Landscape gardening, Herbaceous Perennials of Kashmir Valley for landscape use, Xeriscaping –An approach of landscaping for dry areas, Bonsai, Horti-business through landscaping, Wealth of Bougainvillea, Houseplants, Seasonals, Production of loose flowers, Dried Ornamentals. An option for commercial floricultural enterprises, Design of Protected structures for Indian Conditions, Protected cultivation of flowers crops, future ornamentals.

Singapore is popular for city gardens. The book deals with hanging gardens of Singapore. The book with 20 chapters authored by eminent scientists is a treasure house for gardeners, students and planners. Dr. S.S. Sindhu, Principal Scientist and Head Division of Floriculture and Landscaping, IARI, New Delhi has edited the book under the series of Horticulture Science. I congratulate Dr. Sindhu for the efforts made. Dr. K.V. Peter, Former Vice-Chancellor, KAU and Director, World Noni Research Foundation Chennai is the General Editor of the series. I appreciate the publisher New India Publishing Agency for publishing the book very elegantly and reader friendly.

Dr. Manmohan Attavar
Chairman

Registered & Admin Office : 17th CROSS, II 'A' MAIN, K.R. ROAD, BANASHANKARI II STAGE
BENGALURU-560 070, KARNATAKA, INDIA.
Tel. : +91-080-26760111
Fax : +91-080-26761479

Prof. K.V. Peter

About the General Editor (Series)

Prof. Peter is basically a horticulturist, a plant breeder and a University Professor. He is an acknowledged and decorated scientist and science manager. A post graduate from G.B. Pant University of Agriculture and Technology, he did post doctoral research at BARC Beltsville Maryland USA and worked at Laboratories at AVRDC Taiwan and Guadeloupe(French West Indies). He was associated with development and release of improved and high yielding varieties/hybrids in tomato, brinjal, chilli, bittergourd, melons, amaranths and cowpea. Sources of resistance to bacterial wilt in tomato, chilli and brinjal; aphids in cowpea and viral leaf curl in chilli were located by him and are now used in breeding programes. Prof. Peter provided managerial support to the Indian Institute of Spices Research Calicut to possess the World's largest germplasm of black pepper and cardamom. A prolific writer and academic editor he authored/edited 74 books published both in India and abroad. He received 7 scholarships and fellowships at various stages of education. He is member in the Board of Examiners in 12 Universities, Chairman/member in 130 committees of Government /research institutes, Fellow in 3 National Academies and 3 Scientific societies, member in 18 Scientific societies. His publishers include Taylor and Francis, Elsevier, National Book Trust, I.C.A.R., New India Publishing Agency, Astral International Pvt. Ltd; Studium Press (USA), Universities Press and Kerala Language Institute, Trivandrum.

Prof. Peter is recipient of several awards like Rafi Ahmed Kidwai Award 1996-1998 for outstanding research in Horticulture-ICAR New Delhi; Recognition Award 2000 -National Academy of Agricultural Sciences, New Delhi; Dr. M H Marigowda National Award for the Best Horticulturist-2000; Silver Jubilee Medal for outstanding contributions in Vegetable Research;

Dr Harbhajan Singh Award 1993; Silver Jubilee Memonto; Biotech Product and Process Development Award 2003; Shiva Sakthi-HSI National Award for Life Time Achievement in Horticulture-2008 and Dr K Ramiah Memorial Award- National Academy of Agricultural Sciences, New Delhi for outstanding contribution to Crop Improvement-2009. He was Director when the Best Institution Award was conferred to IISR, Calicut by ICAR, New Delhi. He was Vice-Chancellor when Sardar Patil Award 2003 for the best ICAR institution was conferred to Kerala Agricultural University.

Family: Vimala is wife, Anvar and Ajay sons, Anu and Cynara Daughers-in-law and Antony Ajay Peter, grand son and Anna Vimala Anvar, grand daughter. Parents are Late Kuruppacharil Devassey Varkey and Late Rosa Varkey.

Dr. S. S. Sindhu

About the Editor

Dr. S. S. Sindhu is Professor and Head, Division of Floriculture and Landscape Architecture at Indian Agricultural Research Institute, Pusa, New Delhi. He did his graduation and post graduation from G.B. Pant Univ. of Agri. and Tech., Pantnagar and doctorate degree from Haryana Agricultural University, Hisar (Haryana). He made successful intervention to grow 'off season chrysanthemum' through photo-periodic application and developed three varieties *viz.* Kt-3 of Dahlia, Katrain Sel.-1 in Daffodil and a spontaneous mutant in Gerbera. He also developed two varieties Pusa Kiran and Pusa Shubham in gladiolus, identified six promising lines in annual chrysanthemum, marigold and calendula of the double type and standardized technologies for growing commercial flowers.

He published 70 research papers, 75 popular articles, 25 short communications, 15 books and bulletins and presented 45 lead papers in several conferences and seminars of National and International level. He prepared landscape designs of National Research Institutes (IVRI, NDRI, NCAP and CIFE). He is a life member of several professional societies and Secretary of Bougainvillea Society of India.

He was Project Coordinator (Floriculture) of All India Coordinated Research Project in ICAR, where he made significant contributions in research monitoring, planning and for the up-gradation of the AICRP (Floriculture) to Directorate of Floriculture Research. He was Director of Horticulture at Rashtrapati Bhavan. He guided 07 M.Sc. and 05 Ph. D students. He was honoured with Gold Medal by Delhi Agri Horticulture Society, Dr. H.B. Singh Award, National Leaders Memorial Award and Bharat Jyoti Award for meritorious work in floriculture and landscaping. He visited Japan, Kathmandu (Nepal) and Islamabad (Pakistan). He is recipient of INSA fellowship of Japan under *International Exchange Programme*. He is a Fellow of Horticultural Society of India.

Preface

Ornamental Horticulture is both science and art linking man to nature and its varied blissful manifestations. Garden therapy is an age old system of health treatment against lonliness, mental depression and differentially disabled personal disorders. 'Eden garden' referred to in book of Genesis was created by Divine powers to settle the first human beings in a serene and tranquil environment. 'Kailas' mountain in the epic Mahabharata was blessed by the solemn presence of Lord Shiva and Lordess Parvathi and is the ultimate in gardening in all its dimensions.The Mughal gardens are characterized by symmetry and dominance of paths, edges and hedges. The Rashtrapathi Garden in Delhi and Tajmahal Garden in Agra are classical examples. Japanese gardens are informal gardens with preponderance of water bodies and reflections. Thai, Vietnamese and Chinese gardens are characterized by informality, animations and depiction of nature undisturbed. The natural rain forests maintained in Guateloupe (French West Indies) are worth to be seen to be believed in beauty, diversity and splendidness. With passage of time, there is specialization in gardening-orchid gardens, lotus gardens, foliage gardens, palmatum, arboretum, hanging gardens, vertical gardens, high way gardens and gardens on move. With urbanization and shortage of space container gardens received attention. Flower and foliage forcing through protected cultivation is receiving attention. Industrial revolution raised the purchasing power of people but led to dwindling of natural biodiversity spots. Public gardens are the lungs of a town/city and as such to be maintained and protected.

Space, water, light and human preferences being the factors of gardening; there are a lot of engineering and skill development for gardening. Nursery for raising propagules, soil management, fertigation, light control, integrated pest, disease and nutrient management, pre and post-harvest handling, storage, packing and display require scientific support. Private gardens are in much use consequent to change in life styles and economic status. Possessing an attractive, unique and distinct garden has become a status symbol and in fact it adds to the value of the house. A morning walk, meditation and yoga in the open garden provides peace, happiness and good health to the inmates. The commercial possibilities of gardening and its allied services and tools are to be tapped for nations' growth. India is fortunate to possess two out of 12 worlds' biodiversity spots. Flowers for religious purposes, marriages, festivals and even obituary are now essential part of living. In fact there are no social functions without flowers. With innovations in Information Technology, attempts are made to develop virtual digital gardens and at no stretch of imagination, they can replace natural gardens having life, energy, flavour, colour and vibrations.

The present book Ornamental Horticulture has 20 chapters authored by the best scientists in the field. Suggestions are invited for further improvement.

S S Sindhu

Acknowledgements

The necessity to write this book has been realised based on the feedback received time to time from students, officers working in different fields of horticulture sector and the garden amateurs. Finally, we thought to write this book covering major aspects of Ornamental Horticulture specially landscaping, protected cultivation and green building concept to meet out the day to day demand of students as well as working group of horticulture. We acknowledge all the eminent contributors of 20 chapters for the time, patience and academic inputs which made the book worth reading and refereeing. We acknowledge Dr Manmohan Attavar Padma Sree Awardee and Chairman Indo-American Hybrid Seeds Bangalore for writing the foreword. The support showered on us by our families is appreciated. The smiles of Peters' grand son Antony Ajay Peter and grand daughter Anu Anvar Peter motivated to spend lonely times to constructive academic persuits. K V Peter acknowledges Prof. P I Peter Chairman Noni Biotech Chennai and Dr Kirti Singh Chairperson World Noni Research Foundation Chennai for support and facilities. I also place on record my appreciation to my students, colleagues and particularly seniors to provide me the ideas and outline of this book. I am highly thankful to contributors listed here under in the book and particularly Dr. Sapna Panwar, Scientist, Division of Floriculture and Landscaping, IARI for thorough proof reading and placement of lovely photographs relevant to different chapters. We also acknowledge the staff New India Publishing Agency New Delhi for the care and attention for a user-friendly book. I duly acknowledge the constant efforts and encouragements of my family members my son Bharat Sindhu, Rajat Sindhu and my wife Sunita Sindhu. They kept me spared to devote my time after office hours works to spent on this book. Last but not the least, I appreciate my personal staff Ms. Sunita for providing time to time support, encouragement and editing support to bring out this document upto the level of our expectation.

K V Peter
S S Sindhu

Contents

List of Contributors

Alka Singh
Department of Floriculture and Landscape Architecture, Navasari Agricultural University, Navsari, Gujarat

Arvind Kumar Verma
Division of Floriculture and Landscaping, IARI, New Delhi
Present Address, NRC on Seed Spices, Ajmer, Rajastan

Balaji S. Kulkarni
Department of Floriculture and Landscape Architecture, Kittur Rani Channamma College of Horticulture, (UHS Bagalkot), Arabhavi, Karnataka

B. Hemla Naik
Department of Floriculture and Landscape Architecture, College of Horticulture, (UHS, Bagalkot) Mudigere, Chikmangalur District, Karnataka

B.K. Banerji
Floriculture Section, National Botanical Research Institute, Lucknow, Uttar Pradesh

Chander Prakash
Indian Agricultural Research Institute, Regional Station, IARI, Katrain, (Kullu Valley) Himachal Pradesh

Desh Raj
Department of Vegetable Science and Floriculture, College of Agriculture, Palampur Himachal Pradesh

Gunjeet Kumar
Division of Floriculture and Landscaping, Indian Agricultural Research Institute, New Delhi

Kanwar Pal Singh
Division of Floriculture and Landscapeing, Indian Agricultural Research Institute, New Delhi

M. Hasan
Centre for Protected Cultivation Technology (CPCT) IARI, New Delhi

Mam C. Singh
Centre for Protected Cultivation Technology (CPCT), ICAR-Indian Agricultural Research Institute, New Delhi

M.R. Dhiman
Indian Agricultural Research Institute, Regional Station, Katrain, (Kullu Valley) Himachal Pradesh

Sajad A. Bhat
Division of Floriculture and Landscaping, SKUAST-K, Shalimar, Kashmir Jmmu and Kashmir

Sapna Panwar
Division of Floriculture and Landscapeing, Indian Agricultural Research Institute, New Delhi

Sharat Kumar
National Orchid Garden (NOG), Singapore Botanical Gardens, National Parks Board, (N Parks), 1 Cluny Road, Singapore

S.S. Sindhu
Division of Floriculture, Indian Agricultural Research Institute (IARI)
New Delhi

Sumina Bhat
Division of Floriculture and Landscaping, SKUAST-K, Shalimar, Kashmir Jammu and Kashmir

Sumina Ramzan
Division of Floriculture and Landscaping, SKUAST-K, Shalimar, Kashmir Jammu and Kashmir

Surender Suneja
Department of Urban Landscape, School of Planning and Architecture, IP Estate, New Delhi

S.Y. Chandrashekar
Department of Floriculture and Landscape Architecture, College of Horticulture, (UHS, Bagalkot), Mudigere, Chikmagalur District, Karnataka

T.M. Pal
Division of Floriculture and Landscaping, SKUAST-K, Shalimar, Kashmir Jammu and Kashmir

V.L. Sheela
College of Agriculture and Research Institute, Kerala Agricultural University, PO Vellayani, Thiruvananthapuram, Kerala

Ornamental Horticulture, pp 1-14
© Editor, S.S. Sindhu, 2016
New India Publishing Agency, New Delhi - 110034, India

1

Basic Principles and Elements of Landscape Design

S.S. Sindhu[1] *and Sumina Bhat*[2]

Landscaping combines elements of art and science to create a functional, aesthetically pleasing extension of indoor living to the outdoors. One initial purpose of landscape design is to blend man's technology (house or building) into the natural surroundings. To work towards a desirable landscape design, the landscape horticulturist must have a working knowledge of art elements and design principles.

The "art" is always changing as the plants grow, environmental conditions change and people use the space. For this reason, landscape designers use a design process that systematically considers all aspects of the land, the environment, the growing plants, and needs of user to ensure a visually pleasing, functional, and ecologically healthy design. The design process begins by determining the needs and desires of the user and the conditions of the site. With this information, the designer then organizes the plants and hardscape materials, which are collectively referred to as the features. The features can be physically described by the visual qualities of line, form, color, texture and visual weight— the elements of design. The principles are the fundamental concepts of composition— proportion, order, repetition and unity— that serve as guidelines to arrange or organize the features to create an aesthetically pleasing or beautiful landscape. Knowledge of the elements and principles of

[1] Division of Floriculture and Landscaping, IARI, New Delhi
[2] Division of Floriculture and Landscaping, SKUAST –Kashmir, Jammu and Kashmir

design are essential to designing a landscape and working through the design process.

Basic Principles

It is important to understand that the principles of design must be applied at all the levels that people will experience the landscape. Design principles guide designers in organizing elements for a visually pleasing landscape. A harmonious composition can be achieved through principles of proportion, order, repetition, and unity. All of the principles are related, and applying one principle helps achieve the others. Physical and psychological comforts are two important concepts in design that are achieved through use of these principles.

People feel more psychologically comfortable in a landscape that has order and repetition. Organized landscapes with predictable patterns (signs of human care) are easier to "read" and tend to make people feel at ease. Psychological comfort is also affected by the sense of pleasure that a viewer perceives from a unified or harmonious landscape. Users feel more physically comfortable, function better, and feel more secure in a landscape with proportions compatible to human scale.

The basic principles of landscape design are as under:

1. Proportion

Proportion refers to the size of parts of the design in relation to each other and to the design as a whole. Relative proportion is the size of an object in relation to other objects. Absolute proportion is the scale or size of an object. An important absolute scale in design is the human scale (size of the human body) because the size of other objects is considered relative to humans. Plant material, garden structures and ornaments should be considered relative to human scale. Other important relative proportions include size of the house, yard and area to be planted. Proportion in landscape design usually relates to people and their activities. For example a three-foot pool would be lost in a large open lawn but would fit beautifully into a small private area. And of course, a colossal fountain would dominate a private garden but could enhance a large city plaza.

This gate at Mughal Graden (Verinag Kashmir) is in Perfect Proportion to its Setting

2. Balance and Order

Balance is the concept of equal visual attraction and weight, usually around a real or imaginary central axis. Form, color, size and texture all affect balance. Balance can be symmetrical, asymmetrical or perspective. In formal balance (symmetrical), the mass or weight or numbers of objects on either side of a central axis should be exactly the same. Plants are frequently clipped, lines tend to be straight and edges are clearly defined. For asymmetrical balance, plants should be irregularly placed on either side of an imaginary axis so that the mass or weight on either side of the axis appcars to be balanced. Curved lines, obscure and merging edges and natural contours identify asymmetry in the garden. Perspective balance is concerned with the balance of the foreground, midground and background. When looking at a composition the objects in front usually have greater visual weight because they are closer to the viewer. This can be balanced, if desired, by using larger objects, brighter colors or coarse

Informal Balance.Niagara-on-the-Lake, Ontario

texture in the background. In most cases, either the foreground or background should be dominant.

Order generally refers to the spatial layout or organization of the design and is most often achieved through balance. Order can also be achieved by massing features or elements into distinct groups and arranging them around a central point.

3. Repetition

Repetition refers to the repeated use of features like plants with identical shape, line, form, texture and/or color. Too much repetition create monotony but when used effectively can lead to rhythm, focalization or emphasis. Unity can be achieved better by no other means than repetition. Repetition can be made more interesting by using alternation, which is a minor change in the sequence on a regular basis. An example might be a row of vase-shaped plants and pyramidal plants in an ordered sequence.

Repetition of Diamond Flagstones Creates Movement in Paving Pattern

The Repetitious use of Paving Stone Creates Unity in the Landscape

4. Simplicity

Simplicity goes hand-in-hand with repetition and can be achieved by elimination of unnecessary detail. Too much variety or details create confusion of perception. Simplicity is the reduction of a design to its simplest, functional form which avoids unnecessary cost and maintenance.

5. Unity

Unity means that all parts of the composition or landscape go together; they fit. A natural feeling evolves when each activity area belongs to and blends with the entire landscape. Everything selected for a landscape must complement the central scheme and must, above all, serve some functional purpose. The simplest way to create unity is through use of a design theme or a design style. Design themes and styles have a well-defined set of features that have maintained their popularity over time because they are visually pleasing to many.

6. Transition

Transition is a gradual change. Transition can be obtained by the arrangement of objects with varying textures, forms or sizes in a logical sequential order. For example, coarse to medium to fine textures, round to oval to linear structural forms, or cylindrical to globular to prostrate plants. Transition from shorter to taller plants and from fine to coarse textures would enframe the scene and makes it appear closer, like a painting on a wall. Generally, transition assists in the gradual movement of a viewers' eye to the design and within it.

7. Rhythm

Repetitious use of Sculpture in Landscape Reduces Monotony and Results in the Establishment of Rhythm

Rhythm is achieved when the elements of a design create a feeling of motion which leads the viewers' eye through or even beyond the designed area. Tools like color schemes, line and form can be repeated to attain rhythm in landscape design. Rhythm reduces confusion in the design. Rhythm is expressed through the placement of plants, park furniture, *etc.*, either individually or as group. For example, several benches could be placed at regular indentations along a shrub border. If every other bench was replaced with an attractive piece of sculpture, rhythm would be created that would relieve any monotony from the overuse of one landscape component.

8. Focalization

Focalization involves the leading of visual observation toward a feature by placement of this feature at the vanishing point between radial or approaching lines. Generally, weaker or flowing lines of focalization are desirable in the residential landscape. Focalization can be adjusted by plant materials along the lines to create symmetrical or asymmetrical focalization. Asymmetrical focalization is indirect while symmetrical focalization is more direct, creating stronger focalization. Since focalization can be used to direct attention to a point, traffic in an area is usually directed to that point. Therefore, focalization could be used to direct traffic in a garden area.

Elements

Simplicity is the Essence of Design

Simplicity is the essence of design. How a designer creatively combines plant material and other design components into a simple, unified scheme is always an exciting challenge.

In the landscape palette, the designer is dealing with living plants that are subject to a myriad of weather conditions, different soil types, insect and disease problems, and a host of other environmental and physical circumstances. The landscape artist must deal with plants that celebrate the seasons with the unfurling of leaves in spring, the aroma and visual delight of ephemeral flowers, and the bareness of branches in winter. Change in the landscape is never constant as the seasons come and go. How the designer successfully combines plants and other material components in the landscape planting plan involves paying careful attention to detail, a thorough knowledge of practical horticulture, and a good understanding of the basic principles and elements of design. The designer faces the challenge of creating a plan that is pleasing to the senses and that visually, functionally and aesthetically improves the appearance of the landscape at an affordable cost.

Chaenomeles japonica (Japanese Flowering Quince) Seasonal Design Consideration

Viburnum opulus (European Highbush Cranberry) Winter Fruit is an Important Design Consideration

The elements of landscape design are as under

1. Line

Lines are a powerful tool for the designer because they can be used to create an infinite variety of shapes and forms, and they control movement of the eye and the body. Landscape designers use lines to create patterns, develop spaces, create forms, control movement, establish dominance and create a cohesive theme in a landscape. In curvilinear design, lines should be dramatic, done with a sense of flamboyancy and be very expressive in their shape. Curvilinear lines which have weak, scallopy edges will not be visually interesting or pleasing to the eye. Curvilinear, meandering lines suggest a naturalistic look that invites the user to casually stroll through and experience the landscape.

On the other hand, linear lines such as those found in a straight hedge or the edges of paving materials suggest quick direct movement. Angled lines can create opportunities for creating the "bones or the framework of the landscape". Lines that interconnect at right angles create an opportunity for reflection, stopping or sitting.

Through skillful use of lines in the landscape, the designer is able to direct the attention of the viewer to a focal point.

Effective use of Circumlunar Line form Vancouver Parks Board

Weak, Scallopy Edges Leave a Lot to be Desired

Linear/Curvilinear Lines at Sissinghurst

2. Focal Point or Emphasis

Sculpture / Maze Garden in Japan. Elevated Pieces of Sculpture Create Emphasis in the Landscape

Through the use of emphasis, eye movement is directed towards a center of interest that takes a position of prominence in the landscape. This could be a single tree, a beautifully designed water feature, a piece of sculpture, or a collection of ericaceous plants that automatically draw the eye to this point of interest. Open lawn areas, paths and strategically placed plants can lead the eye to the principal feature without distraction. Plantings should be placed to easily lead the eye to this center of heightened interest.

Secondary features of landscape interest can also be created. In this case, while these components are beneficial in contributing to the unity of the site and tying the total composition of the site together, they have considerably less overall impact than the focal point.

3. Form

Form relates to the natural shape of the plant. Form is found in both hardscape and plants, and it is typically the dominant visual element that spatially organizes the landscape and often determines the style of the garden. The form of structures, plant beds, and garden ornaments also determines the overall form/theme of the garden. For example, a plant that is very fastigiate or upright in its habit of growth is said to have a vertical or aspiring form. Ginkgo biloba 'Princeton Sentry'- Princeton Sentry Ginkgo- is a good example of this form.

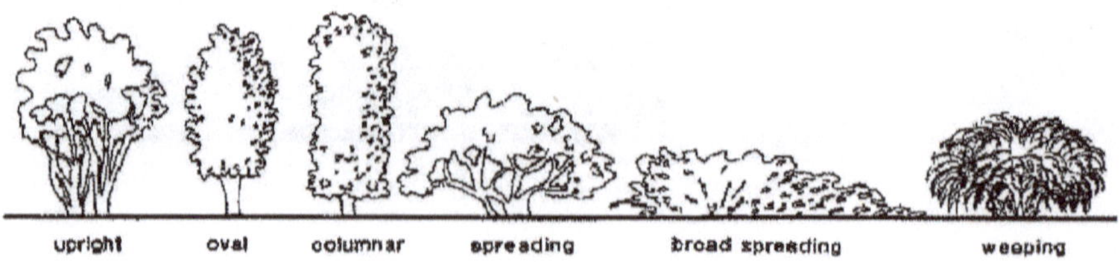

Form Refers to the Shape and Structure of a Plant or Mass of Plants

Other plants that are spreading in their habit of growth are said to have a horizontal or spreading form. A shrub example of this form is *Taxus x media* 'Hillii'-Hills' Yew- and a tree example is *Quercus palustris*- Pin Oak. The Hills' Yew could be effectively used as a hedge to provide special definition between two properties. When horizontal forms are placed together as is the case in the hedge, the individual vertical forms take on a horizontal profile.

Weeping and drooping of pendulous forms can also be used to create softer lines or as interesting accents in the garden. *Fagus sylvatica* 'Purple Fountain" –Purple fountain Beech- is an excellent example of this form.

A Magnificent Example of *Fagus sylyatica* 'Pendula'(European Weeping Beech)

There are also rounded or globular forms that are useful in creating large masses. The majority of shrubs fall into this category.

4. Texture

Texture refers to how coarse or fine the surface of the plant or hardscape material feels and/or looks. Texture is used to provide variety, interest and contrast. The plants' foliage, flowers, bark and overall branching pattern all have texture. The size and shape of the leaves often determine the perceived texture of the plant. A plant can generally be described as having a coarse, medium or fine texture.

The Texture of Plants Differs as the Relationships Between the Leaves, the Twigs and the Branches Differ

Ornamental Grasses and Herbs are Complimentary in Texture

Ostrya virginiana (Ironwood) A Native Tree that Exhibits Excellent Texture

Fagus grandifolia (American Beech) Smooth Bark

When using ornamental grasses for example, a gradation of textures from fine to medium to coarse could be as follows:Texture in the landscape depends upon the distance from which the plant is viewed by the observer. In distant views, the overall mass of the plant is the dominating feature and the fineness or softness of a leaf or branching pattern is lost.

In terms of the overall planting plan, texture must balance in relationship to the axis. Weight on one side should equal the mass on the other side of the axis. For example, much fine texture- as the case would be in using Buxus- is required to balance relatively little coarse texture, as the case would be in the use of *Viburnum rhytidophyllum*, the leatherleaf viburnum. Intermediate plants are recommended to provide the necessary transition from one textural extreme to the other.

5. Colour

Color in plant material and hardscape adds interest and variety to the landscape. Color is the most conspicuous element in the landscape and is usually the focus of most homeowners; however, it is also the most temporary element, usually lasting only a few weeks a year for individual plants. The use of color is guided by color theory (use of the color wheel) to create color schemes. A simple description of the color wheel includes the three primary colors of red, blue and yellow; the three secondary colors (a mix of two primaries) of green, orange and violet; and six tertiary colors (a mix of one adjacent primary and secondary color), such as red-orange.

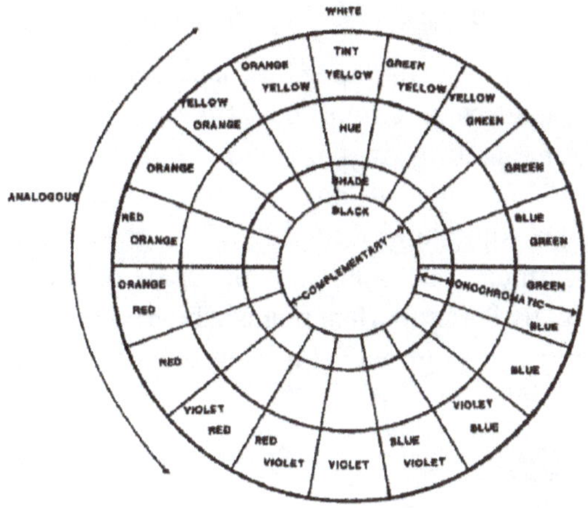

A Color Wheel can be used to Explain Color Variations

Consideration of the use of colour in plantings requires a thorough, practical understanding of the personality of the plants. To vigorously use colour and effective colour combinations requires a thorough knowledge of plants, their colours and seasonal changes with details of twig, leaf, flower and fruit as well as principles of colour.

Vigorously use of Colour in the Landscape. Warm Colours Advance. Salt Lake City, Utah

6. Variety

Variety is the spice of life. In terms of landscape, it is often important to remember that a variety of lines, forms, textures and colours is required in order to achieve an interesting landscape. Without variety in both the use of 'hard' and 'soft' landscape materials, this can lead to unfavourable results.

A Variety of Forms Creates Significant Landscape Interest

7. Grouping

Much greater appeal is achieved when odd numbers of plants are used in the landscape. Groupings of three, five, seven, nine plants etc., will create a strong feeling of mass and a bold landscape statement. Plants should be irregularly spaced and every effort should be made to avoid placement of plants in an equilateral triangle. When grouping, a designer usually starts with a specimen that establishes the scale of the landscape. Around it are grouped slightly less important plants which complement the specimen in colour, texture and habit of growth. Planting one of this and one of that will create a spotty disjointed feeling.

8. Mass

Made up of plants that cannot be seen in their entirety from any one vantage point is mass. Seasonal stability and variety in plant mass is accomplished through a mix of evergreens and deciduous plants. Only rarely should a design consist exclusively of evergreens or deciduous material instead of a mixture of both.

Mass planting of groundcovers for slope stabilization
Seattle, Washington

To create a harmonious effect in any group, a designer should strive to properly fit together plant forms, textures and colours into a harmonious whole or mass. Size of any mass or composition depends upon its location in relationship to other factors such as the need for screening, proximity to other groups, etc. Mass can be any size, but smaller masses or clumps are not normally as effective as larger, bolder mass plantings.

9. Scale

Good proportion and scale have no hard and fast rules. Generally speaking, it is a matter of "does it look right?" Scale usually bears reference to the size of a thing or object that appears to have a pleasing relationship to other things or to the design as a whole. It essentially relates to some finite measure of universal application or a standard of known dimension.

10. Sequence

An Orderly, Sequential Arrangement of Heights

The effective use of sequence is oftentimes employed to create visual movement in the landscape. It is an important consideration to take into account in the development of the overall planting pattern. For example, sequence could be an orderly natural combination of plant material. In this case, low objects would appear in the foreground, intermediate objects in the middle ground, and tall objects in the background.

While it is useful to know the elements and principles of design, it is sometimes difficult to understand how to apply them to your ideas for your yard. Each site presents challenges and opportunities for individual design and expression and requires unique application of the elements and principles. Studying how the elements and principles have been applied in an existing design that appeals to you is a good place to start. The best way to create a good design is to borrow ideas from designs that you find attractive and adapt them to your particular site conditions.

How effectively the designer addresses peoples' needs and the functional requirements of the site, considers ongoing maintenance requirements and the selection of appropriate plants, efficiency and economics will all combine to measure the aesthetic success of the project. All of the above must be carefully woven together to create an outdoor room that is truly pleasurable and enjoyable on a year round basis. Celebrate the seasons in style. Start the most fascinating of the fine arts by developing a garden that takes into account the above principles and elements of design.

Tomb - Itimad-Ud-Dawla 1402 Agra, India

Pavillion on Lake Pichola 1750 Rajasthani, India

Canal between Lakes Dal & Nagin Kashmir, India

Floating Gardens, Kashmir, India

Nishat Bagh, Kashmir, India

Ornamental Horticulture, pp 15-36

New India Publishing Agency, New Delhi - 110034, India

2

Garden Features and Types

S.S. Sindhu[1] *and Arvind Kumar Verma*[2]

Landscape models focus on both the integrated master landscape planning of a home gardening and specific garden design. The people should be educated to realize the importance of gardening, by providing good parks and gardens in cities for a large population to relax and enjoy the beauty of nature. The plants used in the landscape designing of gardens are directly or indirectly involved in increasing the aesthetic value of the city, changing microclimate, social behaviour of the people, checking pollution *etc.* The importance of landscape gardening in human welfare cannot be overlooked. Even in an under-developed country as India, people do not live by bread alone. They also need some finer things of life. The models described in this chapter are very basic features of any garden.

Garden Walls

A garden lover will never like to block the view of his garden by putting a wall along periphery of the garden. A garden as it is enjoyed from inside should also be visible from outside but sometimes, it is necessary from safety and beauty point of view to obscure the ugly sight of an open drain. A compromise is to have a low brick wall of say 60-90 cm height and to put over it some grills, so that the view is not totally obstructed from outside. Alternatively, walls from 1.80 to 5.0 m may also be erected based upon the size of the garden.

[1&2] Division of Floriculture and Landscaping, IARI, New Delhi

Garden Wall

Dry Wall

Garden lay on the wall or walls planted with different plants are termed as dry wall. Plants growing in the crevices of the stones and hanging down on the face of dry wall look beautiful and becoming a common feature of the English gardens. The dry walls are generally retention walls with only one surface available for the gardening or planting while other side is supported against the vertical or near vertical face of two different levels of landscape. Construction of wall in most parts of plains should start just before the rainy season. The aim is to coincide construction with the planting since the rainy season in plains of India is the best time for planting. Before proceeding with the foundation, the slope of land supporting the wall is to be prepared. If the

Dry Wall

total height of the wall do not exceed 75 cm , the wall can be laid vertically and where the height is more, then the wall should be laid in a slanting position with a slope of 1:6.

Plants suitable for the dry wall are: Sweet alyssum, *Bellis perennis*, *Gazania splendens*, *Gypsophila elegans*, *Portulaca* spp., *Zebrina pendula etc.*

Some ferns are very useful as they give natural effect to the wall e.g. *Adiantum caudatum*, *Adiantum venustum*, *Polystichum* spp., *Pteris cretica etc.*

Green Wall

Green wall is a wall, either free-standing or part of a building that is partially or completely covered with vegetation. There are two main categories of green walls: green façades and living walls. Green façades are made up of climbing plants either growing directly on a wall or, more recently, specially designed supporting structures.

Green Wall at the Universidad Del Claustro de Sor Juana, Mexico

The plant shoot system grows up the side of the building while being rooted in the ground. With a living wall the modular panels are often made of stainless steel containers, geotextiles, irrigation systems, a growing medium and vegetation. Most green walls are implemented to create an element of aesthetic diversity that adds to the palette of building materials and the surfaces created. Designers of green walls can work at a great range of scales, from the creation of small private space interactions like intimate garden settings, to the monumental scale of a multi-storey project. Green walls can also be integrated into the entire building site design and utilize multiple systems and forms. Green walls are a key component of living architecture and they will become increasingly important fixtures in our cities in the years to come. Green wall technologies provide a wide range of options for designers who are interested in using the building envelope to accomplish multiple objectives and to provide new free standing design features on the interior and exterior of buildings.

Green walls are found most often in metropolitan cities where the plants reduce overall temperature of the building. Living walls may also be a means for water reuse. The plants may purify slightly polluted water by absorbing the dissolved nutrients. Bacteria mineralize the organic components to make them available to the plants.

Fences

Fences are utility items essentially needed for making the boundary and for protection. Several materials such as wood, bamboo, wire, wire netting and chain like fences may be used. Fences of wood and bamboo get spoilt quicker than the wire fences. Iron grills of various designs can also be used for fencing but the cost of course will be prohibitive.

Wooden Fence

Steps

Steps are the means by which different levels of the gardens can be connected. Steps can be made of various materials like concrete, stone, wood but usually the same material used for the path is used for steps also. The steps in the garden should be different from those in the building. Here, the stepping should be quite broad and the risers low, so that people can have an easier climb.

Stone Steps

Garden Drives

A garden must have a carriage drive leading to the house and the garage besides several other paths or walks leading to different parts of the garden. The main criterion of a good garden drive is to provide

Garden drive at Lodhi Garden, Delhi

pleasant passage for the persons or vehicles mainly. For drives and road a slope 1 : 100 may be allowed for drainage. The width of road of a garden should be about 3-5 m. Gravel drive and concrete drive are mainly used in the garden.

Garden Paths

It links the main features of a garden and provides access to different parts of the garden. A path with graceful and gentle curves will look much better than a straight path. Generally paths are narrower than drives. These garden paths are generally made of gravel, paving stones, crazy paving, bricks, or grass. A garden path should never be less than 60 cm wide but should be preferably between 90 and 120 cm, if sufficient space available. The principles of construction of garden paths are the same as for drives but since these are narrow and not meant for heavy traffic. Types of paths generally used in garden are gravel path, brick path, stone paving, crazy paving and grass paths.

Zigzag Garden Path

Hedge

Shrubs or trees planted at regular intervals to form a continuous screen are called as a hedge. It also includes succulents and cactus. They are useful because they provide protection of an area, demarcate a sector, screen unwanted and ugly sites, acts as windbreak and anti-pollutants and finally beautify the spot.

A) **Trees as hedge:** *Acacia farnesiana, Diospyros embryopteris, Erythrina* spp., *Grevillea robusta, Putranjiva roxburghii, Thevetia nerifolia etc.*

Hedge

B) **Shrubs as hedge:** *Bougainvillea spectabilis, Bougainvillea glabra, Caesalpinia pulcherrima, Clerodendron inerme, Dodonea viscosa, Hamelia patens etc.*

Edging

Lining of borders of flowers, paths, lawn and shrubbery with brick, concrete, living plants *etc*. is known as edging. There are certain plants which are very suitable for edge purpose and they are known as edge plants. Edge plants, unlike hedges, should grow very low as the purpose is not for screening but to provide a lining only for the purpose of decoration of demarcation.

Edging

There are two types of edging.

a) **Formal edging:** It is made of stone, bricks, tiles *etc.*

b) **Informal edging:** It is made of living plants.

Foliage plants for edging: *Alternanthera* spp., *Cineraria maritima, Coleus* spp., *Pilea muscosa* , *Iresene* spp. , *etc.*

Flowering plants for edging: *Allysum* spp., *Amaryllis* spp., *Brachycome* spp., *Gerbera* spp, *Zephyranthes* spp, *Torenia* spp. *etc.*

Shrubbery

Shrubs may be planted in a shrubbery or shrub border, in groups or as accent in the form of a specimen plant. Several kinds of shrubs including tall and dwarf growing, deciduous and evergreen and those flowering in different seasons and having different coloured flowers are planted in a

Shrubbery

shrubbery to give a natural, pleasing and colourful effects almost throughout the year. The common width of a shrubbery is between 3-6 m. and length varies according to the space available for this purpose. The shrubbery was arranged as a walk, ideally a winding one that made a circuit that brought the walker back to the terrace of the house. Its paths were gravel, so that it dried quickly after a rain.

Herbaceous Border

It is an informal arrangement of annuals or biennials and perennials or combinations of both usually to create a dramatic effect through colour and shape on large scale. The term herbaceous border is mostly in use in the United Kingdom and the Commonwealth. In North America, the term perennial border is normally used. Herbaceous borders as they are known today were first popularly used in gardens in the Victorian era. Hybridization and new imported plant species revolutionized the form of British gardens in the 18th and 19th centuries. In addition, the works of Gertrude Jekyll, a British 20th century garden designer and prolific writer, popularized the use of the herbaceous border through a revival of the British cottage garden. The worlds' longest herbaceous border at 215-metres (705 ft), is according to the Guinness Book of Records at Dirleton Castle, East Lothian, Scotland.

Herbaceous Border at East Lothian, Scotland

There are two types of herbaceous border

a) **Annual herbaceous border:** Generally annuals of various kinds are planted in a herbaceous border. The annuals with large variations in plant height, growth habit, colour and form of flower can be utilized advantageously to produce different pleasing effects in an annual herbaceous border.

b) **Perennial herbaceous border:** Perennial plants are ideal for making herbaceous border because of their performance and less care required in cultivation.

Examples: Tall plants: *Solidago* spp., *Rudbeckia* spp, *Tithonia* spp *etc.*

Medium plants: Perennial forms of *Coreopsis* spp, *Phlox* spp, *Gaillardia* spp.

Dwarf plants: *Portulaca* spp , *Gerbera* spp ,*Oxalis* spp and *Verbena* spp.

Flower Beds

A flower bed is a garden area especially prepared for growing flowers. The most essential criterion of a flower bed is to display the flowers in the best way possible. The flower beds can be planted with winter flowering annuals sown during september/october or summer flowering ones sown during february /march and rainy season flowers sown during april/june. The simple designs like circles, squares, rectangles and ellipses are easier to maintain.

Flower Beds

Carpet Bedding

Carpet bedding employs two or more contrasting plant cultivars with a neat, dwarf habit and distinct colouring (of flower or foliage) to create geometric displays. It is often used to form such things as lettering, logos or trademarks, coats of arms or floral clocks. It is an interesting feature in a garden. According to set design, low growing herbaceous plants are planted in beds or series of beds. For contrast and designed effects, plants of one colour are grown as background and other coloured groups are utilized for stripes or lines.

Carpet Bedding

Plants suitable for carpet bedding are *Alternanthera* spp., *Cineraria maritima, Coleus* spp, *Iresene* spp., *Portulaca* spp., *Sedum rupestre etc.*

Ground Covers

Ground covers refer to any plant which forms a low canopy of foliage or spreading type in nature and cover the ground very densely so that weeds cannot grow under them. The low growing and trailing perennial and /or annual plants, both foliage and flowering are ideal as ground covers. Dwarf shrubs are suitable for this purpose which respond to hard pruning and trimming to keep them low growing. It is used to provide protection from erosion and drought and to improve its aesthetic appearance of the landscape.

Ground Cover of *Zebrina pendula*

a) **Upright and spreading type:** These grow upright about 15-30 cm and spreading type of growth which covers the larger area in short period.

   Examples: *Duranta goldiana, Alternanthera versicolour, Iresene lindinii, Cuphea gerlonica, Opheophogon japonicas, Gazania rigens etc.*

b) **Trailing type:** *Ipomoea charyteruse, Ipomoea tricolour, Wadelia trilobata, Zebrina pendula etc.*

Topiary

It is the art of slipping and shearing shrubs and small tress and sometimes even herbaceous perennials into various decorative of geometrical shapes, birds, animals and sculptures and called topiary.

Not to be confused with Bonsai, that is an art creating nature in a miniature form. There is no limit to what you can do. The word derives from

Elephants Shape Topiary at Bang Pa-In Royal Palace, Thailand

the Latin word for an ornamental landscape gardener, *topiarius*, a creator of *topia* or "places". It is common in English, French, Portuguese and Japanese gardens. The plants used in topiary are evergreen, mostly woody, have small leaves or needles, produce dense foliage, and have compact and/or columnar shape and tolerant to heavily training and pruning. Common species commonly used for topiary include *Buxus sempervirens*, *Thuja* spp., *Clerodendron inerme*, *Duranta* spp., *Polyalthia longifolia*, *Putranjiva roxburghii etc.*

Creating Shapes

Wreath: Creating a living wreath isn't difficult. Start with a young plant (ivy or a pliable stemmed plant) and a wire support/frame. Plant two small plants in the container and gently bend each to follow the lines of the frame, securing it with stretch tape or twist tie. If possible, wind the stem around the wire to give it a firmer grasp and camouflage the support. Pinch the growing tip off. As the plant fills in, keep the side shoots clipped to encourage branching.

Stuffed Topiary (animal shapes): These are easier than one-dimensional forms. Acquire chicken wire, make a frame and fill with damp coarse sphagnum moss. Plant directly into the moss with small plants (ivy, baby tears or creeping fig), spacing evenly around the topiary. Keep turning the frame every a few days to expose all sides to light. Keep the inner mosses evenly moist not allowing it to dry out.

Standard Topiary (lollipop): Start out with a plant that has a single stem (fuchsia, coleus, rosemary *etc*) stick a bamboo stake as close to the stem as possible into the soil. Tie the stem to the stake. As it grows, remove all side branches on the bottom 2/3rd of the stem leaving a tuft of growth on top. When it has reached the height you want, pinch the top and let the upper branches grow out to form a crown. All the while, the standard in-training will need constant rotating to make certain that the stem stays straight and the side branches grow out symmetrically.

Spirals and Cones: Usually, the same shaping as the standard, but not removing the bottom branches and pruning to shape into a spiral or a cone.

Lawn

It can be defined as piece of land thickly covered with uniformly green, soft grass. The lawn is the soul of garden or landscape. It creates naturalistic effects and helps to unify the landscape. A patch of green grass in home gardens adds beauty and softness against walls and other concrete structures. It harmonizes with the surrounding and provides a natural green colour. The lawn provides a perfect setting for a flower bed, a shrubbery or a specimen tree or a shrub. A lawn also has its spiritual value. A lawn is the source of charm and pride and reduces tension of the mind after a days' hard work in the materialistic world.

A Beautiful Lawn Around India Gate, New Delhi

Species suitable for lawn

For temperate region: *Agrostis tenuis, Festuca rubra, Lolium perenne, Poa pratensis etc.*

For tropical and sub-tropical region: *Cynodon dactylon , Zoysia japonica , Zoysia matrella, Zoysia tenuifolia etc.*

Arches

The arches are constructed near the gate or over walks. A garden may need some arches for training climbers and ramblers. They are usually made of angle iron and are about 2 m high and 1m broad. Galvanized wire netting may be fixed on the sides of the arched to help the creepers to climb up. The portion of wooden support going underground is painted with coal tar to prevent damage by termites. The angel iron and G.I. pipes are also coated with rust proof paints. The portions above ground should be painted green.

Arches

Pergolas

A series of arches forming a covered walk is called pergola. It is ideal for the display of ornamental climbers and also for screening and partitioning. It is generally contracted over

Pergola

pathways, which add beauty to a garden. Like arched, the support can be made of wooden or stone or brick pillars, angle iron and G.I. pipes. The roof may be made-up of angle iron of different sections with longitudinal and cross rails. The width of pergolas is kept within 2-2.5 m and height is also same but preferably on the higher side.

Bridges

It is a necessary item in a garden to cross a stream. The size, shape and material of bridge depend upon the type of water barrier, it is to transverse. They are made of stone, wood or concrete *etc*. To create a beautiful image, some bridges are built much taller than necessary. This type, often found in East-Asian style gardens, is called a Moon bridge, evoking a rising full moon. A moon bridge is a highly arched pedestrian bridge associated with gardens in China and Japan. The moon bridge originated in China and was later introduced to Japan.

A bridge in the Japanese Tea Garden at Golden Gate Park in San Francisco, California

Gazebo

A gazebo is a pavilion structure, sometimes octagonal, that may be built, in parks, gardens and spacious public areas. Gazebos are free standing or attached to a garden wall, roofed and open on all sides; they provide shade, shelter, ornamental features in a landscape and a place to rest. Some gazebos in public parks are large enough to serve as bandstands or rain shelters.

Gazebo at Will Rogers Gardens, Oklahoma City, USA

Vertical Gardens

Vertical Garden

This new concept of gardening has developed in Switzerland and is not normally found in India. This consists of a wooden/iron frame of thickness from 15 cm to 30 cm depending upon convenience and the height and breadth of the frame and according to available space. The border paces are enclosed with wire frame which serves as the growing medium. It is possible to grow many dwarf flowering and foliage plants in such frames provided the medium is supplied with nutrient and watered regularly. Shallow rooted plants needing very little anchorage grow well. This is very suitable for those areas where no space available for establishment of a garden and people may enjoy the beauty of nature inside the house. Many vertical gardens have been made in Delhi during common wealth games 2010 and the concept has become very popular and being spread in other part of the country.

Roof Gardening

Roof Garden Near Island of Sentosa Adjacent to Singapore

A roof garden is any garden on the roof of a building. In modern times, homes with a compound and lawn especially in cities and towns are becoming rare and skyscrapers are replacing such homes. As a result the private home gardens are vanishing and the only places left for gardening are the roofs of houses and the balcony.

A spacious and well planned roof garden can be a place of joy and recreation. Besides the decorative benefit, roof plantings may provide food, temperature control, hydrological benefits, architectural enhancement, habitats or corridors for wildlife and recreational opportunities. Roof gardens are the most common in urban areas. Plants have the ability to reduce the overall heat absorption of the building which then reduces energy consumption. The primary cause of heat build-up in cities is insulation, the absorption of solar radiation by roads and buildings in the city and the storage of this heat in the building material and its subsequent re-radiation.

Planning

The first thing to check before starting a roof garden is whether the roof surface can bear the weight of soil especially wet soil, secondly the roof should be made water -proof to prevent seepage of water into the room below. The concept of roof gardening itself is artificial; therefore while planning and designing this point should be kept in mind as for example it is not possible to create natural scenery with hills and valleys or natural forest scenery. On a roof, the space available for gardening is limited, therefore it will be a good idea to encourage vertical growth, so this is the reason why a roof garden should use more climbers and trailers in the garden. Hanging baskets also play a major role in a roof gardening

Plant materials suitable for roof gardening

Flowering Annuals: Stock, Pansy, Dog flower, Dahlia, Marigold, Sweet alyssum *etc.*

Herbaceous Perennials: *Pelargonium* spp., *Canna* spp., *Portulaca* spp., *Vinca rosea*, *Verbena* spp *etc*

Shrubs: Many of dwarf and medium shrubs can be grown.

Trees: *Plumeria* spp., *Callistemon* spp., can be grown as specimen plant.

Creepers: *Cobaea scandens*, *Ficus repens*, *Vernonia elaegnifolia*, *Thunbergia alata etc.*

Bahaī Shrine and Gardens, Haifa, Israel

Terrace Gardening

A terrace is a raised space of ground constructed around

a dwelling house or at the corner of a garden or on the sides of hill. When this terrace is used for some sort of gardening, this is known as terrace gardening. This type of garden is generally constructed in front of the house from where a view of whole garden can be obtained. But this can also be constructed in any other corner of the garden from where a clear view can be obtained. A terrace garden is very often referred to as the outdoor living or drawing room.

Planting

Planting in the terrace should be done after much thought and planning. First it is to be decided if the whole area will be paved or some portions will be left out for lawns, beds, planting of one or two specimen trees and creepers for training over the house. As a general principle , the paved surface itself should have very little plant growth .Wider joints of the paving allow more growth of grass which is difficult to clip and requires more time in maintenance than narrow unplanted joints. Only a few low growing creeping plants such as *Portulaca* spp., *Lantana sellowiana* for the tropics and *Thymus serpylllum*, *Veronica repens* for the temperate region can be recommended for the planting in the wider joints. Small specimen dwarf trees or shrubs can be planted amidst grass in the circular, rectangular or square beds left out. Some interesting and attractive potted plants in tubs and bowls can also be arranged. Hanging basket can also be displayed in the terrace. A sundial or bird bath or a stone sculpture may also be constructed in a suitable corner. Chairs and tables may be arranged in places for resting and comfort

Water garden

Planned planting of aquatic and semi- aquatic plants in and around a pool is known as water gardening. Water possesses natural qualities that are both tangible and intangible. Water can be touched, but it also can be heard. It moves, reflects light and sparkles. Water is a strong attraction for people. The reflections of the sky and plants on the water surface, the movement of fishes in water, all give us a very pleasant and happy feeling. Water garden may be natural or artificial. Natural water garden fits well in an informal design and is often

Plants in Water Garden

modeled from an existing water body. An artificial water garden is usually a concrete structure, its form corresponding to style of the garden .The water garden may be formal or informal. In the formal water garden, the pool is situated in a central position and it is in a regular shape like circle, square, rectangle, oblong, oval or any other geometric form with distinct outline. The informal pool is irregular in shape with its outer edges hidden by plants or stones. It blends with any natural landscape. There are three types of pools used in a garden namely concrete pool, container or pre -fabricated pool and polythene lined excavated pool. The concrete pool is durable and can be constructed in any shape and size. Its depth can be variable being deep in the centre and shallow on the margins.

Plants for water garden

a. **Surface flowering aquatics:** *Euryale ferox, Limnanthemum indicum, Nelumbo lutea, Nymphea, Victoria regia etc.*

b. **Oxygenator Plants:** *Callitriche autumnalis, Elodea canadensis, Ranunculus aquatilis, Vallisneria spiralis etc.*

c. **Floating plants:** *Azolla caroliniana, Eichhornia crassipes, Lemma gibba, Trapa natans etc.*

d. **Marginal Plants:** *Acorus calamus, Cyperus alternifolius, Iris* spp. *Sagittaria sagittifolia, Typha etc.*

Rock Garden

The Japanese were the first to recognize the enriching quality of stones in the landscape. In this type of design, large stones are used to suggest mountains *etc.* The site of rockery can be at the end of garden or path, under a tree, in the corner of the garden, against a wall *etc.* The sloppy area is ideal for rockery, but it can also be constructed on a flat ground. The best example of rock garden is in Chandigarh designed by Shri Nek Chand by using the waste materials. A real rock or alpine garden cannot be created at the elevations less than 2,000-2,500 m because most of the true rock or alpine plants do not grow below this altitude

A View of Rock Garden, Chandigarh

Construction

Generally in a garden in cities rockery is constructed on a flat ground. Before construction, the design of the rockery may be marked on the ground using pegs and strings. On a flat ground the soil should be dug to a depth of 30 cm and filled with broken brick rubble or gravel for proper drainage, before forming a mound with soil to desired height. Later big boulders and stones of irregular and varying shape and size are placed in an informal way to give a tiered effect. The slopes of the rockery should be gentle and not sharp so that the soil may not be washed off during rains.

Plants suitable for a rock garden:

Cactus: *Cereus* spp., *Cephalocereus* spp ., *Mammilaria* spp., *Opuntia* spp. *etc*

Succulents: Adenium spp., *Agave* spp., *Aloe* spp., *Euphorbia milli, Furcraea* spp. *etc.*

Flowering shrubs: *Cuphea* spp., *Lantana depressa, Malpighia coccigera, Plumbago auriculata, Russelia juncea etc.*

Flowering annuals: *Dianthus* spp., *Allysum* spp., *Celosia* spp., *Verbena* spp., *Portulaca* spp. *etc.*

Healing Gardens

The "healing garden" is an evolving concept gaining popularity today. What does the word "heal" mean in the context of a garden? If we look in the Webster dictionary the word "heal" is defined as "to make whole and healthy; to cure; to remedy, repair, to grow sound". Healing is defined as "the act or process by which anything is healed or cured; the power to heal. –adj. Tending to cure or heal".

The garden is specifically designed to meet the physical, psychological, social and spiritual needs of the people using the garden. The use of the word healing in the case of a "healing garden" encompasses these definitions to a certain extent, but instead of stressing the idea that they can cure a person, the

Healing Garden at Swedish Covenant Hospital, United States

benefits are related more to the alleviation of stress and the abilities of the space to soothe, to calm, to rejuvenate or to restore one's mental and emotional health. A main role of the space is to provide sanctuary, to allow for meditation or to evoke other qualities desired by the garden user. The World Health Organization defines health as 'a state of complete physical, mental and social well-being and not merely the absence of disease or infirmity'.

The elements of a healing garden

Natural elements: Plants which attract birds and butterflies, ornamental trees, shrubs, perennials, and ornamental grasses, annual plants, sunlight and natural shade, wind precipitation *etc.*

Constructed garden elements: Patio, courtyard paved walking pathways, tables, chairs and benches, landscape lighting, gazebos, pergolas or umbrellas with tables fountain, water garden or other water features, electrical outdoor outlet for music and related activities *etc.*

Clock Garden

The floral clock garden originated with Carolus Linnaeus, an 18th century Swedish botanist. He hypothesized that flowers could accurately predict time based on when they opened and when they closed. He called it specifically the Horologium Florae (lit. 'flower clock'), and proposed the concept in the 1751 publication *Philosophia Botanica.* In fact, many such gardens were planted in the early 19th century using his designs. Many plants exhibit a strong circadian rhythm, and a few have been observed to open at quite a regular time, but the accuracy of such a clock is diminished because flowering time is affected by weather and seasonal effects. The flowering times recorded by Linnaeus are also subject to diffe-rences in daylight due to latitude: his measure-ments are based on flowering times in Uppsala, where he taught and received his university education.

Floral Clock at International Peace Garden, North Dakota, US

The plants suggested for use by Linnaeus are given in a table below, ordered by recorded time. A "-" signifies that data are missing.

Botanical name	Common name	Opening time	Closing time
Tragopogon pratensis	Goat's-Beard	3 a.m.	-
Leontodon hispidum L.	Rough Hawkbit	4 a.m.	-
Helminthotheca echioides L.	Bristly ox-tongue	4-5 a.m.	-
Cichorium intybus L.	Chicory	4-5 a.m.	-
Crepis tectorum L.	Hawks' Beard	4-5 a.m.	-
Reichardia tingitana L.	False Sow thistle	6 a.m.	10 a.m.
Sonchus oleraceus L.	Sow thistle	5 a.m.	12 p.m.
Taraxacum officinale Weber	Dandelion	5 a.m.	8-9 a.m.
Crepis alpina L.	Hawks' beard	5 a.m.	11 a.m.
Tragopogon hybridus L.	Goat's beard	5 a.m.	11 a.m.
Rhagadiolus edulis Gaertner	-	5 a.m.	10 a.m.
Lapsana chondrilloides L.	-	5 a.m.	-
Convolvulus tricolor L.	Morning Glory	5 a.m.	-
Hypochaeris maculata L.	Spotted cat's ear	6 a.m.	4-5 p.m.
Hieracium umbellatum L.	Hawkweed	6 a.m.	5 p.m.
Hieracium murorum L.	Hawkweed	6 a.m.	2 p.m.
Crepis rubra L.	-	6 a.m.	1-2 p.m.
Sonchus arvensis L.	Field milk-thistle	6 a.m.	-
Sonchus palustris L.	Marsh Sow-Thistle	7 a.m.	2 p.m.
Leontodon autumnale L.	Hawkweed	7 a.m.	3 p.m.
Hieracium sabaudum L.	Hawkweed	7 a.m.	1-2 p.m.
Cicerbita alpina (L.) Wallr.	Blue sow-thistle	7 a.m.	12 p.m.
Lactuca sativa L.	Garden Lettuce	7 a.m.	10 a.m
Calendula pluvialis L.	-	7 a.m.	3-4 p.m.
Nymphaea alba L.	White Waterlily	7 a.m.	5 p.m.
Anthericum ramosum L.	St. Bernard's Lily	7 a.m.	-
Hypochaeris achyrophorus L.	-	7-8 a.m.	2 p.m.
Hedypnois rhagadioloides L.	-	7-8 a.m.	2 p.m.
Trichodiadema babrata L.	-	7-8 a.m.	2 p.m.
Hieracium pilosella L.	Mouse-ear Hawkweed	8 a.m.	-
Anagallis arvensis L.	Scarlet pimpernell	8 a.m.	-
Petrorhagia prolifera L.	Proliferous Pink	8 a.m.	1 p.m.
Hypochaeris glabra L.	Smooth cat's-ear	9 a.m.	1 p.m.
Malva caroliniana L.	-	9-10 a.m.	1 p.m.
Spergularia rubra L.	Sand spurrey	9-10 a.m.	2-3 p.m.
Mesembryanthemum crystallinum	Ice-Plant	9-10 a.m.	3-4 p.m.
Cryophytum nodiflorum L.	Ice-Plant	10-11 a.m	3 p.m.
Calendula officinalis L.	Pot marigold	-	3 p.m.
Hieracium aurantiacum	Hawkweed	-	3-4 p.m.
Anthericium ramosum L.	-	-	3-4 p.m.
Alyssum alyssoides L.	-	-	4 p.m.
Papaver nudicaule L.	Iceland poppy	-	7 p.m.
Hemerocallis lilioasphodelus L.	Day-lily	-	7-8 p.m.

Conservatory

It is used for keeping shade or partial shade loving plants. Examples of plants kept in conservatory are ferns, orchids, palms, cacti and succulents, *Alocasia* spp., *Anthurium* spp. and foliage plants like *Dieffenbachia* spp., *Philodendron* spp., *Dracaena* spp., *Maranta* spp. *etc.*

Greenhouse

It is a covered structure for growing plants inside with complete or partial climate control. The modern greenhouse is climate controlled for cultivation of crops under protected environment, particularly when it is not possible to grow the crop in the open due to unfavourable climatic conditions.

Garden adornments

In a garden, several adornments should be provided to make it look more ornamental and useful. For easy understanding, you can see these things in any garden.

Garden seats: These are necessery in any type of garden. The seats should be comfortable, durable, artistic looking, and should not look out of place. Wooden and fabric seats are more comfortable to sit as compared to those built with stone or iron.

Garden Seat

Ornamental tubs, urns and vases: Plants displayed in ornamental tubs or urns at suitable places look beautiful. It can be made of timber, concrete, stone etc. It can be fixed permanently in a place or sometimes it can be movable also.

Vases

Bird baths: These are nothing but large bowl-shaped containers, generally made of concrete, fixed over a pillar or column which is about one meter tall. Water is stored in the bowl for the birds to come and drink or bath in it. Bird baths should be constructed at a quiet corner of the garden. A bird bath is an attraction for many different species of birds to visit gardens, especially during summer and drought periods. Birdbaths which provide a reliable source of water year round add to the popularity and "micro-habitat" support.

Bird Bath

Sundials: A sundial can be used as focal point in a garden and can be placed at the end of lawn and at the junction or termination of a path. In India it can be made with the digits. It should be positioned in a place where the shadow from a tree or building does not fall for a long duration, otherwise the feature becomes meaningless. Before fixing the dial, movement of the sun should be observed and then the dial is screwed to the base. Sundial is used in the garden to know the time.

Sundial in Singapore Botanic Gardens

Fountains: Fountain can also be used as focal point. Unless there is water under pressure, one cannot have any good fountain. Fountains are made to work by circulating the same water contained in a pool. Today various designs of fountain are available i.e. straight upright water jet or finer water jets converging in the form of an umbrella. Musical fountains are gaining popularity.

Musical Fountains at Akshardham Temple New Delhi

Statues: A statue is a sculpture of animate or inanimate objects in the gardens. The statues may be made up of stone, copper, bronze, concrete *etc.* Artistically accenting a garden with extensive selection of statues improves the aesthetic value of the garden. Like in Japanese type of garden we generally found the statue of Lord Buddha. For example Buddha Jayanti Parks, New Delhi have a big statue of Lord Buddha.

The Four Goddesses of the Seasons Statues

Ornamental stones: These stones of various colours and designs can be placed near pools, along or in the midst of various streams and waterfalls. It improves the look of the area to a great extend. Large ornamental piece can also be placed at the corner of the doorway and other suitable places. The ornamental stones are the most common in the Japanese gardens.

Ornamental Stones

References

Arora, J.S. (2004). *Introductory Ornamental Horticulture,* Kalyani Publishers

Bose, T.K., Maiti, R.G., Dhua, R.S. and Das, P. (1999). Floriculture and Landscaping, Naya Prakash, Kolkata, India,

Dubey, K.N. and Singh, S.B. (1985). Ornamental gardens and their layout, *In: Olericulture and Floriculture*, Bharati Bhandar, Baraut, Meerut, pp: 99-112

Furuta, T. (1983). Interior Landscaping, Reston, Virginia, Reston Publishing Company, Inc.

Lancaster, S.P., Bose, T.K. and Mukherjee, D. (1997). *Gardening in India*, Oxford & I BH Publishing Co. Pvt. Ltd., New Delhi.pp 1-15.

Randhawa, G.S. and Mukhopadhyay, A. (1996). *Floriculture In India*, Allied Publishers Ltd, New Delhi, pp 1-631

Swarup, V. (1997). *Ornamental Horticulture*, Macmillan India Ltd, Daryaganj, New Delhi, pp 1-449.

Webster,1990.p257

www.wikipedia.com

www.who.int/abautwho/en/definition.html.

Ornamental Horticulture, pp 37-78
© Editor, S.S. Sindhu, 2016
New India Publishing Agency, New Delhi - 110034, India

3

Temperate Ornamentals in Landscape Planning

M.R. Dhiman[1], S.S. Sindhu[2] and Chander Prakash[3]

Man's progress in this universe is posing a threat to the environment, we live in. If we do not treat the environment well, nature will have its own repercussions and we will inevitably have to bear the consequences. We have already begun to suffer for not taking care of mother earth in the past. The earth is warming up each year and this rise in temperature will have its effects on the earth's inhabitants. Pollution, global warming, acid rain, deforestation, extinction of species, depletion of ozone layer, soil erosion and floods are some of the commonly heard problems of the environment and they are seem to be inter related. The objective of ornamental horticulture is the functional and aesthetic integration of people, building and site, using plants and space as its main tools. The necessity of it in landscape architecture is for positive control of the fast changing landscape for the future. Landscape architecture can help a major way to alleviate the earth's problems. The landscape which includes urban, agricultural forests and other lands becomes a component of nature and thus ensuring that our landscapes are healthy in the first step towards this goal. The landscape gardener is a necessity because very a few people are capable of selecting the right kind of plants for their grounds; they cannot imagine a landscape picture in terms of varieties and groups, the way an expert can. Planning a piece of land with ornamental plants and features transforming it into a beautiful garden requires experience, patience and knowledge about land topography, soils and climatic factors.

[1&3]Indian Agricultural Research Institute, Regional Station, Katrain, (Kullu-Valley), Himachal Pradesh
[2]Division of Floriculture and Landscaping, IARI, New Delhi

The vividity of the garden depends on its design, features and the most vitally on plants, the key elements which make the garden enliven. Selection of variety of ornamentals is done in order to exhibit different form of growth, flowering time, leaf and stem texture, flower colour etc. Therefore trees, shrubs, climbers, annuals, bulbous plants and ornamental grasses usually find place in the garden along with other secondary groups. Plants play a pivotal role in the protection of our environment as well as improving human life. Apart from their physical characteristics, plants aid in cleaning the environment and reducing some of the harsher effects of heat, sound, wind, air and also serves as complementor, attractor, emphasizer, diverter, indicators and provide aesthetic function by creating attractiveness for human activities.

Landscape Planning

A great landscape is the result of planning and vision and without proper plannings a new landscape can be more work than it has to be and can actually take away from the beauty of ones' home and property. The first thing, the landscape architect should do is to put on one sheet of paper a rough and ready plan of the garden to be made from scratch or altered. Before any plan can be studied and drawn, a map or survey of the ground must be secured. As a rule, however, all buildings, building entrances, windows, all walks, roads, paths, all physical features such as vegetation, topography, soil and climate, water sources, rocks, all terraces or other decided irregularities in the ground level, all unnatural features or other features of importance, should be located, named and shown on the map. With the survey and contour maps completed, work on preliminary plan should start. It is the first plan made. In design, it represents a thoroughly worked out scheme or solution of the problem in hand. The essentials of a preliminary plan are: it should offer a practical workable scheme, it should be attractively drawn; and it should present enough information to explain the idea.

A final plan represents a finished scheme, an accepted programme of work. It may or may not be preceded by a preliminary plan. It differs from a preliminary plan in that it should be complete in all respects and presented in a form suitable for its use. In the final plan one can show the adopted scheme, general plan, the planting arrangement, planting plan and the dimension and construction details, working plan, grading plan and drainage plans.

Before preparing the plans for landscape work, the first thing to do is to use our eyes and put our mind to work on the designs of gardens, which we can see around us. From these gardens, we may reach an idea of what sort of layout we would like to have ourselves. Whatever may be the ideas, important considerations for planning a new garden should be the personal taste, practical requirements, size and type of the ground and financial resources. For such interesting ventures, many ideas will naturally crop up in the mind, but by no means would all these be used in any one garden. For the openness, the lawn and paved area are thought of, while for foundation planting, climbers and shrubs which will adopt themselves to walls may be planted against the house together with groups of shrubs. For a riot of colours within a short span of time, herbaceous borders are considered. Beds of roses, individual annuals and bulbs are also very pleasing additions. Water gardens for aquatic plants, rock gardens for cultivation of xerophytic plants or alpine plants; arbours, screens, arches and pergolas for growing many beautiful climbing plants; hedges for screening and decorative purposes; position of bird bath, sundial or a seat are some of the ideas, among many, which will no doubt appear in the mind of the designers. Gardens can be laid out in a severe formal manner or quite informally with no planned balance of line and form, or it may be a combination of these two extremes.

General Principles

A garden is primarily laid out for enjoyment of human beings. In all gardens, it should be the endeavour of designer to secure a reasonable amount of comfort and convenience for those for whom their pleasures are intended. Due respect for natures' teaching, dictates of artistic feeling and ordinary good taste must be priority. The pleasure can be attributed to following points.

All different parts of the garden or various features should exhibit equally in all respects and must be in agreement. Utility is another priority aspect where convenience, comfort and neatness get enough consideration. The lines in the layout must be in order with all correctness and proper finishing. Whatever may be the style of the garden, uniform blend of all features and some sort of symmetry must be achieved in the finished garden. The garden must create a picturesque effect and bring in balance of composition. A well-laid out garden excites and nourishes curiosity and gives a feeling of entanglement. The disposition of objects in the garden should, however, be without any confusion and must be simple in all respects. Abrupt breaks between the objects should be avoided to keep continuity and a link between various parts. Each and every item of the garden must be set apart for a special purpose.

In landscape design, space is utilized for human use, convenience and enjoyment. The basic materials used in the available space are arranged in such a way, to enhance the beauty and as usefulness of each item. Land is the surface on which landscaping is done; this includes different types of soils, mineral materials, organic matter, water, air, rocks, vegetation, *etc.* Land projects and their designs depend on the environment provided by the atmosphere and climate. Variation in grading in the land has a profound influence on the landscape scheme, so a contour map is made to study the plan intelligently and adopt a scheme that will fit with the ground.

Elements of Landscape Design

The components used in a landscape design have a number of characteristics which should be considered. These characteristics are referred to as design elements.

Line: A carefully planned group of lines in a landscape composition will direct the attention of viewer to a focal point or a particular area of interest in composition. Line has a crucial role in controlling movement, either visual or physical, in straight or curved directions. In a design, straight line indicates direct movement without hesitation. Interconnecting straight lines creates points at the intersections for stopping, sitting, changes of views and reflections back to the point of beginning. Sometimes, lines by their juxtaposition make a sharp angle and suggest rigidity, while gentle curves suggest fluidity. Meandering curved lines invite slower movement and are useful in areas that should appear as natural as possible, such as paths through a wild garden.

Form: It is the degree of solidity of forms. In any design, form is very important in creating a particular design. In a tree, the trunk, branches and leaves together create a form. Tall and slender trees have a vertical form. Low and spreading trees have a horizontal form. This form can be modified in proper landscape planning. A group of vertical plant forms, if grouped together in sufficient quantity so that the length of the group is greater than the height, makes the group appear to have a horizontal form.

Texture: It refers to patterning of the components of the landscape. It may be fine or coarse, rough or smooth.

Colour: No garden can be really attractive that has colour clashes. Human response to individual colour varies. In general, reds, oranges and yellows are warm colours and seen to advance towards the viewer. Greens and blues are cooler colours and tend to recede in a composition. Generally, contrasting colours are used.

Variety: It is a critical element in design. Too little variety lead to monotony while too much bring confusion. While choosing plants to develop a landscape plan, it is never a good practice to choose one of everything. Using several of each kind will be much more apt to result in the unity of design. As a matter of fact, a variety of lines, forms, textures and colours are needed to create an orderly, interesting landscape.

Repetition: It gives the element a variety of meanings and expression. Repetition reduces the confusion that may result from excessive variety. It also introduces a sense of order to the viewer of the landscape. Repetition can be achieved by planting individual plants in groups or masses of a single species or variety.

Balance: The visual equilibrium of different garden elements is known as balance in landscape design. Usually, it is possible to perceive a central axis in a landscape composition. It is on the basis of balance that landscapes are judged to be formal and informal, symmetrical and asymmetrical.

Emphasis: The eye is directed to one portion or object of a composition through the use of emphasis. There may, however, be primary or secondary points of emphasis.

Fragrance: One of the greatest delights of a garden is that the fragrance is recognized only in a chance manner. Gardens to be enjoyed primarily after dark could well be planned with this goal in mind. The scent of roses, jasmines, gardenia, *Magnolia grandiflora*, *Cestrum nocturnum*, lilies, hyacinthus etc. are the easiest to recognize and well known to all. It is not only flowers; various kinds of grasses, aromatic herbs, shrubs and trees induce an exhilarating aroma in the garden, which should not be ignored if there is scope to include them in the garden design.

Character: Many plants impart character to a garden setting. It is the pattern of the plant that is important. Pattern is a model that can be found in the natural arrangement of plant parts.

Harmony: The overall plan of the property should be harmonious and functional. One must seek harmony in the design of plant material too. While selecting plants, their ultimate role in the garden, eventual size and the eye level should be thought of.

Garden Features

Thoughtful planning of garden features makes an invaluable contribution to garden design. It is possible to design a garden with a large number of man-made items. Indiscriminate and haphazard use of these items, however, will result in clutter. It should be kept in mind that whatever features may be chosen they must add charm, create an atmosphere of naturalistic design and add character to the quality of design. Some important garden features are: water, rocks, roads, walks, pavements and steps, hedges and edges, arches, pergolas, screens and bridges, lawns, flower beds, borders, carpet bedding and shrubbery. Many other features like stone lanterns, stone basins, statues, towers and wells, sundials, birdbaths, straddle stones, floral clocks, plant containers and raised beds, vertical gardens and plant stands are also introduced in the garden as per need of the landscape design. In some gardens, greenhouses, conservatories and glasshouses are also constructed for comfortable accommodation of some special groups of plants. In modern gardens, artificial light plays an important role. Besides pleasing visual effects, sounds made by wind, water, insects and birds in the garden create inquisitiveness and fill the mind with joy.

Plant Materials

One of the secrets of successful landscape gardening is to choose the right plant for a particular place. Careful selection of the wealth of trees, shrubs, climbers, annuals, bulbs, foliage plants, grass cover, and aquatic ornamentals transform a place into a heaven of beauty.

Vegetation Types in Temperate Regions of India

The Himalayan region is bestowed with varied landscape features which provide multitude of habitats to a diverse array of flora and fauna (Prater 1980). The region covers nearly 11% of India's geographical area. The region is well known for extensive alpine biome, temperate conifers and broadleaf forests, sub-tropical (foot-hill) forests, temperate grassy slopes and various other habitats. The occurrence of 21 forest types (Schweinfurth, 1957; Champion and Seth, 1968; Singh and Singh, 1992), 18,440 species of plants (Singh and Hajra, 1997); 816 tree species; 1748 species of medicinal plants (Samant *et al.*, 1998); 675 species of wild edibles (Samant and Dhar, 1997); 279 species of fodder (Samant, 1998); 155 species of sacred plants (Samant and Pant, 2003) and 118 species of essential oil yielding medicinal plants (Samant and Palni, 2000) throughout the Indian Himalayan region.

The region above natural tree line (*ca.* 3300 –3600 m above mean sea level in the western and north-western and *ca.* 3600 –3800 m in the central and eastern Himalaya) represent alpine habitats, characterized by treeless vegetation, alpine scrub and meadows. The area between *ca.* 3000 m and natural tree line'represents an important ecological belt throughout the Himalaya. The forested habitats in middle elevation rages (1500 - 3000 m) in the western and north-western Himalaya exhibit a great diversity of flora and fauna. The major vegetation types include Himalayan Dry Temperate (conifer), Himalayan

Moist Temperate (broadleaf), and several other categories (Champion & Seth 1968, Singh and Singh 1987). Major forest types in the middle elevation (1500-3000 m) range of Sikkim (Central Himalaya) and Arunachal Pradesh include Temperate Broadleaf Forests, Conifer forests, orchids and bamboo brakes. The temperate belts in the western and north-western Himalaya support an extensive grassland habitat which is largely anthropogenic in nature. These grassy slopes have developed largely on the south facing, steeper slopes which cannot be cultivated.

Native and Naturalized Ornamentals in the Hills

The important hilly areas in India are located in the temperate North West Himalayas (Jammu and Kashmir, Himachal Pradesh, Uttarakhand) and sub tropical East Himalayas (West Bengal, Sikkim, Meghalaya, Arunachal Pradesh); while some are scattered in the Aravali, Vindhya , Satpura ranges (West India) as well as in the Western Ghats, Nilgiris and Eastern Ghats. Although the plant wealth of the hills (especially the Himalayas) is well documented, precious little has been contributed specially on ornamentals. Collette (1921) described the flora of Shimla hills and surrounding areas, mentioned several trees like *Bombax, Cedrela, Aesculus, Acer, Erythrina, Prunus, Rhododendron, Putranjiva, Cupressus* etc. Lele and Misra (1961) added several naturalized ornamentals to the above list, for example *Tagetes minuta, Agave americana, Yucca, Gloriosa, Oenothera, Cytisus, Sambucus, Solanum jasminoides, Opuntia, Passiflora, Lonicera, Wisteria, Lagerstroemia, Callistemon* etc. Rathore (1981) evaluated 150 introduced ornamentals at Shimla, the promising ones being *Acer palmatum*, *Chaenomeles*, *Cotonester*, *Philadelphus*, *Viburnum* and *Weigela*. Polunin and Stainton (1984) described the flora from Nepal to Western Himalaya, with colour photographs of the important native and naturalized attractive species, thus supplementing the earlier work covering this area. Some of these species having beautiful flowers are *Erythrina stricta*, *Paeonia emodi*, *Hypericum*, *Gaultheria*, *Woodfordia*, *Daphne*, *Calanthe*, *Iris*, *Clematis* and *Trichosanthes*. Kohli (1979) described several wild or naturalized bulbous flowers of Kashmir including *Anemone, Fritillaria, Hyacinthus, Narcissus, Tulipa, Iris, Paeonia, Lilium* and *Gagea*.

Gupta (1967) compiled a list of plants flowering during different seasons in Missouri hills. Out of these *Asclepias, Tecoma, Osyris, Woodfordia, Magnolia, Hypericum, Reinwardtia*, and *Caryaopteria* among the spring flowering; *Berberis, Viburnum, Robinia, Cocculus, Vitex, Deutzia* among the summer flowering; *Cryptolepis, Verbascum, Barleria, Inula, Habenaria, Smilax, Crassula* among the rainy season flowering and *Spermadactylon* and *Primula* among the winter flowering species are worth mentioning. Ohashi

(1975) enumerated several beautiful plants of Eastern Himalaya, including *Clematis alternata, Meconopsis horridula, Passiflora nepalensis, Calenthe trulliformis* and *Anoectochilus lanceolatus*. Jana and Mukherjee (1979) listed promising species of bulbous flowers of Shillong area including *Agapanthus, Crocosmia, Hedychium, Hippeastrum, Iris, Lilium, Tigridia* and *Zephyranthus*. Fyson 1974) described the plants of the Nilgiris (including Ooty), the important ones being *Hypericum humifusum, Ilex, Osyris, Prinsepisa, Thunbergia, Cassis tomentosa, Fudchsia fulgens* and *Sarcococca*. Manilal (1988) recorded the flora of Silent Valley (Palghat, Kerala), including *Rhododendron, Ligiustrum, Euonymous, Lagerstroemia microcarpa, Mussaenda, Dendrobium pandburatum, Oberonia* and *Habenaria*.

Landscape planning of temperate/hilly regions is equally important to that of plains. Although there are only a few plant species fit for hills, and most of them are abundant in wild or naturalized, if planted properly they make good skyline and superimpose the beauty of surroundings. Efforts were made hereunder only to give an idea of suitable ornamental plants suitable for hill landscaping with careful consideration of habit, colour, season of flowering, form, cultivation and environmental preferences.

Temperate Ornamental Trees

Trees are essential features of a landscape garden, for roadside planting, public parks, in schools and colleges and some of these can even find a place in small private gardens. Temperate ornamental trees possess an unusually beautiful character due to their attractive flowers, shape and interesting form and also for their wide adaptability. They make wonderful focal points in a landscape and add interest in avenues and gardens. Trees also prevent floods; reduce sound, noise and dust; check the high velocity of wind, condition of weather, provide shade and enhance the beauty of the surroundings. From the floricultural point of view, trees can be classified into two groups, namely shade (foliage) and ornamental and flowering trees. Given below is a list of some temperate trees for use in landscape planning.

Shade or Foliage Trees

Cedrus deodara (C. indica, C. libani deodara): Family: Pinaceae

A large (up to 60 m tall, with a girth of 11m), handsome tree suitable for the hills and is planted singly in large public or botanic gardens as a lawn specimen or as a roadside tree. The timber is used for making railway sleepers and the wood yields a fragrant essential oil. Propagation is from seeds.

Crptomeria japonica: Family: Pinaceae

It is a tall (35-45m), handsome tree of conical shape, good for avenue planting or for growing singly in large parks and gardens. Propagation is from seeds. When planted densely, no vegetation can grow underneath.

Cunninghamia lanceolata (Syn. C. sinensis): Family: Pinaceae

A large tree, attaining a height of 30 m or more. The plant grows well only in temperate climates. Normally propagated from seeds or cuttings of erect shoots. The timber is fragrant and suitable for roadside planting.

Cupressus funebris (Syn. C.pendula) and C. funebris glauca (Syn. C. cashmeriana): Family: Pinaceae

The first species of Chinese origin and the latter of Kashmir or Tibet. The Kashmir cypress is up to 20 m tall and is one of the most beautiful of the cypresses; at the same time it is delicate also. *C. funebris* is a little taller than the above. Propagation is usually from seeds. Hardwood cuttings can be rooted under mist with bottom heat where necessary.

Diospyros embryopteris: Family: Ebenaceae

An evergreen, slow growing tree of medium height, but spreading to a considerable extent. The branches are shining and blackish-brown. The new leaves are very ornamental because of the brilliant red colour. The fruits are decorative and edible, a good shade tree which is propagated from seeds.

Eucommia ulmoides: Family: Eucommiaceae

Eucommia (*Duzhong*) refers to bark of an unusual tree, commonly called the hardy rubber tree originated in central China. It is now grown as an ornamental and shade tree (reaching up to 20 m) everywhere that a temperate climate permits its growth. The tree is monotypic species *Eucommia ulmoides*. Further, this is the only latex-bearing tree that naturally arose in the temperate climate zone. It is also a medicinal tree and propagated form air layering.

Juniperus chinensis: Family: Pinaceae or Cupressaceae

A handsome, erect, dense, pyramidal tree (13-20 m tall), the branches are short and leaves two distinct types, needle-like and arranged in threes on young trees scale-like overlapping in old trees. A hardy tree widely planted for decorative and formal gardening. Propagation is from seeds and by air layering.

Picea smithiana: Family: Pinaceae

A tall (50 m or more) majestic tree with wide-spreading branches and long pendulous branchlets. The dark green leaves are thicker than broad. Young plants may be damaged by frost, planted as a single tree in parks or botanic gardens at high altitudes. In the high hills, this plant looks like a living herbarium as all sorts of orchids and other epiphytes grow on its branches. Propagation is from seeds.

Pinus longifolia (P. roxburghii): Family: Pinaceae

The most commonly grown pine which grows to a height of 30 m or more with a round-topped symmetrical head. The pine is light green, long, pendulous, needle-like leaves and remarkable. It is grown as a specimen in the lawns and can be found in all important gardens in India. Normally propagated from seeds.

Podocarpus macrophylla (*P. longifolia*): Family: Podocarpaceae

A medium-sized, resinous, ornamental and woody plant grown for its foliage. Several variegated forms of this variety are in cultivation. Propagation is form seeds.

Putranjiva roxburghii: Family: Euphorbiaceae

A moderate-sized evergreen tree (10 m tall) with uniform spreading habit. The dense foliage is beautiful, dark green in colour, and gives the plant a weeping appearance. It is a very good avenue tree and can be planted in public places as a shade tree also.

Salix babylonica (*Syn. S. pendula*): Family: Salicaceae

A tree, 10-15 m tall, with spreading habit, the branches are olive-green, very pendulous, and slender, giving the plant a weeping look. In some species branches are spiral shaped and also called spiral willow. This beautiful tree of picturesque habit can be grown on the edge of a bank or an artificial pool. Propagation is from cuttings.

Thuja orientalis (Syn. Biota orientalis): Family: Pinaceae

A pyramidal or bushy tree, up to 12 m in height. A number of erect branches are produced from the base. There are two types- one pyramidal or columnar in habit, and the other dwarfer and rather globular. The branchlets are flattened and arranged at right angles with the branches. The variety *compacta* is dwarf and the best.

Flowering Trees

Aesculus indica: Family: Hippocastanaceae

It is commonly known as horse chestnut tree, a large deciduous tree up to 30 m or more, with very distinctive long-stalked palmate leaves, and with large terminal narrow-pyramidal clusters of white flowers. It flowers during July-October. It is propagated through grafting, seed and cuttings.

Acer campbellii: Family: Aceraceae

A tall tree with light-green foliage and leaves deeply divided into 5-7 more or less equal lobes with fine teeth, and with a long tail-like tip. Flowers cream to reddish borne during April to May. Other important species are *A. pentapomicum, A.caesium, A.sterculiaceum, A.acuminatum* and *A. caudatum*. It is propagated from soft wood cuttings.

Ailanthus altissima: Family: Simaroubaceae

It is also known as tree of heaven. A medium- sized deciduous tree, with grey bark, large pinnate leaves and in the autumn conspicuous with terminal clusters of young reddish winged fruits. Flowers greenish, 5-8 mm across, in erect branched clusters are 10-20 cm length appear during April –May. It is a tall tree suitable for large garden, parks or as a street tree. It is a free suckering tree.

Bauhinia variegata: Family: Leguminosae

A small tree having an irregular shaped crown. The bark is of brown colour. The flowers are large, fragrant, white in colour, but variegated with rose, purple and white. The flowers are borne during February to March when the plant remains in a leafless condition. The tree when in bloom looks very beautiful. It is very useful as an avenue tree, and for parks and open places. It is generally propagated from seeds, but cuttings also root, though with some difficulty.

Callistemon lanceolatus: Family: Myrtaceae

A beautiful shrub or small tree of weeping habit. The crowded scarlet flowers with free stamens are borne in spikes 5-10 cm long on the old branches. A branch with flowers looks like a bottle brush. Flowers between February and November, but the peak season is March –April and August- September. The plants are propagated from seeds or layering. It is specially suited near ponds or lily pools with the weeping branches falling over the pool.

Camellia japonica: Family: Theaceae

A shrub or tree, sometimes growing up to 10 m. The leaves are broadly oval, deep shining green above. The tree is compact in growth and does well at high and medium elevations. The roundish, 7-12 cm across, red, white or pink flowers are double or semi-double and are very beautiful. The flowering season is between February to April. Some species can flower in November also. Propagation is from seeds, cuttings, layers and by inarching. In cutting, a 2.5 cm stem with a leaf attached to it is taken. Camellia is good for planting singly or in groups in parks and gardens.

Cinnamomum camphora: Family: Lauraceae

A medium sized evergreen tree, with elliptic long-pointed 3-veined leaves, which are bright green above and pale and glaucous beneath, and with tiny greenish-yellow flowers in small spreading branched clusters each to 5 cm across. It flowers during April –May. It is propagated from cuttings.

Crataegus songarica: Family: Rosaceae

A deciduous tree to 8 m with grayish branches, deeply and sharply lobed leaves and lax flat-topped clusters of white flowers. Flowers to 1.5 cm across, long-stalked borne during May. Propagation is from cutting and seed.

Dodecandenia grandiflora: Family: Lauraceae

An evergreen tree with silky-haired young leaves and branchlets, with elliptic-lanceolate long-pointed entire leaves, and with solitary or a few pale yellow flowers in axillary clusters 1.5-2 cm. Flowers appear during February – April. It is propagated from seeds and cuttings.

Erythrina arborescens: Family: Leguminosae

A dwarf of tree with a few prickles on stems and leaf-stalks, with long erect spike-like clusters of drooping vivid flowers appearing in autumn with the large trifoliate leaves. It is propagated from long stem cuttings and also from seeds.

Grevillea robusta: Family: Proteaceae

A tree growing up to 30 m, but generally about 10-15 m, it is an evergreen tree with finely divided fern-like twice incised pinnate leaves, the undersurface of which is silvery downy. The plants with golden yellow flowers look very attractive during April-May and flowers attract honey-sucking birds and bees.

The plant is easily propagated from seeds. The plant is very ornamental and can be planted along avenues or as a lawn specimen.

Lagerstroemia indica (Syn.L. chinensis): Family: Lythraceae

A large shrub or a small tree in suitable climatic conditions, the leaves are smaller. The plant looks most handsome from May to August when the sprays of delicately fringed, 5 cm across flowers of pink, rose, mauve or white cover the whole plant. Propagation is from seeds or cuttings.

Magnolia campbellii: Family: Magnoliaceae

A large deciduous tree which may grow up to 50 m in wild, but generally much smaller when in cultivation. The huge lotus like fragrant flowers, pink without and deep rose to crimson within, are borne all over the tree in a leafless condition during February and March or May. Propagation is from layering.

Magnolia grandiflora: Family: Magnoliaceae

An evergreen tree, pyramidal in shape, it may grow up to 25 m, but is generally much smaller in cultivation. The leaves are 15-25 cm long and tapered at both ends. The leaves are thick, glossy dark green above and rusty-felted below. The lotus-shaped flowers are large, 20-25 cm across, and sweet-scented. Propagation is by layering. Veneer grafting or budding may be successful.

Malus baccata: Family: Rosaceae

A small deciduous tree with a short trunk and rounded crown, with elliptic-pointed leaves, and with clusters of white long-stalked flowers. Flowers 2.5-3.7 cm across appear during April –June. It is propagated from seeds.

Michelia champaca: Family*:* Magnoliaceae

A medium sized tree, 10-13 m tall, having long leaves with wavy margins and slightly resembling mango leaves. The 5-7 cm long flowers are golden-coloured, highly scented and silky in appearance. The flowering season is between April and May and September and October. The white flowered and creamy yellow varieties are propagated by grafting on the golden-flowered variety which produces seeds in abundance. Suitable for planting along avenues and in parks and compounds.

Platanus orientalis: Family: Platanaceae

A large deciduous tree with characteristic grey scaling bark, particularly on younger trunks and branches, with broad deeply cut leaves and with long-stalked pendulous globular clusters of numerous small greenish flowers appear during April -May. It is propagated from cuttings.

Prunus campanulata (*P. cerasoides var. campanulata*): Family: Rosaceae

It bears very ornamental flowers in leafless conditions. It is a glabrous tree, 7-10 m tall, and bears pendulous, bell shaped flowers which are the deepest red in colour among the Japanese cherries.

*Prunus sargentii (*Syn. *P. serrulata* var. *sachalinensis, P. serrulata sargentii):* Family: Rosaceae

A large tree with smooth shoots. The flowers are rose-pink borne during March to April.

Prunus subhirtella var. *pendula (*Syn. *P. pendula, P. itosakura, Cerasus pendula, C. itosakura):* Family: Rosaceae

A small tree with drooping and crooked branches. The white to rose-pink flowers are borne during March-April. Most of the ornamental cherries do well when grafted on an easy-to-root cherry rootstock such as *Prunus domestica* or peach, suitable for planting in home gardens or in groups in public parks.

Pyrus pashia; Family: Rosaceae

A small or medium sized deciduous tree, with ovate to broadly lanceolate long-pointed, toothed, quite hairless shining mature leaves. Flowers 2-2.5 cm across appear in March-April. Propagation is from seeds or grafting or also from cuttings.

Rhododendron spp.: Family: Ericaceae

Rhododendrons are mostly shrubs and a few Himalayan species are trees which bear beautiful flowers. The plants are easily raised from seeds in pans containing sandy peat. The seeds can be sown in fresh sphagnum moss also, but in this case the seedlings should be transplanted as soon as they can be handled. Other methods of propagation are grafting, layers or cuttings. Rhododendrons are suitable for planting as specimens in lawns, rock gardens, or in groups in parks backed by dark green foliage of the conifers. The important Himalayan tree Rhododendrons are as follows:

R. arboretum: having blood red, pink, or white, usually spotted flowers are borne in trusses of 20.

R. barbatum: It is a large evergreen tree about 20 m tall. The flowers are packed in a hemispherical truss, 10-12 cm across. The flowers are bell-shaped and blood-red, with five black-crimson blotches at the base.

R. falconeri: It bears bell-shaped flowers of creamy-white to pale -yellow with purple blotches at the base.

R. grande: The flowers are bell-shaped, ivory-white, blotched purple at the base. It is one of the best Himalayan species.

Robinia pseudacacia; Family*:* Leguminosae

A small deciduous tree with deeply longitudinally-fissured bark, with pinnate leaves, and with pendulous clusters of white sweet-scented flowers which appears in April to May. Propagation is form seeds.

Sorbus cuspidata; Family: Rosaceae

A large tree with large elliptic leaves which are white-woolly when young and remain densely woolly beneath when mature. Flower clusters 4-9 cm across, densely white-wooly; flowers white borne during May. It is propagated through cutting.

Temperate Ornamental Shrubs

Shrubs improve the value of the garden by adding colour, texture and fragrant. They offer privacy and can screen unattractive areas. Shrubs can be low growing ground covers or tall as a small tree. Mix different kinds of shrubs for an interesting affect. Evergreen shrubs retain their foliage throughout winter, while deciduous shrubs lose their leaves. Some shrubs flower, while others have unique shapes, berries or unusual bark. From the horticultural point of view shrubs may be classified as 1. Shrubs grown for their ornamental foliage and beautiful form, 2. Flowering shrubs, and 3. Shrubs bearing attractive berries. Selection of shrub for a particular locality depends on several factors like availability of space, sunlight, humidity, temperature, spread, and height of the shrub, the type of the shrub, i.e., whether flowering, foliage, evergreen or deciduous, colour of flower, fragrance *etc.*

Some Important Shrubs

Abutilon spp.: Family: Malvaceae

Abutilon grows better on medium to high elevations and attains a height of 1-1.5m. It is propagated from seeds or cutting. They prefer partial shade and shelter from wind and excessive rains. The flowers are hibiscus-like. In *A. striatum* var. *thompsonii*, the leaves are richly mottled yellow and green. It is grown for its foliage. *A. pictum* is short shrubs with orange-yellow veined crimson flowers appear during spring to summer.

*Abelia x grandiflora (*Syn. *A. chinensis):* Family: Caprifoliaceae

The shrub grows to a height of 1.5 -2m. The white scented flowers are borne in clusters during summer. The persistent brown sepals are attractive. The plant is propagated from cuttings.

Aglaia odorata: Family*:* Meliaceae

A shrub about 2m tall, bearing small, yellow, fragrant flowers during the summer and the rains. It is suitable for shrubbery and as a hedge. Propagated through cuttings.

Agapetes serpens: Family: Ericaceae

A shrub with large swollen woody stem-base and with pendulous bristly-haired branches with two regular ranks of numerous small evergreen leaves. Flowers conspicuous 2-2.5 cm long, axillary, pendulous, scarlet or sometimes yellow appearing during February-June. Easily propagated through cutting.

Aucuba japonica; Family: Cornaceae

Aucuba japonica or Gold Dust Aucuba is an attractive broadleaf evergreen shrub native to eastern Asia and Japan. They are slow to moderate growers and will reach a height of 4-12 feet (1.2-3.7 m) with a spread of 4-6 feet (1.2-1.8 m). Leaves are opposite, elliptical to elliptical-lanceolate 3-7 inches (8-18 cm) long and 1-3 inches (2.5-7.6 cm) wide. Leaf colors on mine are dark green with golden yellow variegation. There are many cultivars with different leaf variegation patterns. They are the best evergreen for deep shade situations and are very drought resistant once established. Flowers are small and inconspicuous. They were planted in deep shade underneath the trees. *Aucuba japonica* is very easily propagated from cuttings.

Azaleas: Family: Ericaceae

Botanically the azaleas are included in the genus Rhododendron. Some important species are *R. indicum, R.kaempferi, R.ledifolium* and *R.poukhanense*. Some other hardy deciduous azaleas are *R. arborescens, R. canadense, R. japonicum, R. nudiflorum, R. occidentale* and *R. rhombicum*. Azaleas do not like lime soils. They need a lot of care in cultivation and should be grown in sheltered position. A light soil with plenty of leaf mold is good for azaleas. The flowers are either single or double and come in white, pink, rose, scarlet red and purple colours during June-August. Azaleas are multiplied by divisions, layering or seeds.

Berberis asiatica: Family: Berberidaceae

A yellow flowering shrub, 2-3 m tall bearing purple berry during early spring to summer. Other species are *B. bhutanensis, B.vulgaris etc*. It is propagated through seeds or cuttings.

Buddleia asiatica: Family: Loganiaceae

The sweet scented flowers are produced in profusion in drooping terminal or axillary panicles during February- April. It is easily propagated by cuttings. Other species are *B. davidii, B. lindleyana, B. officinalis etc*.

Buxus wallichiana: Family: Buxaceae

A dense leafy, evergreen shrub; flowers small, yellowish in dense rounded axillary clusters appear during March- May and propagated by semi-hardwood cuttings.

Camellia japonica: Family: Theaceae

A large evergreen shrub or small tree, with shining evergreen elliptic-lanceolate pointed leaves and erect white almost stalkless flowers, 2-4 cm across during October to November. Species *C. japonica* bears pinkish-red double flower during February–April. It is propagated through semi-hardwood cutting, air layering or by seeds.

*Caryopteris incana (*Syn. *C. mastacanthus, C .sinensis)*: Family: Verbenaceae

The shrub grows to a height of 1.8-2.5 m and produces lavender-blue flowers profusely during winter and spring. The shoots are pruned to 50-70 cm during rains. Propagation is by cutting.

Chaenomeles japonica **(Syn.** ***Cydonia japonica, Pyrus japonica*****): Family: Rosaceae**

A spreading, deciduous shrub that grows 3m high and 6m wide. This plant withstands frost, is fully hardy and can stand to about -2.5°F Celsius degrees. It produces apple-like, edible, aromatic yellow to green or purplish green fruits in autumn. They can also be useful as ground cover or low hedging. Scarlet red flowers are borne in 2-6 clusters during March - April. It is propagated through seeds or semi-ripe cuttings of half-ripe wood or from layering.

Clerodendron **spp.: Family: Verbenaceae**

C. fragrans is a hardy shrub of 1-2 m height, which grows the best in moist and semi-shaded places. The shrub bears usually double, white and highly fragrant flowers in large terminal clusters during summer and rainy seasons. It is propagated from root suckers or cuttings.

Contoneaster frigida: **Family: Rosaceae**

A large, deciduous, spreading shrub grown for their flowers and berries. The numerous white flowers are borne on clusters, followed by abundant scarlet fruits in June. It is one of the most beautiful species when in flowering and fruiting. Propagated by seeds or cuttings. Other ornamental species are: *C .horizontales, C. microphylla etc.*

Cytisus scoparius: **Family: Leguminoseae**

A deciduous, erect shrub, growing up to 2.5 m. The bright shining yellow flowers are solitary and appear in summer. Generally raised from seeds but also from cutting.

Daphne cannabina: **Family: Thymelaeaceae**

An evergreen erect shrub, 2-2.5 m tall. The shrub bears fragrant purplish-pink flowers from January onwards. It is propagated from cuttings and by layering. Other species is *D. surveil*: an erect deciduous shrub, 2-3.5 m tall. The young branches are densely white-downy. The green paling to white, slightly fragrant flowers appear from January onwards.

Datura suaveolens: **Family: Solanaceae**

A large shrub, with large leaves and very large white, pendulous funnel-shaped flowers to 30 cm long, which are very scented at night. It flowers during May- November. It is propagated by cuttings.

Deutzia corymbosa: Family: Saxifragaceae

A vigorous, tall (3m) shrub, suitable for growing at medium to high elevations. The pure white, small and fragrant flowers appear in dense panicle or corymb during summer. It is easily propagated from cuttings.

Euonymus japonica: Family: Celastraceae

A group of shrubs which are not grown for any beauty of their flowers, but many of them are extremely fine in fruit and develop good colouring. It is easily propagated from cuttings.

Euphorbia pulcherima: Family: Euphorbiaceae

A deciduous, soft-wooded shrub to 4m tall, distinguished by its vermilion bracts surrounding the yellow flower heads. It flowers during November-January. Propagation is from soft-wood cuttings.

Forsythia spp.: Family: Oleaceae

A group of shrubs which grows in any soil provided it is well drained. The shrubs are cut back after flowering in March-April. Commonly propagated from cuttings. Some ornamental species include *F.* ×*intermedia* var. *densiflora*: the flowers are pale yellow and rather crowded. *F.* ×*intermedia* var. *spectabilis* (*F. spectabilis*): the flowers are bright yellow and crowded. A beautiful shrub which flowers in a leafless condition during February- April.

Gardenia jasminoides: Family: Rubiaceae

It is a very popular shrub and grown for its highly fragrant flowers. The large, white, double, highly fragrant flowers cover the plants during summer. It is propagated by cuttings or by air layering.

Hamamelis mollis: Family: Hamamelidaceae

Hamamelis (Witch hazel) species and varieties are large shrubs or small trees which bring flower and leaf, golden colours to temperate landscapes when gardeners least expect color and fragrance. Their diversity of bloom times and leaf colors, mature shapes, ease of maintenance and lack of serious disease and insect pests provide an extensive woody plant collection that can strengthen landscape designs. It bears blue, red or yellow flowers in late winter during January –March. Flowers are set close to bare branches. The leaves turn bright yellow in late fall. Grows well in light shade. It is propagated by layering and grafting on *H. virginiana*.

Hibiscus **spp.: Family: Malvaceae**

A group of flowering shrubs consisting of several species and hybrids. Important temperate species are: *H. rosa-sinensis*: a large, 1-2 m tall, bushy shrub having numerous branches. The broadly-ovate, sharp-toothed green leaves are ornamental. The fine-petalled large, single or double, rose-scarlet coloured flowers are borne terminally on the branches. It is propagated by air layering, cuttings or by budding. *H. syricus*: a deciduous shrub, colour of flowers varies from white, light pink or red to purple or blue. Flowering time is often from end of July till the first part of October. *H. paramutabilis*: flowers are larger, more and more vigorous.

Holmskioldia sanguinea; **Family: Labiateae**

A straggling climbing shrub with broadly ovate leaves and with scarlet tubular flowers with large conspicuous red cup-shaped calyx, borne in crowded axillary clusters during August –December. It is easily propagated by cuttings.

Hydrangea paniculata**: Family: Saxifragaeceae**

A group of shrub which can be cultivated at medium and high altitudes. The leaf is handsome, large, oval to roundish in shape and toothed at margins except in one species. The flowers are borne in large (20-30 cm in dia.) round heads of white, pink or blue colours during summer. It is propagated from cuttings.

Hypericum hookerianuml: **Family: Guttiferae**

A small branched shrub to 1.5 m tall. It bears a cluster of few yellow flowers during each 3-4 cm across during September to October. It is propagated through cuttings and seeds. Other temperate ornamental species are *H. oblongifolium, H. choisianum* and *H. uralum*.

Indigofera cylindracea**: Family: Leguminosae**

Usually a spreading shrub with ascending hairless branches. It bears lax spike-like clusters of pink or red flowers during May- August. It is propagated through cuttings. Other species are *I.dosua* and *I.pulchella, I. gerardiana etc.*

Inula cappa**: Family: Compositae**

An aromatic shrub with wooly or silky-haired branches, leaves and inflorescence, with oblong-lanceolate leaves and with numerous small yellow flower heads in large terminal domed clusters 8 cm or more across appearing during August- February. Propagation is from seeds.

Juniperus communis: Family: Cupressaceae

A dense erect, evergreen shrub to 1.5 m tall, having needle like sharp pointed, 6-13 mm long, with a broad bluish white band above, in whorls of three. Propagation is by seeds. Other species are *J. recurva, J. squamata, J. indica*, and *J. macropoda*.

Kalmia latifolia: Family: Ericaceae

It is a broadleaf, evergreen shrub, 3-6 m tall. Its flowers are round, ranging from light pink to white occurring in clusters in May to June. It is propagated from seed, cutting and tissue culture.

Lagerstroemia indica (Syn. *L. chinensis*): Family: Lythraceae

Commonly known as crap myrtle, it is *a* pretty deciduous shrub or small tree, which never grows taller than 2.5 m if pruned. The flowers are borne in profusion, terminally or sub-terminally, from the branches during summer and the rains. It is propagated by cuttings.

Lonicera rupicola: Family: Caprifoliaceae

A dense low growing shrub with stiff interlacing branches which become somewhat spiny, with tiny oblong leaves and with almost stalk-less clusters of pale-pink to purple flowers appearing during June-July. It is propagated by cuttings.

Luculia gratissima: Family: Rubiaceae

One of thc most attractive autumn flowering evergreen shrubs with large pink **fragrant** flowers in large terminal domed clusters, 8-10 cm across flowering during September- October. It is propagated by cuttings.

Magnolia stellata: Family: Magnoliaceae

It is a deciduous shrub, grows 3 m tall, flowers during March -April. Flowers white, purple or creamy, 6-0 cm across, fragrant, many petalled, freely produced. It flowers whilst only quite young. Propagation is by layering in spring or by seed sown in February.

Mahonia napaulensis: Family: Berberidaceae

An erect evergreen shrub up to 3m tall. Flowers yellow, in terminal clusters of several or dense flowered spikes, 10-25 cm long, flowering during October –April. It is propagated by cuttings.

***Nerium indicum: Family*: Apocynaceae**

An evergreen shrub bearing large terminal branched clusters of fragrant red, pink or white funnel-shaped flowers during April-May. It is propagated easily by cuttings.

***Olea fragrans* (Syn. *Osmanthus fragrans*): Family: Oleaceae**

A delicate shrub or small tree, up to 9 m tall, which bears small, **sweet-scented** creamy-white flowers several times in a year but mainly during June – August. Propagated by layering.

***Osbeckia nepalensis* (Syn. *O. speciosa*): Family: Melastomaceae**

A dwarf (45 cm), much- branched shrub, having lanceolate leaves which are often spotted brown. The purplish rose flowers appear during June. It is propagated through cutting or seed. Other species is *O. stellata*, a shrub often hanging from rocks and banks, with large pinkish-purple flowers in terminal clusters during July –November.

***Osmanthus suavis*: Family: Oleacae**

An evergreen shrub or small tree to 7m, with leathery narrow-elliptic acute, finely toothed or small leaves, and with short axillary clusters of small **scented** white tubular 4-lobed flowers appearing during early spring. Propagation is by tip cuttings.

***Paeonia saffruticosa* (Syn. *P. montan, P. arborea*): Family: Ranunculaceae**

The shrub is about 2 m tall. The flowers have a close resemblance to rose and may be single, semi-double, and double, in colours varying from white to rose-pink and red flowers during May to June. The plant thrives the best at medium to high altitudes. It is propagated by air- layering or grafting.

***Philadelphus coronarius (*Syn. *P. pallidus, P. coronarius nivalis)*: Family: Saxifragaceae or Hydrangeaceae**

It is a hardy, tall and deciduous, spreading shrub, densely covered with leaves and flowers. The flowers are creamy white, strongly scented and are borne during spring and early summer. Other species are *P.* × *lemoinei* and *P. pubescens* (Syn. *P. latifolius*). It is propagated by cutting or seed.

Potentilla polyphylla: Family: Rosaceae

Distinguished by its leaves which have many elliptic toothed leaflets, the larger 1.3-4 cm, alternating with much smaller more rounded leaflets. Flowers yellow, rather small 8-15mm across, appear during June –July. It is propagated by layering or soft wood cuttings or by seed. Other species include *P. fruticosa, P.coriandrifolia,* and *P. atrosanguinea.*

Prunus jacquemontii: Family: Rosaceae

A small straggling shrub. It bears solitary nearly stalk less pink flowers during May –June. It is propagated through cuttings.

Pyracantha angustifolia: Family: Rosaceae

An evergreen, thorny, protective and large shrub. It is suitable for town garden. They like to be planted in shade or semi-shade. They have white or yellow blossom in June, but these are insignificant compared to the bright orange or red berries that appear during winter. Propagated by seeds or cuttings.

Reinwardtia indica: Family: Linaceae

A spreading or erect branched shrub upto 1 m tall, with showy axillary bright yellow flowers bearing during November to May. It is propagated through cutting.

Rhododendron spp.: Family: Rubiaceae

Important species include: *R. hookeri*: bears blood red flowers, *R.keleticum*: bears purplish-crimson flowers suitable for rock gardens, *R. lindleyi*: bears white, fragrant flowers, *R. schlippenbachii*: having pale pink to soft rose spotted with red-brown flowers, *R. thomsonii* : blood red flowers. It is propagated through cuttings.

Rosa sericea: Family: Rosaceae

A stiff erect shrub with usually hairless often unarmed twigs, with leaves with small leaflets, and with solitary white or cream, axillary somewhat drooping flowers during May -August. It is propagated through cuttings. Other important temperate species are *R. moschata, R. macrophylla, R. webbiana, R. foeida* etc.

Sambucus nigra: Family: Caprifoliaceae

A large shrub or small tree, mainly grown for the handsome foliage. The flowers are yellowish or dull white and are borne almost round the year. Propagated from seeds and hard-green wood cuttings.

***Sarcococca coriacea*: Family: Buxaceae**

A small evergreen shrub to 1 m, with green stems, and with shinning leathery lanceolate entire leaves. Flowers white, in short dense axillary clusters 6-10 mm, during September –May. Propagation is by cuttings.

***Skimmia anquetilia*: Family: Rutaceae**

A strongly aromatic evergreen shrub or small tree, creeping or erect to 150 cm tall. It bears very compact terminal clusters of yellow flowers during April –May. It is propagated from cuttings or seed.

***Spiraea betulifolia*: Family: Rosaceae**

The white flowers are borne in dense panicles during June - July. The brown-tipped anthers provide a good contrast against the white corolla. It is propagated from seeds, cuttings or layering.

***Syringa vulgaris*: Family: Oleaceae**

It is known as lilac and a popular shrub of temperate region. The flowers and foliage are the most attractive. The flowers are lilac, fragrant and borne profusely in dense terminal spikes during spring. The heart-shaped leaves are attractive. It is propagated by grafting, seeds, semi-hardwood cuttings and layering.

***Thespesia campas*: Family: Malvaceae**

A shrub upto 2 m tall with large yellow bell shaped flowers with dark crimson centers, and with usually bi-lobed leaves, flowering during August – October. It is propagated through semi-hard wood cuttings.

***Tibouchina elegans* (Syn. *Pleroma elegans*): Family: Melastomaceae**

A shrub, 150 cm tall. The flowers are small, 3.6 cm across, solitary, purple in colour and produced in abundance during May and continues up to October. It is propagated through semi-hard wood cuttings. In other ornamental species *T. semidecandra* (Syn. *Lasiandra macrantha*) flowers are rich purple, very large, 8-13 cm across in size, and are borne solitary or in clusters of a few flowers throughout the year in glass but mainly in outdoors during autumn.

***Vaccinium corymbosum*: Family: Ericaceae**

Commonly known as blueberry. A tall (1.2 to 3.5 m) shrub, worthy for its white-tinted flowers and blue-black globose berries which are edible. It flowers during March-April. The leaves turn rich-red in autumn. Propagated by layering or cuttings.

***Vaccinium vacciniaceum*: Family: Ericaceae**

An evergreen shrub, flowers greenish, pale yellow, white or pinkish, borne in one or several stalked one sided clusters at the ends of leafy branches during April –June. It is propagated from seed.

***Viburnum grandiflorum*: Family: Sambucaceae**

It is a large shrub, 3 m tall, distinguished by its dense nearly stalkless clusters of relatively large pink fragrant flowers which appear on bare stems during April –May. Propagated through cuttings. Other important species are *V. cylindricum, V. erubescens, V. cotinifolium, V. nervosum* and *V. mullaha*.

***Vitex negundo*: Family: Verbenaceae**

A large deciduous shrub with whitish hairy branches. Flowers are small pale mauve colours, in clusters on a long terminal branch during April October. It is easily propagated through cuttings.

***Weigela florida*: Family: Caprifoliaceae**

This deciduous shrub is grown for its showy bell-shaped flowers and spreading and arching growth habit. Leaves are dark green, oval and tapered to 7.5 cm long. Bell or funnel shaped dark pink flowers with pale pink-white interiors are produced in corymbs on short lateral twigs on previous years' growth in late spring to early summer. Flowers are attractive to humming birds. Very nice in a mixed shrub border. Propagated by softwood cuttings in late spring through early summer or hardwood cuttings in mid-summer.

Temperate Ornamental Climbers

Botanically a climber is defined as a plant which possesses special structures to climb over a support. These special structures may be hook-like thorns, tendrils, rootlets, or modified leafstalks. The climbing plants which do not have such structures but climb over a support or a plant by twining themselves spirally around such supports are called twiners. There are still other plants which fail in their attempt to climb but somehow manage to support themselves over the trunks, stems, or branches of other plants; these are termed ramblers and stragglers. Creepers are those plants which are unable to climb vertically on their own because of their weak stems. They are important ornamental plants and if properly selected and planted in suitable situations; enhance the beauty of a garden to a great extent. Climbers can be used for the purpose of screening to maintain privacy from the adjacent houses. They may also be grown to be trained on trellises, over specially constructed pergolas, arches, arbours, and against pillars or similar structures.

Climbers can be classified according to their growth habit and the purpose for which they are grown.

1. **Showy flowering climbers**: Climbers of this group generally display a show of colour with their flowers at a certain time of the year and are more appealing to the common man.
2. **Climbers with scented flowers**: A climber may be attractive because of its beautiful flowers and foliage or the scent of its flowers.
3. **Climbers with attractive foliage**: Certain climbers are mainly grown for their foliage and the flowers of such plants are insignificant or inconspicuous.
4. **Shade-loving climbers**: Though most of the climbers thrive the best under the full sun, there are a few which grow well under light or partial shade.

Some Important Climbers

Ampelocissus rugosa: Family: Vitaceae

A large woody vine like tendril-climber, with large ovate-heart-shaped finely toothed leaves which are conspicuously red-woolly beneath. Flowers reddish to yellow-green, in large branched clusters arising opposite the leaves during May -June. It is propagated from cuttings.

Apios carnea: Family: Leguminosae

A tall climbing perennial, with large pinnate leaves, and long-stalked clusters of reddish flowers distinguished by their narrow incurved keels which are longer than the standard and shorter wings. Flower clusters 15-25 cm long; flowers 2-2.5 cm long appear in August -September. Propagation is by tuber

Aristolochia griffithii: Family: Aristolochiaceae

A tall climber with large heart -shaped stalked leaves, and with very distinctive brown-purple, spotted, flowers, with curved trumpet-shaped corolla with a large rounded mouth. Propagated from seeds or by layering. Cuttings are taken with a heel, root in propagating frame.

Campsis grandiflora Syn. C. chinensis, Tecoma grandiflora : Family: Bignoniaceae

A climbing shrub with or without aerial rootlets with the help of which the creeper climbs. The deep orange, red and purple flowers are funnel-shaped, opening into a trumpet-shaped mouth. The creeper flowers during rains in to 6-12 flowered panicles. Propagated easily from root suckers and cuttings. The creeper is pruned during winter when it is deciduous.

Clematis x jackmannii: Family: Ranunculaceae

This genus consists of a large number of species, of which many are good climbers. The creepers produce flowers only on new growth and hence these are pruned to five or six buds on current seasons' growth. It bears violet- purple coloured flowers during July - September. Propagation is from layering or inarching on hardier species or by cuttings.

Hedera nepalensis: Family: Araliaceae

A woody climber to 30 m, climbing by aerial roots, with simple lanceolate to ovate, entire or variously lobed, leathery, dark green glossy leaves. Flowers tiny, many, yellowish-green, in stalked globular umbels 1.5-2 cm across which are arranged in domed clusters to 8 cm across, flowers during September - October or during February -March. It is easily propagated from cuttings.

Ipomoea nil: Family: Convolvulaceae

A hairy climbing annual, with ovate-heart-shaped 3-lobed leaves, and with large purple funnel-shaped flowers with pale bases and often tinged with pink appear during June to September. It is easily propagated from seeds.

Jasminum dispermum: Family: Oleaceae

A woody climber, with many fragrant pinkish-white flowers in terminal and axillary pendulous clusters, and with pinnate leaves with leathery leaflets. Flowers during April to May. It is propagated from cuttings.

Lonicera japonica: Family: Caprifoliaceae

A quick growing climber, very popular for its sweet-scented flowers which appear in large number during the winter. The tubular fragrant flowers are white, tinged with purple but turn to yellow with age. It is easily propagated from cuttings and by layering. Other species is *L. glabrata.*

***Passiflora coerulea*: Family: Passifloraceae**

These are tendril climbers which can be trained over trellises or walls. The plants need pruning after flowering which not only keep them in shape but often help in producing more flowers. The flowering bracts are large, green and the petals are pale pink. Flowers during July- September. Propagation by seed or from cuttings of partly-matured shoots in July or August.

***Parthenocissus himalayan*: Family: Vitaceae**

A very large deciduous climber up to 18 m, with trifoliate leaves with 3 ovate long-pointed, sharply toothed, stalked leaflets, which are shining dark green above and pale beneath, the lateral leaflets asymmetrical. Flowers yellow-green, in spreading flat-topped clusters appear during April –May. It is propagated from aerial cuttings.

***Rosa brunonii*: Family: Rosaceae**

A stout climber with hooked prickles, and terminal clusters of many fragrant white flowers each 2.5-4 cm across. Flowering occurs during April –June. It is propagated from cutting or grafting.

***Schizandra chinensis*: Family: Magnoliaceae**

It is a hardy deciduous climber, grows 6 m. tall. Flowers during April – May. Flowers pale pink, fragrant, on long stalks, singly or in clusters. Fruits are red berries in a grape-like cluster. Propagation is from cuttings of half-matured shoots in August.

***Schisandra grandiflora*: Family: Schisandraceae**

Hairless woody deciduous climbers with drooping globular pinkish-white sweet-scented flowers 2-3 cm across are borne during late spring and early summer (April –June). Fruits of numerous scarlet carpels borne on a long cylindrical fleshy axis. Propagation from seed and cuttings of half-ripe wood.

***Senecio scandens*: Family: Compositae**

A climber with zig-zag, grooved branches, with arrow-shaped long-pointed leaves, and with many lax terminal and lateral domed branched clusters of bright yellow flower-heads. Flowering occurs during September –December. It is propagated by cuttings.

Solanum jasminoides or S. album: Family: Solanaceae

A hardy semi-evergreen climber known as the potato vine. It is fast grower and good bloomer, relatively frost-hardy. It grows up to 4 m. tall, vigorously to form a graceful intertwined climber. Flowers pale blue, white, borne in many-flowered sprays during June to the first frosts. Propagation is from cuttings in July or August.

Trachelospermum lucidum: Family: Apocynaceae

Usually a tall woody evergreen climber, sometimes creeping over rocks, with shining elliptic pointed leaves, and with small fragrant white tubular flowers in loosely 3-branched clusters appear during April –July. Its main distinguishing feature is that it has both pinnate and trifoliate leaves on the same plant. It is propagated from cutting and by layering during rainy season.

Vallaris heynei: Family: Apocynaceae

An evergreen, strong-growing climber with pretty foliage. The climber bears white fragrant flowers from February to April. Propagation is by layering.

Wisteria chinensis: Family: Leguminosae

It is the best climber for cool areas. The fragrant mauve or deep lilac flowers appear in their thousands during summer (April-May), when they are objects of great beauty. It is a heavy, woody climber, up to 30m long in extreme cases. The creeper may be trained on walls, porches or trees. To control growth, severe pruning may be necessary. Propagation is by layering.

Plants Suitable for Herbaceous Borders

The following plants are suitable for herbaceous borders on the high hills and temperate regions:

1. **Dwarf:** *Dianthus allwoodii, Dianthus chinensis, Campanula carpatica, Linum narbonense, Sedum spectabile, Statice latifolia, Viola cornuta, ornamental cabbage* and *kale.*

2. **Medium:** *Achillea elegans, Achillea eupatorium, Anemone hupehensis, Aster amellus, Campanula persicifolia, Chrysanthemum maximum, Coreopsis, Delphinium belladonna, Gaillardia, Galega, Gypsophila elegans, G. paniculata, Helenium pumilum, Lupins, Lythrum, Oenothera, Papaver orientale, Phlox decussata, Rudbeckia newmanii, Scabiosa caucasica* and *Tradescantia.*

3. **Tall:** *Anchusa azurea, Chrysanthemum uliginosum, Delphinium, Helianthus annuus*, Hollyhocks, Peony, *Rudbeckia* (Tall).

Plants Suitable for Rocky Areas

The following plants are suitable for rocky areas on the high hills and temperate regions:

Adonis vernalis, Androsace lanuginosa Anemone spp., *Campanula* species, *Cotyledon oppositifolia, Cyclamen* (hardy types), *Draba* species, *Gentiana* species, *Iris* species (dwarf), *Lilium longiflorum, Lilium tigrinum, etc., Meconopsis* spp., *Narcissus, Penstemon, Potentilla* spp., *Primula* spp., *Rhododendron* spp., *Azalea* spp., *Selaginella* spp. and *Tulipa* spp.

Temperate Orchids for Landscape Planning

Cool-climate orchids grow naturally in areas from as far a field as China until Northern Australia. The orchid plant that comes to mind immediately is the *Cymbidium* orchids. *Cymbidium* orchids make up the largest of the potted orchid selections which can be found.

In the Andes the orchids are cooled by moist clouds in late afternoon. The examples here include *Odontoglossum* and *Masdevallia* orchids. It is found that *Odontoglossum* orchids hybridize easily with the *Oncidium* orchid, a warmer-growing orchid. In South Africa, especially on Table Mountain and in Madagascar the *Disa* orchid grows in the cool winds all year round with cold rain and occasional snow falls in winter. They experience high summer temperature, but mountain streams keep the orchid plant roots cool and make the weather conditions tolerable.

The following examples are the naturally cool-climate orchids: *Cymbidium* orchids, the *Odontoglossum alliance* and the different types of *Odontoglossum* orchids, the *Disa uniflora* and its hybrids, the *Pleione* orchid, the *Zygopetalum* orchid and the *Dendrobium* orchid, the *Masdevallia*, *Brassia* and of course the *Coelogyne* orchids.

There are also Terrestrial orchids which resort under the cool-climate orchids. They are the *Cypripedium* orchid, *Oriental Cymbidium* orchids, the *Diuris* and *Pterostylis* orchids, the so-called Jewel Orchid (*Goodyera* and *Ludisia*), and the *Epipactus* and *Spiranthes* orchids.

Palms for Temperate Gardens

Rhapidophyllum hystrix, Sabal minor McCurtain', Sabal minor Emerald Island Giant,' Sabal minor Bear Creek,' Sabal minor Castor Dwarf, Sabal minor Chipola Dwarf,' Sabal minor Blountstown Dwarf,' Sabal minor McCurtain,' Sabal minor Oriental Giant,' Sabal minor Woodville,' Sabal x texensis Brazoria,' Sabal minor var. louisiana, Sabal Birmingham,' Sabal etonia , Sabal palmetto Bald Head Island,' Sabal palmetto Mt. Holly,' Sabal palmetto Rock Hill,' Sabal palmetto Tifton Hardy,' Sabal spp. Tamaulipas.

Trachycarpus fortunei Bulgaria,' *Trachycarpus fortunei* Charlotte,' *Trachycarpus fortunei* Greensboro,' *Trachycarpus fortunei* Hayes Stiffie,' *Trachycarpus fortunei* Nanital,' *Trachycarpus fortunei* Norfolk,' *Trachycarpus fortunei* Taylor Form, *Trachycarpus fortunei* Tennessee Form, *Trachycarpus fortunei* Wagnerianus', *Trachycarpus nanus.*

Butia capitata, Chamaerops humilis var. *cerifera, Chamaerops humilis, Guihaia argyrata, Jubaea chilensis, Nannorrhops ritchieana,Erythea armata, E. edulis,* Phoenix *dactylifera, P. sylvestris, Serenoa serrulata, Washingtonia gracilis, W. sonorae.*

Ornamental Bulbous Flowers

Horticulturally, bulbous plants include the plants with true bulbs, corms, rhizomes, tubers and tuberous roots. Bulbous plants are usually herbaceous perennials which play an important role in the beautification of our landscape. They add beauty and interest to the landscape. There are a large number of different types of bulbs, offering variation in form, fragrance, colour and lasting display. These plants can be used in formal and informal landscape in beds, borders, pots, etc. Many of them survive untended in the open field and blossom year after year. Majority of these plants are grown for their attractive blossom and only a few are grown for their foliage value. Thus, they need a suitable background in the form of shrubs or a wall. They look much better when comfortably grouped with other plants. Naturalization of bulbs intensifies beauty of landscape and also helps the bulbs to mature and multiply in a given environment without any additional care. Some of the important bulbous plants found suitable for temperate landscape planning are listed below with their characteristics. Some important bulbous ornamentals are:

Agapanthus umbellatus, Anemone coronaria, Bergenia ligulata, Begonia evansiana, Belamcanda chinensis, Convallaria majalis, Crinum amoenum, Crocus sativus, Eremurus himalaicus, Fritillaria imperialis, Galanthus nivalis, Hemerocallis, Hosta plantaginea, Hyacinthus orientalis, Iris species, Kniphofia uvaria, Narcissus tazetta, Muscari botryoides, Nerine sarniensis, Ornithogalum umbellatum, Paeonia lactiflora, Platycodon grandiflorum, Ranunculus asiaticus, Sternbergia lutea, Tulipa species, Zantedeschia aethiopica, Lilium species and Hybrids, *Alstroemeria*, and *Zephyranthus species etc.*

Temperate Ornamental Bamboos

Fargesia robusta: *Fargesia robusta* has become one of the most important new bamboos in western horticulture. It was becoming apparent that true Clumping Bamboo has great value in small, urban gardens where, because of limited space, it is sometimes not feasible to grow the better known, spreading timber-type bamboos. It is able to withstand cold winters and windy conditions showing minimal leaf stress. Its versatility, growing well in sun or shade, makes it ever useful in the landscape. It matures to about 5 m tall and sends up two crops of new shoots in a single growing season. The first occurs in early spring. The second flush, though much smaller, in August through September. It can be easily propagated from division of clumps.

Fargesia sp. Rufa:' Among the coldest hardy of the true Clumpers, Rufa' is becoming well recognized as versatile, non-invasive bamboo, used to create a medium sized hedge or screen. Rufa'shows minimal sign of stress when exposed to very cold winters, or even full sun in mild climates. The canopy is gently weeping; if the root mass is one meter in diameter, the canopy is usually about twice as wide and 8 feet tall at the center. Its versatility and non-invasive nature make it an important new bamboo for the landscape and small, urban gardens across the country.

Temperate Ornamental Grasses

Stipa brachytricha, S. arundinacea, S. splendens, S. calamagrostis, S. gigantea and *S. pulchra, Calamagrostis x acutiflora Karl Foerster,'* ***Helictotrichon sempervirens*** (Blue Oat Grass), *Festuca arundinace*a (good for lawn and for golf courses), *Melica ciliata, Achnatherum calamagrostis, Lolium perenne, Agrostis vinealis, A. capillaris, A. alba, Phleum bertolonii, Dactylis glomerata* and *Poa pratensis.*

References

Bhattacharjee, S.K. (2006). In: Advance in Ornamental Horticulture, Vol. 5. Agarwal Printing Press, Jaipur, India,

Bose, T.K., Dhua, R.S. and Maity, R.C. (1999). Floriculture and Landscaping. Naya Prakash, Kolkata, India.

Carpenter, P.L., Walker, D.T. and Lanphear, F.O. (1975). Plants in the Landscape, W.H. Freeman and Co., San Francisco, USA.

Champion, H.G. and Seth, S.K. (1968). A Revised Classification of Forest Types of India. Manager of Publications. New Delhi.

Collette, H. (1921). Flora Simlensis. Thacker Spink and Co., London.

Fyson, P.K. 1974. Flora of the Nilgiri and Pulney Hill-Tops. Bishan Singh and Mahinder Pal Singh, Dehradun.

Gupta, R.K. (1967). Seasonal flowers of the Indian summer resorts-Missouri hills. Nav Yug Traders, New Delhi.

Jana, V.K. and Mukherjee, A. (1979). Shillong- a Treasure house of bulbous plants. *Indian Horticulture,* 24: 17-22.

Jhon, A.Q., Rather, Z.A., Paul, T.M. and Neelofar. (2007). Evaluation of ornamental bulbous plants for landscaping in temperate regions. *J.Ornamental Hort.*, 10(1): 1-8.

Kohli, P.N. (1979). Beautiful bulbous wild flowers of Kashmir. *Indian Horticulture,* 24: 23-27.

Lele, V.C. and Misra, D.P. (1961). Some unreported wild naturalized flowering plants of the Shimla Hills. *Himachal Horticulture,* 2: 5-8.

Manilal, K.S. (1988). Flora of Silent valley. The Mathurbhumi Press, Calicut.

Miller, G.T. (1992). Envirnomental Science, Fourth edition, 1992.

Ohashi, H. (1975). The flora of Eastern Himalaya. University of Tokyo Press, Japan.

Polunin, O. and Stainton, A. (1984). Flowers of the Himalaya. Oxford University Press, New Delhi. 580 pp.

Prater, S.H. (1980). The book of Indian animals. Bombay Natural History Society, Oxford University Press. 324 pp.

Randhawa, G.S. and Mukhopadhyay, A. (1986). Floriculture in India. Allied Publishers Limited, New Delhi, India.

Rathore, D.S. (1981). Advances in ornamentals in hills. *Indian Horticulture,* 26: 19-26.

Roy, R.K. (2006). House plants, Economic Botany Information Service, National Botanical Research Institute, Lucknow, India.

Samant, S.S. and Palni, L.M.S. (2000). Diversity, distribution and indigenous uses of essential oil yielding medicinal plants of Indian Himalayan region. *J. Med. Arom. Plant Sci.*, 22: 671-684.

Samant, S.S. and Pant, S. (2003). Diversity, distribution pattern and traditional knowledge of Sacred Plants in Indian Himalayan Region. *Indian Journal of Forestry* 26 (3): 201-213.

Samant, S.S. and Dhar, U. (1997). Diversity, endemism and economic potential of wild edible plants of Indian Himalaya. *Int. J. Sustain. Dev. & World Ecology* 4:179-191.

Samant, S.S. (1998). Diversity, distribution and conservation of fodder resource of west Himalaya, India. In: *Proceedings of the Third Temperate Pasture and Fodder Network (TAPAFON).* Edited by B. Misri. Pokhra, Nepal, sponsored by F.A.O. Rome, pp. 109-128.

Samant, S.S., Dhar, U. and Palni, L.M.S. (1998). Medicinal Plants of Indian Himalaya: Diversity, Distribution Potential Values. HIMAVIKAS Publ. No. 13, Gyan. Prakash. Nainital.

Saqua, V.O.(1986). Drought and Desertification in: Ecological Disasters in Nigeria, pp.251-258. Federal Ministry of Science and Technology, Lagos, Nigeria.

Schwell-Cooper, W.E. (1961). The A.B.C. of flowering shrubs. The English Universities Press Ltd., LONDON, E.C.I.

Singh, D.K. and Hajra, P.K. (1997). Floristic diversity. In: *Biodiversity status in the Himalaya.* Edited by Gujral. British Council, New Delhi, pp. 23-38.

Singh, J.S. and Singh, S.P. (1992). Forests of the Himalaya: Structure, Functioning and Impact of Man. Gyan. Prakash. Nainital.

Singh, J.S. and Singh, S.P. (1987). Forest Vegetation of the Himalaya. *Botanical Review* 53:80-192.

Temperate Ornamental Trees

Magnolia campbellii

Grevillea robusta

Ailanthus altissima

Magnolia grandiflora

Michelia champaca

Aesculus indica

Camellia japonica

Callistemon lanceolatus

Malus baccata

Temperate Ornamental Trees

Prunus campanulata

Pyrus pashia

Rhododendron

Salix babylonica

Sorbus cuspidata

Podocarpus macrophylla

Platanus orientalis

Cupressus funebris

Cedrus deodara

Temperate Ornamental Trees

Cryptomeria japonica

Bauhinia variegata

Cinnamomum camphora

Robinia pseudacacia

Prunus subhirtella

Prunus sargentii

Temperate Ornamental Shrubs

Pyracantha

Kalmia latifolia

Lagerstroemia indica

Aucuba

Abutilon

Lilac

Luculina

Osbeckia napalensis

Paeonia suffruticosa

Philadelphus coronarius

Potentilla

Quince

Temperate Ornamental Shrubs

Skimmia

Spirea

Weigela florida

Camellia japonica

Vaccinium corymbosum

Forsythia

Hydrangea

Rhododendron

Azaleas

Temperate Ornamental Shrubs

Pyrus

Magnolia stellata

Junipers

Temperate Ornamental Climbers

Schisandra grandiflora

Parthenocissus himalayan

Rosa brunonii

Hedera nepalensis

Lonicera japonica

Campsis grandiflora

Temperate Ornamental Climbers

Vallaris heynei

Jasminum dispermum

Senecio scandens

Clematis x jackmannii

Wisteria chinensis

Solanum jasminoides

Important Bulbous Plants for Hill Landscaping

Convallaria majalis

Eremurus himalaicus

Fritillaria imperialis

Galanthus nivalis

Hosta plantaginea

Kniphofia uvaria

Important Bulbous Plants for Hill Landscaping

Muscari botryoides

Nerine sarniensis

Platycodon grandiflorum

Sternbergia lutea

Narcissus

Tulipa

Ornamental Horticulture, pp 79-98
© Editor, S.S. Sindhu, 2016
New India Publishing Agency, New Delhi - 110034, India

4

Green Building Concept-the Needs of 21st Century

S.S. Sindhu[1], Surinder Suneja[2] and Arvind Kumar Verma[3]

The green buildings are the needs of present day in India as well in the world. The easy availability of most of the green materials and equipments in the country has made it easier for designers to adopt local materials to a very large extent in the construction of green buildings. Now, there is an imminent need for service providers, who would be required in large number, not in hundreds but in thousands, as the movement is heading to reach greater heights. The green building movement is here to stay for the benefit of individuals, society and the country at large in the era of climate change.

With increase in the urbanization and industrialization, there is more pressure on available land and natural resources. In coming era, there will be only limited space available for city dwellers. In buildings one can grow a wide range of flowering and foliage plants. Plants not necessarily one confined to the outdoor only but bed room, kitchen, staircases, patios, windows and other locations can be made green with suitable plants. Indian Council of Green Building (ICGB) has developed the rating system.

Green Building in Santiago, Chile

A large amount of materials are used and energy consumed in construction and operation of average building. Many steps have been taken to reduce consumption of energy by developing new types of vehicle, energy sources,

[1&3] Division of Floriculture and Landscaping, IARI, New Delhi
[2]Department of Landscape Architecture, School of Planning Architecture, New Delhi

recycled materials and designing environment friendly buildings. These buildings are also known as green buildings or green construction or sustainable building.

A View of Green Building

Rapid increase across the world in construction of green building that reduced the building impacts on human health and the environment. Green buildings ranged from residential complexes, exhibition centres, hospitals and educational institutions to laboratories, IT parks, airports, government buildings and corporate offices. Well designing of plants in green buildings reduces the energy requirement and purifies the air.

Interior Vertical Wall

Vertical garden inside a building and greenery inside a building

The human may be benefited in several ways by using different types of plants in decoration of green buildings.

Green building concept involves a building, which is designed, built, operated, maintained or reused with objectives to protect occupants health, improve employee productivity, use natural resources and reduce the environmental impact. In construction of a green building, environmental issues should be taken into consideration at every stage of building construction like designing and development efficiency, energy and water-use efficiency, natural resource, indoor environmental quality, building-owner maintenance and buildings' overall impact on environment.

Significance of Green Buildings

During the last 30-40 years, we are facing the bitter experience of global warming, ozone depletion, resource depletion, energy scarcity, ecological toxicity and imbalance, human toxicity, acid rains *etc*. These have alarmed, rather compelled the mankind to change the way they operate on earth. Though we cannot avoid affecting the environment, green buildings will aim and contribute towards minimising the environmental impact. It should also be emphasized that green buildings do not only contribute towards a sustainable construction and environment but it also brings a lot of benefits and advantages to building owners and users. It contributes towards lower development costs, lower operating costs, increased comforts, healthier indoor environment quality and enhanced durability and less maintenance costs.

The buildings which are constructed today in the world are still a kind of steamship, polluting, contaminating and depleting the surrounding environment and relaying on scare amounts of natural light and fresh air for breathing. Imagine, instead, a building as a kind of tree. It would purify air, accrue solar radiation, produce more energy than it consumes, creates shade and habitat, and changes with the reasons.

Energy Use Efficiency

Energy Efficient Backyard

The high energy use efficiency should be the main objective of a green building. Direction of windows and insulation in walls, building envelope air sealing (the barrier between conditioned and unconditioned space), duct sealing, proper placement of air and vapour barriers, use of solar powered heating/cooling systems will contribute towards high energy efficient buildings. Designers orient windows and walls and place awnings, porches and trees to shade windows and roofs during summer while maximizing solar gain in the winter. Onsite generation of renewable energy through solar power, wind power hydro power, or biomass can significantly reduce the environmental impact of the building. Power generation is generally the most expensive feature to add to a building thus, plants used in green buildings minimise the demand of energy. Now a day large amount of planting materials are being used by the corporate landscaper to save the energy.

Plants can be successfully used to mitigate heat. Plant leaves intercept, reflect, absorb and transmit solar radiation. Trees are called natures' air conditioners, because they lower air temperature, shade buildings in summer and block winter winds. Thus, trees can reduce building energy use and cooling costs. A single large tree can produce the cooling effect of 10 room-sized air conditioners operating 24 hours/day. Dew and frost are less common under trees, because less radiant energy is released from the soil in those areas at night.

Indoor Environmental Quality Enhancement

An increase in respiratory ailments and allergies with use of chemicals which give off gas from deodourants have greatly contributed to sensitive awareness of the air we breathe inside our homes. Buildings rely on a properly designed ventilation system to provide adequate ventilation of cleaner air from outdoors, filtered air as well as isolated operations (kitchens, dry cleaners, etc.) from other occupancies. Indoor plants can be used in bed room, kitchen, staircases, patios, dining table and other locations which give fresh air and enhance indoor environment quality.

Effects of Green Buildings

Green buildings increase the economic stability by attracting people and shop longer along tree lined streets, and apartments/ offices in wooded areas rent more quickly and have higher occupancy rates. Property values are higher in well-planned areas lined with well planed indoor and outdoor plants. Studies attribute this response to stress relief, the association of nature with positive emotions through memory and natures' ameliorative effect on mental fatigue. One study showed that hospital patients with a view of trees from their windows had a more positive recovery course than those without. They had shorter hospital stays by 10 percent. Less violence occurs in urban public housing, where there are plants due to nature-linked decrease in stress and irritability. Parks and open spaces provide opportunities for active and passive recreation to the people, enhancing interaction among them and building a sense of community.

Houseplants which Clean Indoor Air

Indoor plants are able to purify indoor air to some degree through their normal photosynthesis processes. But some were more beneficial than others in removing harmful household toxins, even removing 90% of chemicals in the air in only 24 hours. Carcinogenic chemicals are used in the manufacture of synthetic substances and materials and are offgases from new materials. These types of houseplants may just decrease risk of cancers, asthma, allergies, auto-immune disorders and other diseases.

Aglaonema: Emits high oxygen content and purifies indoor air by removing chemicals like as formaldehyde, benzene and other toxins.

Aloe-vera: Not only can be used for burns on the skin, it is known to remove formaldehyde from the air.

Baby rubber plant: This houseplant cleans the air by emitting high oxygen content and purifies indoor air by removing chemicals like formaldehyde or other toxins.

Boston fern: The ferm acts as a natural air humidifier, removes formaldehyde and is a general air purifier.

Chrysanthemum: Very effective in removing benzene, trichloroethylene, formaldehyde and ammonia.

Dendrobium and Phaeleonopsis orchids: They help to rid off the air of xylene, a pollutant found in many glues and paints. They make wonderful housewarming gifts for anyone recently moved into or renovated a new space. Unlike some other plants, orchids also respire and give off oxygen at night – so they are great for the bedroom.

Dracaena: These plants absorb undesirables like formaldehyde, toluene and xylene.

English ivy: It is known for removing the benzene.

Ficus plant: Known to remove common airborne toxins and increases oxygen levels.

Gerbera: This plant is fantastic at removing benzene, a known carcenogenie. It also absorbs carbon dioxide and gives off oxygen overnight improve our sleep.

Golden pothos: Great for removing formaldehyde. It is also known for removing carbon monoxide and increasing general indoor air quality.

Ornamental annuals: e.g. *Calendula officinalis, Althaea rosea* and *Impatiens balsamina* had higher tolerance and accumulation ability to Cadmium (Cd) and Lead (Pb).

Palms: It is the best known for being natural air purifiers. Palms specifically target and remove formaldehyde, benzene and carbon monoxide.

Peace lilies: It removes harmful toxins like acetone, ammonia, benzene, ethyl acetate, formaldehyde, trichloroethylene and xylene.

Philodendron: Removes formaldehyde

Schefflera: Removes benzene from the air.

Snake plant: It absorbs toxins like nitrogen oxides and formaldehyde.

Spider plant: Known to remove formaldehyde also removes carbon monoxide and other toxins or impurities.

Selection of Plants for Green Buildings

Easy to Manage Plants: Aglaonema, Impatiens, Araucaria, Maranta, Aspidistra, Monstera, Chlorophytum, Philodondron, Dieffenbachia, Sansevieria, Dracaena, Scindapsus, Ficus, Tradescantia *etc.*

Plants for Dark Corner: Araucaria, Sansevieria, Aspidistra, Scindapsus, Maranta, Selaginella, Monstera, Tradescantia, Philodendron, Zebrina *etc*

Plants for North Window: Aglaonema, Ferns, Araucaria, Hedra, Monstera, Begonia, Philodendron, Sansevieria, Chlorophytum, Tradescantia, Dieffenbachia, Flowering plants *etc.*

Plants for South Window: Acalypha, Lantana, Amaryllis, Cacti, Pelargonium, Cycas, Poinsettia, Coleus, Euphorbia, Zephyranthus *etc.*

Plants for East and West Windows: Anthurium, Palms, Araucaria, Begonia, Hedera, Beloperone, Impatiens, Bromeliads, Cycas, Dieffenbachia, Tradescantia, Dracaena, Pandanus, Ferns *etc.*

Creeping, Trailing and Hanging Plants: Asparagus, Portulaca, Begonia, Zebrina, Vernonia, Antigonon, Bougainvillea, Passion flower, Ipomoea, Rangoon creeper, Wisteria, *Ficus repens*, Indian Ivy *etc.*

Energy Conservation

Courtyard Conserving Energy

Energy conservation is an essential component of energy policy in most of the developed countries of the world. Energy conservation is the most economical solution to energy shortages being faced by every country in the world. The potential of landscaping features to improve the energy efficiency of residential colonies, business houses and public buildings can be exploited. Proper use of trees, shrubs, climbers and man-made structures can modify the climate around homes and public buildings to reduce heat gains in summer and heat losses in winter. There is as much as a 30% reduction in cooling and heating costs through careful landscape planning.

In todays' scenario of rising energy costs, everyone is looking for ways to conserve energy in residential colonies, business houses and public buildings. Energy conservation is an essential component of energy policy in many developed countries. The energy can be conserved either through efficient energy use, in which use of energy is minimized while achieving a similar outcome or by reducing energy consumption. The energy conservation may result in increased financial capital, ecological balance, national security, personal security and human comfort. Energy conservation is the most economical solution to energy shortages and is a more environmentally benign alternative to increased energy production. Today, the ability of landscaping features to improve the energy efficiency of residential colonies, business houses and public buildings has gained increased attention in the scientific and professional communities. Energy conservation landscaping is a new and popular concept in the field of landscaping. In this type of landscaping, main emphasis is given to reduce energy demands in a building. A well planned landscape reduces energy demand in a building by lowering temperature during summers and preventing from chilling winds during winters. A well planned landscape can make a significant difference in the amount of energy needed to maintain a comfortable home. A well planned use of trees, shrubs, vines, ground covers and man-made structures can modify the climate around homes and public buildings to reduce heat gains in summer and heat losses in winter. It is not possible to fully control temperature, wind and other weather elements, but certain landscape practices can modify the climate in and around homes significantly. There is as much as 30% reduction in cooling and heating costs through careful landscape planning. It is necessary to analyze the area, before starting landscaping, for existing plants at the building site to identify those plants which can play a role in energy conserving landscape. Because the established plants will require less effort to maintain and will generally be of a larger size and better established than new plantings. The landscaping needs are determined by orientation of building, wind direction and water. Nearby buildings, walls, trees and bodies of water can produce significant effect on climate, which should also be considered while developing a plan. A proper study of existing propertys' features is important to plan an energy-efficient landscaping plan. The energy is consumed for cooling during summers and for heating during winters.

Protection Against Sun

In summer, the solar radiations are very intense and cause very high temperature inside buildings due to direct exposure of roof surfaces to sun. Trees serve as an important component in energy conserving landscape. The trees can reduce temperature inside the home by as much as 8 to 10 ^{0}F. The trees canopy provides shade to roof during hot summers, thus reducing cooling costs and increasing comfort. The type of trees to be planted and direction of planting are decided by its habit i.e. evergreen, deciduous. The evergreen trees should preferably be planted on the north and west sides to get protection from the setting summer sun and cold winter winds. The plant species namely, *Ficus religiosa, Azadirachta indica, Alstonia scholaris, Ficus infectoria* and *Morus alba* were more efficient to modify micro-environment for betterment. *Ficus infectoria* reduced the average summer noon temperature beneath the canopy by 3°C and by 5.6 °C on an extremely hot summer noon. *Morus alba* increased the winter morning temperature by 0.8°C. Overall, *Morus alba* was the best species for both summer and winter. The deciduous trees with high, spreading crowns (i.e., leaves and branches) give adequate shade in the summer when its leaves are in full bloom and warm the home in winter by letting low-angle winter sun filter through its bare branches. Such deciduous trees are the best suited for south and east sides of a house. The trees should have moderate growth rate because moderate growing trees often have deeper roots, stronger branches, more drought resistance and are less prone to breakage by windstorms or heavy snow loads. The size of the tree determines the distance from house wall, where it is to be planted. Trees, shrubs and groundcover plants can also shade the ground and pavement around the home. This reduces heat radiation and cools the air before it reaches houses' walls and windows. The groundcover reduces the temperature a few cms above turf than above asphalt or concrete surfaces. The summer temperatures can be reduced by using turf and groundcover plants between homes and paved areas. A trellis can be used for climbing vines to shade a patio area. In addition to shading roof areas, vines can protect walls from heat and cold. Some vines such as English Ivy have specialized roots which cling to wall surfaces. Deciduous vines like *Wisteria* allow the sun to penetrate during winter. If a quick effect is desired, annual vines such as morning glories and moon vines can be planted in early spring and will provide the needed shade for the hottest summer weather.

Protection Against Wind

During winters, flow of winds accelerates air exchange between house and surrounding environment. These winds increase the heating costs in a house. The use of tree as windbreaks shows energy savings up to 23 percent. Basically, a windbreak lowers the wind chill near home. Wind chill occurs when wind speed lowers outside temperature. A windbreak will reduce wind speed for a distance of as much as 30 times the windbreaks' height. Dense evergreen trees and home shrubs planted to the north and northwest of the area are the most common type of windbreak. Windbreaks obstruct and redirect the flow of wind. As wind strikes an obstruction, it can move over, around or through it.

Windbreaks

Impenetrable windbreaks create a strong vacuum on the protected or leeward side which reduces the protection. Windbreaks composed of living plants allow some of the wind to penetrate which makes them more effective. In addition to traditional windbreaks, shrubs can also be used closer to the home for protection against winds. This is more useful for small areas and subdivision lots where space does not allow use of conventional windbreaks. For this type of protection, dense evergreen plants are the most appropriate. They should be planted close enough to form a solid wall and minimum of about 1.25 to 1.75 m away from the house to create a dead air space. This relatively still or dead air has much less cooling power than moving air which can decrease the loss of heat through walls. Another way to moderate the temperature inside the home is to use rows of shrubs on northwest side to protect it from cold winter winds as well as to direct summer breezes around it.

Water Conservation

A good landscape plan should also conserve water. The local plant species should be used in landscape plan. The use of turf grass should be minimized as it requires a lot of supplemental watering. The soil should be improved for better water absorption and deeper roots. The efficient irrigation method should be adopted and mulches should be used. The water needed depends on plants' watering requirements and evapo-transpiration. The plants should be watered in early morning when evaporation rates are low.

Use of Solar Energy

Energy conservation can also be achieved by using solar energy during winter for solar water heater, photo-voltaic arrays and cooking food, instead of using conventional non-renewable petroleum resources. The plant material is designed in such a manner that there is no obstruction to winter sun on the roof.

Energy conservation can be an additional benefit of landscaping to conserve our natural resources as well as fulfilling, to some extent, shortage of energy.

Concept of energy conservation landscaping should be promoted to save non-renewable energy resources and to maintain ecological balance for our future generations.

Micro-Climate and Vegetation: Design for Comfort

To design an open space for maximum comfort throughout the year, the first pre-requisite is to have understanding of the relationship of human physiology to the environmental conditions like temperature, wind and humidity. A careful analysis of orientation, sun position, topography and site factors are essential. Finally one must have the knowledge and ability to make the best possible use of plant and innovative architectural materials and others devices to improve the comfort conditions within the open spaces.

A general understanding of the processes of heat loss from the body is necessary for improving the climatic conditions of specific outdoor spaces in question.

Micro-Climate Efficient Building

During warm weather in order to remain comfortable, the body must lose heat through sweating. While in cold weather, retention of the body heat is necessary to keep warm. Heat is lost from the body by radiation, conduction, convection and evaporation. At body surface temperatures above 85° F, lot of heat are carried away from the body by the evaporation of perspiration, but if the air temp-erature are higher than the skin temperature, the heat is transferred to the clothing and the exposed skin by convection. If the surfaces (i.e., walls and paving) are above skin temperature, heat will also be transferred to the clothing and expose the skin surface by radiation.

At air temperature below 78F, the temperature of the skin is below that of the body (interior), the secretion of perspiration is minimum and the loss of heat is largely due to radiation and convection. Heat is also carried away from the body by the moisture and evaporates into the exhaled air.

Comfort is not a constant factor among individuals but may vary widely between people of different age groups, health, activity *etc.*

Some feel comfortable at lower temperature, other at higher temperature. An increase in the velocity of air movement permits comfort at a higher temperature because the convective and evaporative heat loss from the body increases for a given temperature difference. A decrease in the velocity of air permits comforts at a lower temperature because the convective and evaporative heat loss decreases for a given temperature difference.

Before deciding upon a given open space, thorough investigations on the specific microclimate problems need to be considered.

Topography plays pivotal influence on the local climate, because at night the cold wind flows downhill. Sites exposed to the winter winds and valley bottom are usually less desirable because of extra coldness at night. Frost occurs at lower elevations. In summer, the bottom land may be the warmest because it may be cut off from the cooling breezes.

The best location is probably part way up a slope but the site at the top of hill may be desirable in warm areas. The lee side of the hill well below the crest is out of the area of strong winds and is usually much more comfortable.

South or South East exposure on slope generally makes the most desirable site. The South facing side offers the best possibility for taking advantage of the winter suns' natural heating qualities. In climate where cool living areas are desired in winter and in summer, it may be more desirable to have a north or north east exposure.

Soil conditions and natural plants cover have a great effect on the total climate of the open spaces. Soils of different texture, colour and composition have varying effects. For example, moist soils have lower temperature than do dry sandy or gravel soil.

Nearness to water bodies is also important. The winter climate near bodies of water is always mild than in other areas. They raise winter temperature, lower the summer temperature and always raise the humidity. The change is more pronounced near larger water body and less in near the small one. The lee ward side has a less extreme climate than the wind ward side, offshore breeze lower the temperature more than 10°C. Thus the topography, buildings and tree along the shore may be limiting factor depending upon their size and extent.

It is well known that wind flows nearly the same as water and therefore, it should be handled in much the same manner. The wind can be troublesome in open countryside. This can be partially overcome by proper design of shelter belt and use of hedges, screen and wind break. Hedges and other plant material may be more effective against the wind than the wall or fence since the foliage tends to diffuse the wind rather than push it over the top.

Plants in addition to shading effect, absorb and re-radiate far less heat than inorganic material such as wall, walks, terraces and buildings.

Summer temperature has been recorded 25-30°lower over the surface of grass than the black asphalt. Surfaces light in colour reflect heat but produce more glare. Large paved shall be designed with large opening left in for trees and other plants to produce shade, relieve the glare and reduce the amount of heat absorption.

Vegetation and Climate

Extent of the impact of vegetation upon the climate should not be underestimated. Vegetation may absorb over 90% of light falling upon it, reduce wind speeds in an area to less than 10% of that in the open, reduce day time temperature by as much as 15°F and on certain situations raise night time temperature. The principal uses of vegetation of special importance to the energy conscious designer are:

- Large and small trees and shrubs to check undesirable winds, coniferous plant should be used to control wind speed.
- Trees to channelise wind and to increase ventilation in specific areas.
- Vegetation, especially needle-leave trees to capture fog and this increased sunlight reaching the ground.
- Deciduous trees to screen sunlight during summer to reduce required heating load.

Temperature Control through Vegetative Devices

Solar Radiation Control

Sun is the ultimate source of all the earths' light and heat. It immense effects on climate factors, trees and forest are among the best solar radiation control devices. The major functional uses of forest are both in tropical climates where solar radiation is oppressive requiring year round control and in temperate climates where solar radiation the most oppressive in summer. Trees and forest play a critical role in controlling excess or unwanted solar radiation. They provide this control in at least four ways, these are:

- Absorption
- Reflection
- Radiation
- Transmission

In the forest, the tree and the leaf absorb all solar radiation throughout the day. The tree in the forest receives different amount of radiation and is continually changing area of its foliage presented to sun. The leaf on the tree, continually moving and receiving radiation from sun where the position is constantly changing above shifting cloud cover.

In the absence of wind and with restricted convection, leaf temperature may be as much as 13° F higher than air temperature.

Street level microclimate can be separated into three broad classes. The first included areas with extensive evaporating or transpiring surfaces –parks and gardens. The second includes wide tree less streets. Square and parking lot open to the sky are dry. The third includes narrow streets and courtyards surrounded by relatively tall buildings showed diurnal variation in temperature. This can be amply demonstrated by following example, i.e., the broad streets and squares which do not have trees are very hot at noon, but cool off more noticeably in the evening. Avenues and squares containing trees are cooler and have a narrower daily range in variation. The narrow streets, however, are distinguished by much greater coolness during noon and after noon hours, amounting to from 5 to 6°C lower than the temperature of the surrounding area. Even the orientation plays an important role in the narrow street. North-South street will not be shaded from intense noon-time solar radiation. They will have higher temperature than narrow east-west street.

The importance of evapo-transpiration is quite obvious in the parks which stands out as 15°F cooler than the surrounding city.

Wind Control

One of the ways in which vegetation affects the microclimate, hence the comfort of man, is through its control of flow of wind. C.A. Federe in his paper 'Effects of Trees in Modifying Urban Microclimate' has the following to say concerning the relationship of wind alteration and vegetation to human comfort. The most significant effect of wind is due to increase of convective and evaporative cooling. Trees reduce wind speed by increasing resistance to wind flow within the crown of a single tree. Under the forest canopy wind is light and almost unrelated to the external wind. An isolated tree or stand of tree acts almost like a solid barrier to the wind forcing the wind over or around. The wind speed can be increased at the edges of a tree. Deciduous trees have an adverse seasonal effects on wind partially compensated for their beneficial shading characteristics. They decrease wind speed mostly in summer and the least in winter.

Vegetation, trees, shrubs and wind break are extensively studied to ascertain their effectiveness in controlling velocity, movement and direction of wind near the earth surface.

Effectiveness of Various Forms of Vegetation

Effectiveness of various forms of vegetation is well documented. Extensive literature to delineate the effectiveness of trees and forests in improving the quality of environment through their ability to control winds is available. A design wind break reduces a 30 miles per hour (48 kmph) wind to 10 miles per hour (16 to 8 kmph). The spacing and size of trees and the relationship to wind velocity are quoted 'Narrow spacing is naturally intended to produce a dense and relatively high screen when the conditions of soil and climate are favourable. Age of the tree has at least a two fold influence, it affects the height and width of screen, secondly, the influence on density is the number of rows, and even so, the density is not strictly proportionate to rows which are planted.

Deflection of wind over trees or shrubs is another method of wind control. Plants of varying height, width, species and composition planted either individually or in rows having varying degree of effects on wind deflection, conifers, ever green trees are generally the most effective year round for wind control. Deciduous shrubs and trees are the most effective in summer. Wind velocity is cut from 15-25% of the open field velocity directly on leeward of a dense screen planting.

Shelter Belts

Studies show that shelterbelts and wind breaks are the most effective when placed perpendicular to the prevailing wind velocity which may be reduced by 50% for a distance of 10-20 times the tree height, down ward of a shelter belt, and the degree of protection and wind reduction depend upon the height, width and penetrability of plants used.

Nearly a moderately dense shelter belt, at the end of wind break wind speed is increased more than 10% of the open field velocity prior to its interception. The leeward sheltered zone is not confined within lines drawn perpendicular to the end of the barrier, but is broader than length of barrier.

Wind speed is also affected within or on the wind ward side of wind break. The wind speed is reduced for a distance of 100 yards on the wind ward or front side of a 30 feet high shelter belt, and is reduced for a distance of 300 yards downwind or behind the shelter belt.

Accordingly to Gorshenin (1934) the influence trees have on reducing wind velocity extends from 30-40 times the height of trees leeward of the wind break, and the sharp reduction in the wind velocity extends from 10-15 times height of the trees. Commenting on the same subject, Logginov (1941) stated that maximum shelter belt from the wind is obtained at three to five times the height on the leeward side of trees. Accordingly to Lizukee (1950) wind velocities recorded behind an artificial wind break in Japan at distance of 10, 20 and 30 times the height of the wind break were respectively 66, 44, 69, 33 and 44% of open wind speed.

The influence of forests on vegetation and climate was noted from the earliest time. Neither, the exact records of the nature and extent of changes in climate are available nor would it be practicable to determine what is attributable to deforestation. It is known, however, that vegetation has an ameliorating effect of the local climate.

General Effect of Vegetation on Microclimate

The most noticeable effect of a forest on microclimate occurs within the forest itself, in an adjacent open land, the effect is less marked. In contrast, the important effect of narrow shelter belts on microclimate is on the adjacent open areas which they protect.

Every vegetative type develops a distinctive microclimate and there can be a wide variation with different kinds of forest. There is a marked contrast between the effect of dense forest and open forest of broad-leaves species (i.e, *Eucalyptus*). With evergreen species, the effect on microclimate tends to be fairly steady throughout the year, but with deciduous species, the microclimate is modified less during winter.

The principle climatic factors influenced by the tree growth are wind, temperature, relative humidity and evaporation.

Condition within Forest Vegetation

The most noticeable effect of forest is on movement of wind which may be reduced within the forest by 50-70% or even more, as compared with open spaces. The reduction is the greatest near the ground and noticeable at all velocities. In case of velocities below 5 miles/hour in the open, condition of calm may exist within the forest. The pattern of turbulence is affected by presence of vegetation, whilst local wind direction may be changed, particularly in hill country.

Air Temperature

Temperature gradients occur within vegetation and the greatest extreme usually being experienced at top of crowns. The mean annual temperature at 412 feet above the ground tends to be 1 to 3° F lower than in the open. This difference is the most marked in summer but may be absent in winter. The amptitude of the air temperature range may be as much as 8°F lower in the forest than in the open.

Soil Temperature

The mean annual temperature at soil depth at 6 -18" is slightly lower for soils within the forest than in the open. The difference is the most marked in summer and may be negligible in winter. Variation of surface soil temperature is considerably less in forests than in open.

Frost

The mean air temperature is lower under forest cover than in open. The diurnal variation is smaller because less direct heat is received from the sun at ground level and less heat is lost by radiation at night. This reduction of heat loss by radiation results in formation of fewer frosts on ground vegetation under the forest cover than in the open. Heat loss by radiation from the ground is reduced not only by the protective cover of frost but also by the higher humidity of the forest microclimate.

Precipitation

Some investigators have come to conclusion that forest increases rainfall and a few that they have a slight reducing effects, the bulk of evidence is that forests do not result in significant change in total precipitation.

Only a proportion of the total rainfall received by a forest reaches the ground. The actual amount varies considerably with the type of forest and nature of precipitation. The percentage of rain intercepted by dense coniferous forest is much greater than that held by any other type of forests and is inversely related to both intensity and direction of rainfall.

Relative Humidity

The relative humidity is 3-10% higher within the forest than in the open, and that the amptitude of the range is considerably less within the forest.

Evaporation from a Free Water Surface

Evaporation is a product of temperature, wind and humidity and forests have marked effect on wind. It follows that there is a large reduction in evaporation within a forest as compared to within that occurring in the open. This reduction ranges 25-75%.

Conditions in Open Space Protected by Shelterbelt

The ameliorating effect of shelter belts upon the microclimate of protected areas was investigated both in the field and by means of models in wind tunnel experiments. Shelter belts bring about substantial change in microclimate, particularly by reducing wind movement. As a result of these changes, it was found that in cool climate, grass production may be increased by as much as 10% and in potato crops by up to 20%.

Techniques for Establishment of Shelter Belts

The principal method of reducing surface velocity of wind, upon which depends the abrasive and transportation capacity of wind, is by vegetation measures (Rama Rao 1974).

Vegetation methods are the most effective in the form of wind breaks and shelterbelts. In aerodynamic terms, these provide protection as follows (Konda Reddy 1979):

i) Sheltered zone on the leeward side extends to approximately 15-30 times the height of belt.

ii) A dense belt provides greater shelter immediately to leeward but the sheltered area is not as extensive as when a more permeable zone of vegetation is provided.

iii) Porosity is important in the effectiveness of shelterbelt and proper selection of tree species is necessary. Porosity near ground level is desirable.

iv) Effectiveness of shelter planting depends more on height and permeability on width. The width influences the general microclimate but above a certain minimum width, it does not effect greater reduction in wind velocity.

v) A belt which rises and falls abruptly on windward and leeward sides is more effective. Smaller trees and shrubs should occupy inter-spaces between tall trees. Some authorities maintain that a triangular section of shelter belt planting is more effective.

vi) Depth of shelterbelt should be approximately ten times its height. This is, however, only a thumb rule. Much lesser widths of 20 –30m are also useful in particular situations; 15m should be considered as minimum width.

vii) Apart from factors like climate, soil and fast rate of growth, one of the more significant considerations in choosing species for shelter planting is the possibility of a particular species serving the dual role of wood-production (for fuel, fodder) and shelter.

viii) Spacing of plants in windbreaks and shelterbelts:

Windbreaks: Usually consist of a single or double rows of trees planted at 0.7 to 1.5 m according to species. Normally, one year old trees are used (Leloup 1955). If tractor is used for cultivation then 3 to 4 m should be distance between rows. Otherwise the distance between rows can be 2 m –2.5 m. Species suggested for various Indian situations: *Casuarina equisetifolia and Tamarix articulate.*

Eucalyptus is not recommended because of the strong competition exerted by the spreading root system. Usually evergreens are preferred, but deciduous trees like poplar and *Dalbergia sissoo* can also be used. Since the roots of trees extend for some distance beyond the rows in which they are planted, fruit trees and garden crops should not be grown closer than 4 to 5 metres to the windbreak. Trees with long roots are not suited for windbreaks as their roots compete with the crops they are supposed to protect.

Shelterbelts Design: The most common layout where shelter planting is part of an extensive planned programme, is that of tree belts arranged in a chessboard pattern, each field being protected from every side. This pattern gives full protection to all the fields, provided that right distance between the fields is chosen. Efficient protection is achieved, if belts are separated by a distance of not more than 20 times the height of trees.

A considerable mixture of species is recommended to compensate different rates of growth and also to achieve in crowns. Under arid conditions choice of species is limited. Within shelter belts, close spacing of trees is the general practice. Recommended spacing for shrub is 1m and for trees 1.5 –2m. Spacing between rows should be 2.5- 3m for mechanized cultivation. Five rows of trees and shrubs are considered necessary for protection. A soil improving crop of legumes may be grown between rows during first years, to foster growth of belt.

ix) Management: Shelterbelts should be regarded as living groups of trees to be managed in perpetuity. Cutting is done individually by single tree selection method of selection.

Continuous cultivation may be required in areas with scanty rainfall.

If individual trees do not survive, then they should be replaced immediately so that no gaps remain in vegetation belt. The shelterbelts should be protected from cattle, either by fencing or other means, especially in early stages.

The following examples, from India and abroad illustrate experiences of use, function and techniques in planting of shelter vegetation:

i) Without giving the tag 'Shelterbelts', the forest Department in Andhra Pradesh is planting Casuaria all along the coast from Vishkhapatnam to Nellore since 1920. Eucalyptus and cashew plantation on leeward side of the casuaria belt, provided reinforcement necessary and met demand of since 1958, fuel and small timber.

ii) Shelter planting on a massive scale was done on Eastern bank of Hazari between 1946-53 to prevent sand drift from the wide sand bed of the river and protect fields and villages from being overrun by moving sand and dunes. Prosopis was used as the land was arid, with clayey soil. These shelterbelts are worked on a simple coppice system to provide feed for cattle and small timber to local population.

iii) In arid districts of Andhra Pradesh, shelterbelt planting is being done along river Penna in Anantapur and Cuddapah and along river Cheyyer. Species chosen are: *Dalbergia sissoo, Eucalyptus, Terminalia, Tecoma* and results are satisfactory.

iv) Mangrove forests: Rich and extensive mangrove forests occurred originally in the estuary of rivers Krishna and Godavari. Mangrove forests can persist in areas where salt water of sea and fresh river water alternate. As the rivers deposit the silt when they are in spates, the salt water cannot reach the vegetation and the raised mud flats gradually get desalinated by river water and rain. Mangrove forest gives way to general vegetation, pioneered by

Terminalia species

Thespesia and *Prospis*. Farmers in the delta bund raised the mud flats and flood then with river water; gypsum is added to assist in the desalination process and practice agriculture on land so available. Dry evergreen forest and estuarine forest are gradually dwindling, and the natural shelterbelt on East Coast has almost disappeared.

Shelter Planting: Storms and tempests hitting the coast lift sands from ridges and dunes and blow them inland resulting in sand drift which covers several area planted with casuarinas. Shelter planting suffered comparatively less.

References

Bhattacharjee, S.K. (2006). In: *Advance in Ornamental Horticulture*, Vol. 5. Agarwal Printing Press, Jaipur, India.

Carpenter, P.L., Walker, D.T. and Lanphear F.O. (1975). Plants in Landscape, W.H. Freeman and Co., San Francisco, USA.

Fyson, P.K. (1974). Flora of the Nilgiri and Pulney Hill-Tops. Bishan Singh and Mahinder Pal Singh, Dehradun.

Gupta, R.K. (1967). Seasonal flowers of the Indian summer resorts-Missouri hills. Nav Yug Traders, New Delhi.

Jana, V.K. and Mukherjee, A. (1979). Shillong- a Treasure house of bulbous plants. *Indian Horticulture,* 24: 17-22.

Jhon, A.Q., Rather, Z.A., Paul, T.M. and Neelofar. (2007). Evaluation of ornamental bulbous plants for landscaping in temperate regions. *J.Ornamental Hort*., 10 (1): 1-8.

Lele, V.C. and Misra, D.P. (1961). Some unreported wild naturalized flowering plants of the Shimla Hills. *Himachal Horticulture*, 2:5-8.

Miller, G.T. (1992). Environmental Science, Fourth edition, 1992.

Ohashi, H. (1975). The flora of Eastern Himalaya. University of Tokyo Press, Japan.

Polunin, O. and Stainton, A. (1984). Flowers of the Himalaya. Oxford University Press, New Delhi. 580 pp.

Prater, S.H. (1980). The book of Indian animals. Bombay Natural History Society, Oxford University Press. 324 pp.

http//www.igbc.in

http//www.greenlandscape.in

Ornamental Horticulture, pp 99-130
© Editor, S.S. Sindhu, 2016
New India Publishing Agency, New Delhi - 110034, India

5

Hanging Gardens- A Softer Touch to Urban Landscapes for Economic Development

Sharat Kumar

Cities or urban conglomerations are fast developing and where the rate of land availability has *infinitesimally* small; vertical expansion and taller buildings have emerged and are defining the city skyline. Though this may speak well of the economic development, but the increase in hard paved areas and roads may spell trouble for the environment as urban heat island (UHI) effect. The causes of Urban Heat Island are primarily due to the absorption of solar radiation (Short wave) by building/urban materials that is subsequently re-radiated (long wave) to the surroundings. In addition, heat generated from the use of air-conditioning coupled with the greenhouse effect of pollutants also contribute to the increase in temperature. Such increase in temperature with the presence of air pollutants can result in the accumulation of smog, damages the natural environment and jeopardizes human health. It also costs consumers more money because it takes more energy to cool buildings. Who can forget the regular news of Beijing repeatedly topping the list of most polluted cities and fresh air is available in cans!

Fresh air cans are in no means a firm solution. Plants are atmosphere cleansers, can serve to filter carbon dioxide and other toxins in environment, cool the air by transpiration and also provide cooling shade. Research work shows the benefits of using light colour roofs and roof gardens to minimize the heat gain inside the buildings. This would in turn reduce the energy consumption

National Orchid Garden (NOG), Singapore Botanical Gardens, National Parks Board, (N Parks), 1 Cluny Road, Singapore

of air conditioning in buildings. Rooftops, balconies and terraces have often been referred to as the "last urban frontier"of cities, in the sense that they represent the last significant remaining space to be exploited to create a green oasis and an area of relative peace and privacy away from the urban bustle. Given the social, economic and environmental contributions of urban greenery to the quality of life in cities and the realization that cities more often than not lack sufficient greenery, greening of rooftops, balconies and terraces is therefore a logical, effective and high impact way to re-inject the much needed greenery in urban settings thus reinventing the concept of hanging gardens.

History- Beginning of Green Architecture

People have strong influence on nature. In every moment, from every side we are strongly connected with our environment. We take part in it. With our architecture we try to copy nature. We are building awesome buildings for instance pyramids and towers. With them we imitate mountains and create new landscapes. We attempt to make buildings adapted to nature. Looking in history of architecture we can find one amazing construction. People still feel delighted with edifice –"Hanging Gardens of Babylon"; is one of the old seven wonders of the World. Hanging garden is not a modern phenomenon. Archaeologists have, for instance, suggested that trees were already planted in ziggurats or pyramid-shaped stepped towers in ancient Mesopotamia more than four thousand years ago, apparently to provide relief against the scorching sun of the Babylonian plains. The famous Hanging Gardens of Babylon, built around 600 B.C, were said to be a series of stone terraces on which a multitude of plants, including big trees, were planted to create a magnificent garden setting. Sod or grass roofs were also used for centuries in Scandinavian countries before the advent of modern-day heating to insulate buildings against heat loss during long harsh winters.

Starting in 1960s, advances in horticulture knowledge, building sciences and in new building materials, have enabled the development of contemporary green technology to the extent that the placement of vegetation layer in places defying gravity are becoming widespread across the world. Perhaps hanging gardens have 2500-year old history but reached maturity with contributions from contemporary multi-discipline subjects. Today, hanging gardens have spread around the world, especially in developed countries and regions in Europe, North America and Asia.

Historical Evidences Around the World

The earlier recorded and the most famous example quoted in many literature of a successful attempt to incorporate natural functions in a dense urban environment was Assyrian king Nebuchadnezzars' Hanging Gardens of Babylon built in 6th Century BC. However, the 'true' green roof concept originated from Iceland, where sod roofs were used for hundreds of years. Sod roofs soon became popular in Scandinavia and throughout history all over Europe, as well as, by ancient tribes in America. They took advantage of limited resources and placed layers of turf on their roofs for insulation and waterproofing. In France, Le Corbusier in his famous 'Five points of New Architecture,' written in 1923, roof gardens came second. 'Roof gardens' he said 'provide the city with the means to win back the whole of its developed area.'

Some impressive examples of 20th Century highlights are:

The Casino Patio in Bern, is a large roof greening project, created as a roof garden as early as 1936. It consists of a thick layer of nutrient-rich earth that makes it almost unnecessary to fertilize and irrigate the garden. Still standing, today the garden of over 78 years has 20 huge chestnut trees with trunk diameters of over 170 cm.

The roof garden of the former Derry and Toms department store in Kensington High Street, London is approximately 6,000 m^2, also planted 19. Even today this is an impressive garden landscape, located around 35 m above a shopping street in the middle of London. This green roof concept "sunk into obscurity" for many centuries was revived in European cities since 1960s' green movement. Therewith, green roof industry takes off in large strides. Rooftop greening is the most established in Switzerland and Germany. By 1996 over 3.2 million square feet of green roofs were constructed in Germany alone. Many Swiss cities now require constructions relocate to the rooftop the area of greenspace covered by the building and one in 10 flat roofs in German cities is now vegetated (PortPhilipEcocentre).

In Europe, FLL published 'Principles for Roof Greening,' which provided guidelines for roof greening in 1982. In 1984 it was supplemented by a new examination procedure for checking root resistance of root protection membranes. Energy saving and environmentally friendly construction were also examined with regard to roof greening.

The City of Esslingen in Germany has a by-law, which requires that flat and sloping roofs (up to 15 degrees) must be vegetated. Similarly, in Mannheim, West Germany, declining air quality prompted the City Council to impose a by-law in 1988, which required all central business district buildings to be vegetated.

What has caused roof greening to move from the fringe into the present over the past a few decades?

Increasingly, urban problems such as increased storm water runoff, the urban heat island effect, deterioration of air and water quality, and loss of habitat and biodiversity are becoming disturbing. In the 1970's and 80's, rooftop greening began to gain prominence as a viable strategy for improving the urban environment. Green roofs are well received in developed countries for they address the urgent ecologically demanding issues of air and water quality and storm water management. Upon 'rediscovered,' advancement in technology has made improvements in materials and techniques, updated variation of green roof system thus become enormously popular again in developed countries. Motivated by same objectives like improving access to green open space, improving viewscapes, controlling storm water and moderating climate, Germany and Switzerland lead the field in promoting and encouraging rooftop greening. 'Green roof' technology is practiced for at least 30 years in Europe, where it is embraced as a way to beautify a building while reducing harmful storm water runoff.

Hanging Garden as a Feature

Definition: Sky gardens or hanging gardens refer to the green spaces on the intermediate floors of high-rise buildings (podium gardens) or on rooftop (roof gardens). Osmundson (1999) defined sky gardens as 'any planted open spaces that may be below or above the ground level, intended to provide human enjoyment or environmental enhancement that is separated from the earth by a building or other structure'. Sky garden is a technical term in the high rise world. It maximizes the scope of greenery on buildings and can be planted on raised planting beds, walls, railings, window boxes, aisles, balcony, rooftops and underground or above-ground car parks.

Hanging gardens provide invaluable ecological, economic and social benefits. They can increase the life expectancy of roofs and podiums by protecting construction materials from climatic extremes, ultraviolet (UV) exposure and mechanical damage. They alleviate congestion and provide open spaces required by local municipal land-use ordinances in less costly ways, and they lower energy costs and urban heat island effects. They extend the longevity of the residents by bestowing leisure and open spaces, visual aesthetic value, health and therapeutic value and environmental values such as absorbing air and sound pollutants. Furthermore, they provide peaceful places and a natural ambience for the residents of the urban jungle, preserve historically or culturally significant places and soften hard portions of planting structures. Hanging gardens increase access to private outdoor green spaces at home or at work, which increase housing prices for owners and tenants and create job opportunities in the fields of research, design, gardening, landscaping, health and food production. Near to the greened parts of hanging gardens are always amenity areas for social activities or children pleasures. These areas can provide social services. Variety of soil depth, vegetation type and character collectively determine the biodiversity and wildlife in the hanging garden.

In summary, hanging gardens improve living and working environments and solve the serious problems of limited ground level green spaces in the built-up areas of compact cities. Hanging garden has a major advantage over gardens that are at ground level; they do not need a lot of space, but provide many benefits analogous to the green spaces on the ground. Thus, hanging gardens are important in realizing sustainable cities by combining energy efficiency and urban ecology through plants, water and landscaping. As previously mentioned based upon the vegetation requirement and growing habit led to the different typologies as follows.

Typologies

A typical sustainable development involves the creation of vegetated space integrated structurally on top of built or man -made structures. The vegetated space may be below, at, or above ground, located on a podium deck, a hanging garden on an intermediate floor level, or at the very top level of the building, but in all cases the plants are not planted on the ground. Different typologies of hanging gardens are garden at roof, terraces, balconies, vertical greenery, green facades, planters on wall and greenery on other parts of buildings.

1. Rooftop gardens

The terms ģreen roofànd ŕooftop gardenhave become a byword used to describe various architectural and landscape treatments for roofs. These range from freestanding planters to utilitarian structures such as paved recreational courts and swimming pools ornamented by vegetation atop a roof. Green roof development involves the creation of contained green space on top of a human-made structure. There are two types of green roof systems, extensive and intensive, and their difference lies (Table 1) in the desired function of the roof space.

Extensive Green Roofs

Extensive green roofs are not designed for public use and are mainly developed for aesthetic and ecological benefits. They are distinguished by being low cost, lightweight (50-150 kg/m^2) and with thin mineral substrates of up to 100mm. Minimal maintenance is required and inspection is performed one to two times per year. Plants selected tend to be of low maintenance and self-generative.

Intensive Green Roofs

Intensive green roofs are usually called 'roof gardens.' They are a greater load of more than 150 kg/ m^2 and have more than 100 mm of substrate with a higher amount of organic material than extensive systems. They are developed to be accessible for people and are used as parks or building amenities (Scholz-Barth 2001). Hence, they usually occupy areas of paving and seating. The added weight, higher capital cost, intensive planting and higher maintenance requirements characterize intensive green roofs. The plant selection, ranging from ornamental lawn to shrubs and trees, consequently affects the weight, build-up heights and cost of the roof garden. Further, regular garden maintenance such as mowing, fertilizing, watering and weeding are required for intensive green roof landscapes.

Table 1: Differences between extensive and intensive green roof

Details	Extensive green roof	Intensive green roof
Maintenance	Low	High
Irrigation	little or no irrigation	Often
Plant density	Low	High
Plant group	Grasses, Ground covers, Moss	Grasses, Perennials, Shrubs, Trees
System build up height	Up to 150 mm	Up to 1500 mm
System build up weight	50 to 150 kg per m^2	290 to 970 kg per m^2
Attractive	Less	Very attractive
Costs	Low	High

2. Vertical greenery/ green walls

Vertical greening or facade greening is a technique of covering walls of buildings. This can either be self-regeneration of the plants or they can be supported by a trellis or steel cables. Planting on balconies and planter boxes can also be considered as one of the techniques of vertical greening.

3. Green facades

Green facades refer to a guidance structure that provides support for vines or climbing plants to grow up the side of a wall. Most systems use wire fencing, modular trellis systems, or stainless steel cables as support systems. Wall mounted facades are either flush or 7.5-45 cm away for the wall. Green facades are attached with brackets or mounting clips and a waterproof membrane should be installed between the wall and the facade in most arrangements. When using cables or wires, turnbuckles and anchors need to be mounted and attached to each wire for possible adjustments needed as the plants grow and expand.

Since green facades contain continuously growing plants, the support systems are essential for directional guidance and load bearing capabilities. There are three main types of support; a trellis of vertical and horizontal elements, just vertical supports, or just horizontal supports. Each system can be made of different materials like wire or wood. Different types of supports are used for different kinds of plants, and there are four main types of climbing plants; self-clinging climbers, vines-climbing plants, leaf-twining climbers and tendril climbers, and ramblers and scrambling plants (Table 2).

When climbing plants are utilized for high vertical landscaping in green facades, intense weather is a common dilemma; therefore, resilient types of plants should be chosen. Depending on the area, climbers with a high tolerance for frost, wind or heat should be selected. In addition to planning for the right type of plant, considerations must also be made for a proper schedule. Some climbers may take up to 3-5 years to completely cover the entire area, depending on the height.

4. Gardens on terraces

Terraces are literally semi open voids carved out of buildings at intermediate floors usually with landscaped areas incorporated to create green spaces within the buildings for easy access by the occupants.

5. Gardens on bridges

In an urban setting, a Skyway or sky bridge is a type of pedway consisting of an enclosed or covered bridge between two buildings. They are built and owned separately by the city government, connecting between railway stations or other transport with their own footbridges, and run many kilometres are the potential usable space to establish greenery among multi-storey buildings above the ground-level.

6. Hanging baskets

Hanging baskets play a major role with most bungalows in metro cities going the apartment way and garden space becoming a luxury. Hanging baskets can bring in that little bit of green, that some of us hanker for. They do not take too much space, they are easy to maintain and can fill them with whatever plants your heart desires. Gardening is difficult when you live in an apartment but the concept of hanging baskets makes things easier. All one needs is a lightweight flower pot, some cord and a flowering plant. It does not take up too much space and can also add to the decor of a balcony.

Various types of basket either containing flowering or foliage plants can be displayed with great advantage at suitable places. One can also display with advantage the collection of bonsai and a few alpine plants. One or a few vertical gardens can also be displayed in a planned manner in the hanging garden. It is possible to grow a wide range of plant materials in a hanging garden excluding, of course, the large shrubs and big trees.

7. Greenery on other parts of buildings

This includes greening of other parts of buildings, such as the façade with planting on balconies, planter boxes, creepers on walls and mid-level planting on intermediate levels and terraces.

Design is a subtle business. Good design should combine strength of purpose, compatibility with the surroundings and, above all, simplicity. Over complication is the antithesis of good design and nowhere is this more apparent than in a garden. Hanging garden needs careful thought, especially planning and designing are imperative. One advantage of hanging gardens is that they invariably have wide views - it's a bit like gardening on top of a hill or nature slopes - but it will also have similar problems with wind at roof top height and temperatures more extreme than at ground level too. It is a great opportunity to make a private green oasis that looks intriguing to all who get a glimpse of it - but only a few will enjoy a closer look. In hanging gardens especially one should never copy exactly a scheme intended for a different garden; it may look fine where it is, but will, without doubt, prove unworkable for you.

Structural Considerations

The structures are visible from the ground or those which alter the roofline of a building. The structural strength of the roof must be gauged. It has to be capable of supporting the additional weight of the garden components, and planting media are very heavy even without any planting. The bulk of extra weight in hanging gardens is usually concentrated around the sides, where it is

Table 2: Climbing plants for Green facades

Name	Description	Examples	Considerations
Self-clinging Climbers	They use lateral shoots with glandular discs that adhere to any surface and have strong roots that grow over long distances without any auxiliary support.	Boston Ivy (*Parthenocissus tricuspidata)* Climbing Hydrangeas *(Hydrangea petiolaris)*	May cause harm to buildings by damaging sidings and mortar with their invasive adhesive pads and roots, so they should only be used with a support system and proper maintenance to avoid structural damage
Vine Climbers	A strong affinity to grow upwards by winding around the support with circular movements of their stem tips.	Wisteria (*Wisteria sinensis)* Honeysuckles (*Lonicera* sp.) Hops (*Humulus lupulu)* Morning glory (Ipomoea) Ground Ivy *(Glechoma hederacea*)	Vertical supports might be needed in most cases, but spacing between vertical supports varies depending on the species.
Leaf-twining and Tendril Climbers	Use their leaves to hold on. Primarily deciduous, but they hold on to the system with their brittle stems in the winter	Clematis and Nasturtiums (*Tropaeolum* sp.) Passion flowers *(Passiflora* sp.) Grape vines *(Vitis* sp.) Ampelopsis (*Ampelopsis brevipedunculata)*	Vertical and horizontal supports recommended for leaf-twiners and tendrils.
Ramblers and Scrambling Plants	Holds on to the supports with thorns, bristles and prickles. Can be used over stretches of horizontal growth with a small vertical range.	*Rosa* sp.	Should not be used where people can easily come into contact with them because they are so jagged. Not ideal for vertical growth.

supported by load-bearing walls. The load bearing capacity of a roof or balcony will determine where beds, containers and other heavy features can be safely placed. The climate pattern will dictate what provision is needed for shelter from winds or hot sun.

Style Considerations

Good architects and designers create spaces which respect and are compatible with their surroundings. A problem often arises with the popularity of a style, whether it is cottage, formal, informal, Japanese or something else. While any of these may be appropriate, and indeed superb in the right setting, they can simply be a fashionable clichéin others. Many people see the spaces inside and outside as two separate entities, when they should be linked as strongly as possible and carry on to use their immediate surroundings and more distant views to their advantage. In this way we can achieve a progression from inside to outside and beyond, drawing together materials, colour schemes, furnishings and plants into an all- embracing whole that will help to create a feeling of space at any point in the overall composition. The linkage between buildings and the immediate garden area outside should be as seamless as possible.

Colour, often under valued and under used, is another theme that can draw built structures and garden together. If you have decorative scheme inside and an adjoin wall outside, why not paint the outside wall in the same shade? Awnings offer an ideal opportunity to pick up a fabric pattern from an interior scheme. Incidental features such as pots, containers and furniture can be chosen in much the same way, with a coherent and simple theme in mind.

Gardens come in all shapes and sizes. The patterns in which the various elements of the different areas are laid out have an enormous impact on how the space works in both visual and practical terms. An obvious parallel can be drawn between how decorative effects are used indoors and how they are used in roof gardens, terraces and balconies. For example, indoors, with a large, bold pattern tends to emphasize a surface, drawing the eye towards it and making the room feel smaller. On the other hand, a delicate, small scale pattern, or a sponged or rag-rolled finish, tends to create a feeling of greater space. The elements of garden can be selected and positioned to manipulate the space visually.

Plants - Planning and Selection Criteria

Plants are an integral part of garden design. They define the visual impact of each part of the garden; their colour, shape, size and texture can be used to bring an area to attention and attraction or to make it fade into the background. Which plant species you choose will, ofcourse depend on the style and size of your garden, as well as the climate. Although the choice of plants is largely dependent on the microclimate of the hanging garden, the types of plants that can be grown increase with the amount of maintenance funds available. For highly intensive gardens, the choice of plants is unlimited. Landscaping design considerations also influence the selection of plants. For example, if shade is highly desired in the garden landscape, trees rather than palms become the plant of choice, but this would imply a higher maintenance cost.

a) Criteria for plant selection

1. Native plants - creating a more resilient landscape

Having evolved under local environmental and biological conditions, native plants are more likely to attract native animals, such as birds, butterflies, and other insects. Specifically, native plants will provide shelter and food, such as nectar, pollen, seeds and leaves for herbivores, thereby creating an ecosystem that is more resilient towards pest and disease outbreaks. Native plants are also likely to be more adapted to the local environmental conditions. Such characteristics of native plants could lead to beneficial impacts on the longevity and maintenance requirements of green landscapes. Through the introduction of native plants into the urban landscapes, such as rooftop gardens, home gardens on terraces and balconies, the populations and genetic diversity of our native plants will be maintained. This is also important in safeguarding the genetic resources of our plants, from a biodiversity conservation perspective.

2. Drought tolerance

Under local climatic conditions, gardens on roof, balconies and terraces are generally exposed to constant high light and temperature, which lead to high evapotranspiration rates and the demand for water by plants. In elevated rooftop gardens, high evapotranspiration rates can be exacerbated by strong winds or proximity to hot and dry air from air conditioning or ventilation vents. The selection of drought resistant plants, woody shrubs and succulents rather than herbaceous plants cuts down on irrigation requirement. Given that water is the key limiting factor of growth, a useful starting point for plant selection is knowledge of the physiological behaviour of plants towards water use. Plants which close the stomata during daytime (Crassulacean Acid Metabolism -

CAM), prevent evaporative loss of water from the leaves during day and keep stomata open at night to absorb carbon dioxide. CAM plants should thus offer many potential candidates for hanging garden plants. Fortunately, CAM plants are very widespread and are represented in thirty-three families encompassing 18,900 plant species. The huge diversity of CAM plants can be found in families such as Aizoceae, Agavaceae, Asclepiadaceae, Bromeliaceae, Cactaceae, Commelinaceae, Crassulaceae, Dracaenaceae, Lamiaceae, Orchidaceae, Piperaceae, Portulaceae, etc. For roofs which are less exposed, many tropical epiphytes and sclerophyllous plants in the families Orchidaceae, Bromeliaceae, Asclepiadeceae and Piperaceae will offer numerous possibilities. Undoubtedly, continuous research and exploration of plants will also lead to the discovery of non-CAM plants suitable for roofs. Indeed, the tropics and sub tropics are blessed with such great plant diversity that there are exciting opportunities to create hanging gardens featuring tropical look which distinctively different from gardens commonly seen in temperate climates.

3. Plants for all year round

Considering potential of a garden in winter is crucial since it can be several degrees warmer on a roof than at a ground level. Meanwhile several higher degrees of temperature in summer are taken while plant groupings. Groupings largely depend on individual beauty of a plant, broader appeal of form, shape and texture are valid choices. Flowering is a seasonal bonus rather than the prime object of a species; as blooms can last only a short time, the rest of the plant has to provide interest for the remainder of the year.

4. Compact growth form

Shrubs which assume a compact form naturally can minimize the pruning required to keep them aesthetically pleasing. Large trees cannot be practically grown in rooftop gardens because of limited root space. Although in theory, a limited root space can limit large trees from attaining their full size, regular attention will be required to keep them safe and healthy.

5. Pest and disease resistance/incidence

The relationship between a plant and a pest is very complex. The ability of a pest to cause disease in a plant depends on environmental conditions, the properties of the organism itself and the capacity of the plant to defend itself. Under different climatic conditions, the interaction between the same plant and pest may have different outcomes. In this view, selection of plants for hanging gardens need to consider susceptibility and resistance of plants to a number of diseases and insect pests. Plants may survive without a basic pest control program, but they may not be very attractive for hanging gardens. A pest control program starts with proper site selection, good soil preparation, good drainage, proper spacing, resistant plant or cultivar selection, and plant maintenance. These all foster healthy and attractive plants for hanging gardens.

b) Making a plant selection

Plants are the most challenging factor for successful garden installations. Difficulty relies on horticulturist to deploy a right plant at a right place in consideration with a role. Efforts are made to summarize below based on type of plants available and just what they can do for you. In addition, the plant selection should go hand in hand from the construction itself as the durability itself is a major determining factor in plant selection.

Trees

The largest plants are trees which provide both vertical emphasis and shelter from a prevailing wind, as well as shade. Trees can be grown for their foliage, which can change dramatically from spring to autumn, as well as blossom, fruit, berries, or nuts. Small foliage with variegated species are often more appropriate for hanging gardens as their canopies cast lighter shade and the leaves are less likely to be buffeted by strong winds. Trees occupy the highest storey of planting. They will need to be secured to prevent from being worked loose by the wind. Shapes vary from fastigiate, or upright, to weeping, flattoped, contorted or rounded, all of which have a different character and visual emphasis.

Shrubs

Shrubs have an enormous range of shape, size and characteristics. They can be evergreen, deciduous or semi-deciduous, and can be grown for their leaves, foliage; they can also form a natural framework into which smaller and more colourful material can be interlaced. Many shrubs flower prolifically at different times of the year, including winter. While the foliage can provide enormous interest, particularly when contrasted with other species, many deciduous species have an attractive branch structure that comes into its own during the winter, especially if it is set against a suitable background. Many shrubs can successfully grow in pots and containers, which are valuable bonus on any forms of hanging gardens like balcony or terraces.

Climbers and creepers

This group of plants is especially useful in those areas partially or fully enclosed by walls, fences or screens. Most climbers are shrubby; they retain their branch structure in winter and they can be evergreen or deciduous. Although there is a climber for every situation, sunny or shady, some are grown for their beautiful, often fragrant flowers and some for their foliages which varies enormously (Table 3). There are two main types of climber- those that attach themselves (climbers) and those that need to be supported (creepers). The support can be some kind of trellis fixed to the surface, which can also be decorative element of hanging garden. Climbers are very important ornamental plants and are used commonly on walls, arches and pergolas but in urban landscape their utility is increased for the purpose of screening the premises from adjacent houses, maintaining privacy and they largely take up vertical space in hanging gardens to create floating greenery. According to the growth habit, appearance, suitability and uses, and the least observed but the most important is the final weight of the climbers are selected for hanging gardens:

Table 3: Compendium of climbers and creepers for hanging gardens

Category	Botanical name	Common name
Annual climbers	*Clitoria ternatea*	Butterfly or Blue pea
	Lathyrus odoratus	Sweet pea, Kasthuri Dal
	Thunbergia alata	Black -eyed Susan
	Ipomoea rubrocaerulea	Sweet pea, morning glory
For screening Climbers which grow quickly and show a thick and neat growth are mostly used for screening.	*Antigonon leptopus*	Coral vine, Bee bush Lovers' chain
	Clerodendrum splendens	Flaming glorybower, Pagoda Flower, Bleeding Heart Vine
	Thunbergia laurifolia	Laurel clock vine or Blue trumpet vine,
	Pyrostegia venusta	flame vine, golden shower, Chinese cracker flower,
	Vallaris solanacea	Bread flower
	Bougainvillea spectabilis	Paper flower
	Passiflora edulis	Passion flower
	Clerodendrum thomsoniae	Bleeding-heart vine, Glory bower
	Vernonia elaeagnifolia	Vernonia, Curtain Creeper
For low walls/ trellis Light type of climbers and annual climbers are selected for low walls and trellis to avoid live loads on structures.	*Lonicera japonica*	Japanese honey suckle
	Solanum seaforthianum	Blue potatocreeper, Brazilian Nightshade, Italian Jasmine
	Tristellateia australasiae	Australian gold vine
	Pandorea jasminoides	Bower Vine
	Jacquemontia pentanthos	Skyblue Clustervine, Pentantha
	Clerodendrum thomsoniae	Bleeding-heart vine, Glory bower
	Thunbergia alata	Black -eyed Susan

(Contd.)

Category	Botanical name	Common name
For pergola Usually heavy climbers are grown for long pergolas in a public garden.	*Quisqualis indica* *Petrea volubilis* *Mansoa alliacea* *Allamanda cathartica* *Thunbergia mysorensis*	Chinese honeysuckle , Rangoon creeper Queens' Wreath, Purple wreath Garlic vine Golden trumpet, Yellow allamanda Mysore trumpet vine or Indian clock vine
For porches Selection of a good flowering climber on a sunny porch is important for a colorful display.	*Pyrostegia venusta* *Petreavolubilis* *Clerodendrum splendens* *Bougainvillea sp.* *Jasminum sp.*	Flame vine, Golden shower, Chinese cracker flower, Queens' Wreath, Purple wreath Flaming glory bower, Pagoda Flower, Bleeding Heart Vine Paper flower Jasmine
Flowering in partial shade	*Aristolochia sp.* *Quisqualis indica* *Clerodendrum splendens* *Trachelospermum jasminoides*	Birthwort, Pipevine or Dutchmans' pipe Chinese honeysuckle, Rangoon creeper Flaming glory bower, Pagoda Flower, Bleeding Heart Vine Confederate jasmine, Chinese ivy
For attractive Foliage in a shade, flowering climber fails to grow or flower, those with attractive foliage& luxuriant growth can be selected for both indoor and outdoor decoration.	*Jacquemontia pentanthos* *Syngonium podophyllum* *Scindapsus aureus* *Philodendron* sp. *Monstera deliciosa*	Skyblue Clustervine, Pentantha Arrowhead vine DevilsIvy Philodendron Monstera
For pot culture For growing in pots, light and free-flowering climbers should be selected.	*Tristellateia australasiae* *Mansoa alliacea* *Clitoria ternatea* *Bignonia binata*	Australian gold vine Garlic vine Butterfly or Blue pea Bignonia

Annuals

Annuals are a group of herbaceous plants which grow from seeds, produce flowers, set seeds and complete their life-cycle within one year or one season. They exude the joy of life, putting forth a great profusion of flowers in a spectacular range of colours. They provide a beautiful display of colours in the hanging garden. Therefore, more people prefer annuals to other plants for growing in pots, beds, borders, window boxes and hanging baskets of hanging gardens. Irrespective of any garden, it is incomplete without beds of annual flowers. They enhance the decorative value of a garden within a short span of time. At their blooming time, one feels elated when a rain of beauty drizzles in the garden. A wide range of annuals are available of selection can be made considering temperature, location and utility. *Acroclinium, African Marigold, Ageratum, Agrostemma, Amaranthus, Anchusa, Antirrhinum, Arctotis, Balsam, Bells-of-Ireland, Brachycome, Calceolaria, Calendula, Campanula, Candytuft, Carnation (annual), Celosia, China aster, Chrysanthemum (annual), Cineraria, Clarkia, Coreopsis, Cornflower, Cosmos, Daisy, Delphinium, Dimorphotheca, French marigold, Gaillardia, Garden Poppy, Gazania, Godetia, Gomphrena, Gypsophila, Helichrysum, Hollyhock, Larkspur, Limonium, Linaria, Linum, Lupin, Matricaria, Mimulus, Myosotis, Nasturtium, Nemesia, Nemophila, Nicotiana, Nigella, Pansy, Petunia, Phlox, Pimpinella, Poppy, Portulaca, Primula, Rudbeckia, Salvia, Saponaria, Scabiosa, Schizanthus, Stock, Sunflower, Sweet alyssum, Sweet pea, Sweet sultan, Sweet william, verbena, Viola, Wall flower* and *Zinnia*.

Palms

Palms are the princess of vegetative kingdom and are mainly tropical plants distinguished in the wild for their tall, unbranched trunks topped by a crown of fan-shaped or feathery fronds which are prominent in the plant world for their graceful appearance and sculpturist in Mother Nature. Most palms have upright and straight stem and large leaves at the terminal portion of the trunk. Some of them are bushy giving rise to suckers and a few are naturally branched. Several species are sturdy to grow well in shade or semi shade, though most of the species prefer sunlight as they come from sunny places. Palms less than 6.00 m are considered to be small. Small Palm Trees, also called Dwarf Palm Trees, are great for the front yard of the house, balconies or terraces because they are easier to maintain and they do not obstruct the view. Also, small palm trees shorter than 6.0 m are good for hanging gardens. There are thousands of palm tree species to choose from, but efforts are made to list only the most popular small palm trees which are easy in maintenance.

Table 4: Compendium of palms for hanging gardens

Botanical name	Common Name	Appearance
Beaucarnea recurvata	Ponytail Palm, Bottle Palm, Elephant-Foot Tree, Elephant Foot	It is well-known because of its stunning trunk that is greatly swollen at the base making this palm look like a giant onion. This slow growing palm is great for small gardens. This palm can tolerate low temperature. Young Ponytail Palm has no trunk, but as the palm matures it develops a brownish-grey bark that is greatly swollen at the base, reaching up to 4 m in diameter. When the palm is only a few years old it has a single smooth trunk topped with crown of dark green fronds, but eventually develops a few branches.Ribbon-like leaves grow in clustered at the tips of branches. Thin leaves are 6ft long and only 3 cm wide. They emerge in a fountain-like fashion, curling downward. The cascading nature of the leaves gives much the appearance of a ponys' tail, hence the name Ponytail Palm.
Brahea armata	Mexican Blue Palm	It is a tall palm but can grow it as a small tree for its growth rate is naturally much slower. That said the plant has a beauty and charm at a young age that surpasses the mature state, the young trunk forms a triangle which is mirrored in opposite by the fronds. This tree requires superb drainage and sun higher to flourish.
Butia capitata	Pindo Palm, Jelly palm, Wine Palm	It is one of the most popular palms in the world because of its stunning appearance, cold hardiness and bright yellow fruit, that can be made into a jelly. Its graceful appearances with blue-green fronds make it great for pool-side plantings, and also for container use. It has a heavy gray trunk covered with old leaf bases. Trunk is around 5 m tall and 30-45 cm in diameter. Leaves are pinnate, or feather-like, ranging in color from green to bluish gray, about 1-5-3.0 m cm long, with 80-150 leaflets that are about 60-75cm long. They are supported by 1.1.25 cm long petioles that have spines along both edges.
Caryota mitis	Clustered Fishtail Palm	Native to Southeast Asia and India. This palm is cold hardy with multi stem grey

Botanical name	Common Name	Appearance
		trunk. Each stem has 8 to 10 bipinnate leaves that can be up to 3 m long. The leaflets are shaped like fishtails therefore the palms' common name is Fishtail Palm.
Chamaerops humilis	European Fan Palm, Mediterranean Fan Palm	Grows in multiple trunk clumps. Growing from the base sucklings' with a shrubby look. It can take 10-15 years for it to achieve 2 m. It has triangular, palmate, or fan shaped, leaves that range from blue green to silvery grey in colour. Leaves grow outward then upward. They grow up 0.8 m long and 0.8 m wide, supported by 1.1m stems. Stems are covered with sharp thorns, so you need to be careful when pruning them.Its toughness and wind resistant foliage make it one the most useful plants to use on exposed roof terraces.
Chambeyronia macrocarpa	Red Feather Palm, Red Leaf Palm, Blushing Palm, Watermelon Palm	It is an exotic palm tree with spectacular bright red leaflet that emerges among dark green feathery fronds and lasts for about 2 weeks. Because of the red leaflet this palm has a commonly called Flame Thrower Palm or Blushing Palm. Flame Thrower Palm Tree is a great choice for shady conditions or indoors. It has clear single trunk with 3-4 m long pinnate leaves which are growing from a long yellow crownshaft.
Chrysalidocarpus lutescens	Yellow Palm, Butterfly Palm, Cane Palm, Golden Feather Palm Areca palm	It is one of the most popular palms in tropical and subtropical climates because of its beautiful appearance and low maintenance. This palm can be grown indoors as well as outdoors.Palm has smooth silver-green trunks, which are topped with arching feather shaped fronds. It grows in clusters forming thick clumps of many stems.There are usually around six to eight yellow-green leaves on long petioles curve upwards creating a butterfly look. That is why Areca Palm is often referred to as a Butterfly Palm. It has pinnate type leaves that can grow up to 1.5-2.5 m long.

Cacti and Succulents

Practically every discussion on cacti invariably includes the word "succulent," which literally means "juicy plant."It is a reference to the ability of these plants to retain vast stores of moisture, which enables them to survive harsh and dry climates. While all cacti are succulents, not all succulents are cacti. These are group of plants which have developed a special capacity to store water in thick fleshy leaves or stems. They thrive the best in sunny situations and are light loving. They need little care except when actively growing. All the cacti are succulents on account of storing water but all the succulents are not cacti. There is a clear cut distinction between both. Cactus is characterized by the presence of areoles, which often look like woolly cushions carrying spines, hairs and the flowers arise from or near the areoles. The spines in cactus are modified leaves which provide shade against scorching sun and help in conservation of moisture.

Cacti are members of a huge family that includes more than 1,300 species. Most cacti have virtually no leaves. Instead, their swollen stems have evolved to take the place of leaves. Without a lot of exposed leaf surface, cacti do not lose moisture as rapidly as other plants. Spines on various cacti are the most versatile and effective design features in the entire plant world. They serve many functions, the most obvious of which is physical protection. The spines also absorb and reflect sunlight so that the internal temperature of the plant remains fairly constant both day and night. Because cacti have adapted well to adverse conditions, they make great houseplants and landscape objects in any form. Where other plants might fail in the hot, dry atmosphere of the typical living room or barely survive, but cacti often thrive.

Cacti and succulents require a lot of light, and a southern or south-eastern exposure is ideal for most species. A sun porch is ideal for growing cacti and hanging garden site with a sunny window will work fine. Hanged glazed pots with cacti and succulents are attractive which add on impressive beauty to roof terraces as well as gardens, courtyards and patios. Century plant (*Agave americana*), Christmas cactus (*Schlumberger abridgesii*), Easter cactus (*Rhipsalidopsis gaertneri*), Flowering Kalanchoes (*Kalanchoe blossfeldiana*), Hens and Chicks (*Sempervivum* sp.), Jade plant (*Crassulaar borescens*), Medicine plant (*Aloe barbadensis*), Moon cactus (*Gymnocaly-ium mihanovichii*), Golden Barrel Cactus *(Echinocactus grusonii),* Old man cactus (*Cephalocereus senilis),* Old lady cactus (*Mammillaria hahniana),* Sedums (*Sedum* sp.), Silver or wooly torch (*Cleistocactuss trausii*), Snake plant (*Sansevieria trifasciata*) and Thanksgiving cactus (*Schlumbergera truncata*), are the smart choice of plants for hanging gardens.

Ornamental bamboos

Ornamental bamboos are among the most fascinating of plants. Their unusual life cycle, tremendous power of growth and extraordinary diversity of size and form give them a special significance in hanging gardens. Bamboo is a very interesting grass whose graceful culms, branches and leaves lend peace and serenity to any landscape. Landscape gardens of all sizes can accommodate these elegant plants, which are cold and warm -resistant and surprisingly easy to grow. Some species of bamboos make impressive specimens for the border, others form a fast-growing hedge or screen, and short forms provide a leafy groundcover in pots and containers of hanging garden. Bamboos are highly coloured, eye catching specimens often the best appreciated for ornamental value. Stripped of their lower branches, and maintained in contained areas, these make an impressive, highly visible focal point in any landscape. Bamboos are considered as the best material for the development of urban landscapes. More than 1350 botanical species of bamboos are scattered all over Asia, South America and Africa in areas between latitudes of 0°to 45°and up to altitudes of 3000 meters. Over 60% of the species known are distributed in tropical and subtropical Asia with good assortment of coloured mutants. The differences in size and architecture make ornamental bamboos an interesting element for interior as well as for exterior landscape of hanging gardens. Thus, bamboo is becoming a favourite and splendid feature in all types of gardens as ground cover, live fence, hedges, or as a focal point in a garden landscape. Planting bamboos in hanging garden or other forms of landscape needs a little thought and planning. Thoughtful planning is necessary to maximize the ecological and aesthetic impact of a greening strategy for urban landscape, and the choice of the suitable vegetation for particular purposes is crucial.

Planning involves choice of the bamboo species as well as the mode of planting. Minimal care involves one or two rounds of maintenance per year to improve the ornamental value of the bamboo and make sure it does not grow too vigorously in the limited space of hanging garden. Bamboos are grasses, and like most grasses they tend to expand and spread. Clumping bamboos pose no problems of spreading and excellent for the patios, pots and containers.

Bamboo for hedges

Bamboos are used as hedges. It can be planted as hedges along pathways. It can also serve as dividers or partitions. Bamboo used as hedges is maintained by pruning the leaves usually at the end of summer. Species commonly used for this purpose is *Bambusa multiplex*. Other species which belong to genus Pleioblastus may also be used.

Bamboo as live fence

Bamboo used as "live fence"serves as a beautiful private screen. It gives the area privacy and serenity. It is maintained by employing top pruning which may be done at least once a year. Removal of unwanted and damaged old culms may also be done every two to three years. Species appropriate for this purpose are: *Thyrsostachys siamensis* and *Bambusa multiplex.*

Bamboo for windbreak

Tall bamboos are ideal for windbreaks at high rise buildings where there is a threat of high wind blown on the roof and can add protective ornamental components for hanging gardens. Species good for this purpose are *Bambusa blumeana* and *Bambusa bambos*.

Bamboo in containers

Many kinds of bamboo make wonderful indoor and outdoor companions for hanging gardens. These bamboos can also be grown in containers as indoor plants or as decorative plants inside the house and even in small space of patios. Medium-sized bamboos are the best kind for this purpose. When growing bamboos in containers, it is important to put pebbles (2-3 cm in size) at the bottom of the pot to provide good drainage. Keeping the plants moist but not waterlogged is necessary in maintaining the good condition of the plants. *Phyllostachys aurea, Phyllostachys nigra* and *Bambusa multiplex* ssp. *variegata* are some of the good bamboo species that can be grown for this purpose.

Bamboo as solitary specimen plants

Bamboos are multi-purpose landscaping plants. They can be used as solitary specimen plants, the extensive assortment of bamboos of different sizes and habit preferences make the plant suitable for virtually every application of landscaping, Bamboos are quite simply beautiful plants that can be combined with trees or flowering plants. Excellent choice of species for solitary specimen plants are; *Fargesia rufa*, *Fargesia nitida*, *Fargesia robusta, Phyllostachys nigra* and *Semiarundinaria fastuosa.*

Ferns

The tropical and warm temperate hills are the natural home of most of the ferns. Most ferns are decorative and beautiful foliages. They are often compact and of graceful appearance. Since they do not need adequate light and can grow in a wide range of environment, many types of ferns are useful as house plants and epiphytic nature is added advantage to make them suitable for hanging gardens to create serenic beauty with lush of greenery. Ferns are grown as house plants, specimen plants, in groups and also with other house plants. Ferns are foliage plants distinguished from almost all other plants in that they do not produce flower and seeds but they reproduce by means of spores. The flowerless plants belong to several different families and yet they have many common characteristics which they can logically converse as a single group.

Horticulturist always looks for a plant that requires little care, which thrives in low-light conditions fulfilling visual interests. In such instances ferns are always the best answer which has foliage ranging in appearance from delicate to dramatic, depending on plant family.

Ferns are (relatively) delicate plants which only grow in areas where there are suitable moist conditions. They favour sheltered areas under the forest canopy, along creeks and streams and other sources of permanent moisture. This nature of ferns helps them to establish successfully in shady areas of hanging gardens, where rest of vascular plants find it difficult to establish. Ferns reproduce differently from conifers and flowering plants. It all has to do with moisture. Not just the moisture that allows the plant to live where it does, but the moisture that allows it to reproduce there.

Hanging gardens with a secluded pond, sunny rock wall or a serene woodland setting are the best place for assortment of ferns to add beauty. Ferns come in an amazing range of texture, colour, sizes and shapes. Their ease and versatility make them an essential part of any well rounded garden.

Ferns for hanging gardens are selected based on locality, temperature, moisture level, colour, texture, size and considering clients.

Table 5: Compendium of Ferns for hanging gardens

Category	Botanical name	Common name
Evergreen ferns	*Adiantum venustum*	Himalayan Maidenhair
	Arachniodes simplicior var. *variegata*	East Indian Holly Fern
	Asplenium scolopendrium	Harts' Tongue Fern
	Asplenium trichomanes	Maidenhair Spleenwort
	Blechnum spicant	Deer Fern
	Cyrtomium spp.	Holly Fern
	Dryopteris championii	Champions' Wood Fern
	Dryopteris erythrosora	Autumn Fern
	Dryopteris formosana	Formosan Wood Fern
	Dryopteris lepidopoda	Sunset Fern
	Polystichum neolobatum	Long-eared Holly Fern

(Contd.)

Category	Botanical name	Common name
Sun tolerant ferns	*Asplenium trichomanes*	Maidenhair Spleenwort
	Athyrium filix-femina	Lady Fern
	Blechnum penna-marina	Alpine Water Fern
	Ceterach officinarum	Rusty-back Fern
	Dryopteris affinis	Scaly Male Fern
	Dryopteris erythrosora	Autumn Fern
	Osmunda regalis	Royal fern
	Polypodium glycyrrhiza	Licorice fern
	Polystichum munitum	Sword fern
Ferns for deep shade	*Adiantum aleuticum*	Western Maidenhair Fern
	Asplenium scolopendrium	Harts' Tongue Fern
	Blechnum spicant	Deer Fern
	Cyrtomium (all species)	Holly Ferns
	Dryopteris dilatata	Broad Wood Fern
	Gymnocarpium dryopteris	Oak Fern
	Polystichum acrostichoides	Christmas Fern
	Polystichum munitum	Sword Fern
	Polystichum setiferum	Soft Shield Fern
Drought-tolerant ferns	*Blechnum penna-marina*	Alpine Water Fern
	Dryopteris crassirhizoma	Thick Stemmed Wood Fern
	Dryopteris filix-mas	Male Fern
	Polystichum munitum	Sword Fern
Ferns for containers	*Adiantum aleuticum*	Western Maidenhair Fern
	Arachniodes simplicior var. *variegata*	East Indian Holly Fern
	Asplenium scolopendrium	Harts' Tongue Fern
	Cyrtomium spp	Holly Fern
	Dryopteris affinis	Scaly Male Fern
	Dryopteris erythrosora	Autumn Fern
	Dryopteris lepidopoda	Sunset Fern
	Osmunda regalis	Royal Fern
	Polystichum tsus-simense	Korean Rock Fern
Ferns for color	*Adiantum hispidulum*	Five-Fingered Jack
	Athyrium niponicum	Japanese Painted Fern
	Athyrium otophorum	Auriculate Lady Fern
	Dryopteris erythrosora	Autumn Fern, Japanese Wood Fern, Copper Shield Fern
	Dryopteris lepidopoda	Sunset Fern
	Dryopteris wallichiana	Alpine Wood Fern
	Onoclea sensibilis	Bead Fern
	Woodwardia orientalis var. formosana	Oriental Chain Fern
	Woodwardia unigemmata	Chain Fern
	Woodwardia virginica	Virginia Chain Fern

Orchids

Orchids are the largest of flowering plants, with an estimated 30,000 naturally occurring species worldwide. The world of orchids is tremendously vast and diverse with many incredibly beautiful flowers. An epiphytic nature of orchids is like extra bonanza for orchid enthusiasts to grow them on balconies, terraces, hanging pots, mounted on tree bark in any forms of hanging gardens in the most unpropitious city sites. Growing orchids as balcony plants or as roof plants has always been attempted by enthusiasts. Orchids can hang off on trees, on wooden plank and dont' need to be specially planted, their thick exposed roots (velamen roots) gather all of the water and nutrients they need right from air, making them considerate to grow on trees and plants at balconies.

The plants which grow the best on balconies are those that stand drying out well, and have intermediate sunlight demands. Many Oncidium, Dendrobium and Cattleya, Phalaenopsis plants fill this bill. Monopodials such as Ascocendas or Vandas are excellent choice to grow under more sunlight.

Tree branches are the natural hosts for many orchids, so it seems logical to use tree branches as mounts. In nature, however, many orchid species prefer particular tree species and it follows that these orchids may not do as well on some mounts as they do on others. Generally the branches of rough-barked hardwood trees, such as tree fern, osmunda fibre, oak, casuarinas, peach, apricot etc., are the most suitable. Orchid roots seem to cling better to rough bark than to smooth, and rot-resistant hardwood is more durable than many types of softwood, such as pine.

Orchid mounts can choose from many natural materials like as tree bark, cork bark, or wood slats. The mounting process is very simple, with a small amount of sphagnum moss evenly wrapped around the roots of the orchid and tied with thin wire or fishing line around the bottom of orchid to attach it to the mount. Orchids mount needs watering more frequently than potted orchids since they tend to dry out fairly quickly. Orchid mounts can be used to create an incredible natural displays on limited space available on balconies of city scape.

Orchid enthusiasts will be tempted to try growing orchids on mounts to craft a beautiful and naturalized way of displaying orchids and will provide the grower with an increasing range of viable window and balcony plants with beautiful orchid display in hanging gardens. Orchids growing successfully on mounts include Lady of the Night (*Brassovola nodosa*), *Schomburgika tibicinus, Dendrobium loddigesii, Dendrobium schneiderae, Maxillaria sophronites, Encyclia polybulbon, Baptistonia echinata, Zygostates grandiflora* and *Polystachya pubescens.*

c) Arranging the plants

Planting in small spaces, including roof balconies and terraces, is particularly important; too much and the whole area can be swamped, too little and the result can be meagre, emphasizing the harshness of the surroundings. Roof gardens and terraces each have their own particular problems which are largely to do with the severe weather they may experience. However shelter the space may be, wind will almost certainly prove troublesome at some point. This combined with a hot sun, can make conditions difficult to deal with.

Plants that are drought-tolerant thus need to be used. Obviously, if irrigation is provided for the plants on built structures, the type of plants that can be used increases tremendously. The use of shallow substrate also imposes another limitation, in that because of limited anchorage, ground cover or low shrubs, rather than tall shrubs need to be used to prevent toppling of plants.

Hanging Garden as a Part of Environmental Serenity

From an aesthetic stand point, hanging gardens clearly help to improve the environment. But adding such green space can do more than simply improve the looks of a space. When developed on greater scale, hanging gardens can play a significant role in maintaining healthy ecosystem, especially in heavily built-up areas. Forests, farm and vegetation can accomplish this task in suburban, rural and undeveloped areas. Ironically, in the city, where air pollution is the heaviest due to traffic and industry, the limited land available for plants helps compound an already difficult problem. By making optimal use of available space for plantings, hanging gardens can play an important role in maintaining air quality in urban areas.

Hanging gardens can also help to moderate the climate of urban areas. Cities do not only seem hotter than outlying areas do in summer, they actually are warmer, by as much as 7°C at night and 3°C during day. This is due to most of land area is covered by materials like asphalt and concrete, which have a low albedo. Consequently, built up areas get hotter and stay hotter than outlying areas, where plant cover reflects more sunlight. An extensive network of gardens on roof in an urban area could reduce the temperature by several degrees by covering heat-retaining spaces with more reflective plant cover, well as by increasing the evaporation of moisture, which helps to cool air. In addition to making urban areas more comfortable, such a temperature reduction could result in a significant reduction in energy usage, as less energy would be needed for air conditioning.

Any well designed open space amid the chaos of a modern city can provide some respite from the noise and bustle all around. But a hanging garden, particularly a garden located at a roof above ground level, is like a peaceful island within the urban concrete jungle. A feeling of isolation from the traffic, noise, dust and general confusion of the typical downtown city street can be sensed in most roof gardens above ground level. It is one of their major attributes, and one which downtown park at street level can rarely achieve. One of the most notable qualities in almost all above-grade hanging gardens is their quietness.

One of the most common reactions of those first visiting a hanging garden is astonishment that such a quiet, natural place can exist in a busy city. This initial feeling of pleasure seems to remain as they continue to enjoy the garden over time.

Hanging gardens give a sense of isolation from the urban environment and they promote community within a city. Residential gardens atop apartment buildings bring together neighbours whose only other contact might be a brief nod in an elevator. Hanging gardens associated with office buildings provide a place for employees to mingle in more relaxed scenery. Public hanging gardens aid a myriad of community functions, as spaces to meet, to socialize, to attend special event. In the suburbs, ground-level gardens and plazas could fill such needs, but in built-up cities, roofs are often the only space available.

Hanging gardens can also enable city dwellers to maintain a connection with nature that might not otherwise be possible. Over the last fifty years, efforts have been made to insert nature into the urban landscape by placing potted plants on sidewalks, planting street trees and median strips of streets.

Hanging gardens offer an opportunity to provide natural habitat for resident and migratory birds and insects. This is achieved through the provision of one or more of birds' basic needs like food, shelter, water and a place to breed. Although individual vegetated roofs or walls of hanging gardens may not serve all of the above mentioned, but they may act as stepping stones for creating links between larger habitats offered by nearby natural green areas, hillsides or parks. Suspended greenery, with its high potential to link greenery at roofs to greenery at ground, can provide a natural extension of habitats. Hanging gardens serve as a transit route for wildlife between the ground habitat and those established on greenery at roofs. The enhanced biodiversity can enrich the ecological quality and health of the environment as well as additional emotional, intellectual, social and physical benefits to humans who bustle with city life.

Roof spaces offer opportunity for growing certain types of shallow rooted vegetables, flowers and green herbs. Albeit the roofscape gardening is mainly confined to ornamental crops still a small window for fresh and organic self-grown crops is available. Hanging gardens with Balcony baskets offer herbs to grow in them. Mint, rosemary, coriander and basil are some of the herbs which grow comfortably in gardens on balconies.

Institutes Specializing in Hanging Gardens

Hanging garden concept is popping up on buildings all over the world, and has seen an especially big surge in popularity in Europe, America and Asia. The city of Chicago's green-roofed city hall, an environmentally friendly roof space is just a one of the most prominent example, but there are a few hundreds more in cities both big and small as both private and public organizations look for ways to be more conscious about their environmental impact. Universities and colleges have not missed out on this trend, adapting current and new buildings on their campuses so that green roofs now grace college-owned buildings all over. Here is the highlight of a few colleges, institutes (in no particular order) leading the way in using green roofs, vertical greenery, doing research, and building the greener, more sustainable college campus of the future.

University of Illinois, Chicago

UIC has a number of LEED-certified buildings on campus, but even those that are not certified are getting an environmentally friendly update via green roofs. The campus boasts two buildings on campus that are employing green roof technology to save energy, extend the life of the roof, and retain rainwater. Students can currently enjoy heading out on to a green roof at the school's Behavioural Sciences Building or the Education, Performing Arts and Social Work Building. Additional green roofs may be popping up in the future as the school aims to help Chicago maintain its reputation as a green roof leader.

National Parks Board, Singapore

NParks has introduced the Skyrise Greenery Incentive Scheme (SGIS). Under the Scheme, NParks will fund up to 50% of installation costs of green roofs and vertical greenery. As more and more of Singapore land is taken up by buildings, there is need to ensure that greenery lost on the ground is replaced in the urban environment. To reinforce Singapore's 'City in a Garden' aspiration an incentive scheme was started to achieve.

- Encourage the installation of skyrise greenery on existing buildings across Singapore.
- Create a distinctive image of the city in the tropical climate through extensive greenery adoring building facades and skyrise levels.
- Bring about environmental benefits such as mitigating the urban island heat effect and improving the air quality through the plants' transpiration and filtration of dust particles.

Augustenborg Botanical Roof Garden, Sweden

The Augustenborg Botanical Roof Garden, Sweden is the world's first botanical roof garden and located atop the Scandinavian Green Roof Institute. Situated within a residential neighbourhood atop 9,500 m^2 of industrial buildings, the extensive roof garden is designed as both demonstration and research garden, testing various plants, engineered soils, and water runoff. The research is ongoing and test gardens reflect different greenroof systems as well as different designers.

Center for Green Roof Research, Pennsylvania State University

The Center, the only location in North America with small test green roofs on replicated buildings, characterizes and quantifies the performance of green roofs and promotes their discoveries through education and outreach. Some of the cutting edge research performed in the Penn State Center has included studies in storm water runoff quality and quantity, building energy consumption and insulation value, chemical and physical characteristics of media plant selection, plant water use, waterproofing resistance to root penetration and drainage materials.

National Aeronautics and Space Administration (NASA)

NASA is also involved in research use and improvement upon the concept of "green roofs"- urban rooftops covered with vegetation. Although the idea is centuries old, it has great relevance in addressing today's challenges. With over half of the entire human population living in cities, it is important to understand how cities can be affected by climate change, and to devise strategies to employ adaptation strategies to deal with possible impacts.

NASA plays an increasingly important role in research into climate change and its potential impacts on the quality of life on earth. This includes several thrusts of research into the impact of climate change on urban life. One such effort at NASA involves utilizing green (a.k.a. vegetated) roofs. NASA's research reveals that roofs covered with vegetation provide a cleaner environment, energy savings and increased insulation. Additionally, utilization of green roofs can dramatically reduce the surface temperature of urban environments, something critically important for elderly and other sensitive populations.

Colorado State University

Colorado State University has an extensive green roof program. The school not only builds green roofs but also uses them in its environmental and agricultural studies programs, which are working to determine which plants are the best for their region and what soil mixes are the most productive. Their research is instrumental in helping green roofs all over the nation maximize their benefits. While the campus does have a small, 400-square-foot green roof on site, they are planning much larger green roof projects in the future.

Some of the other institutes associated with green roof and vegetative roof research are:

- University of Guelph and Vineland
- Michigan State University
- Green Roof Centre at Southern Illinois University, USA
- Texas A&M Interdisciplinary Green Roof Research Programme
- University of Toronto Green Roof Research, etc.

Conclusion

Hanging gardens can be delightful substitutes for natural landscaped areas at ground level, with all the amenities of gardens on the ground. Indeed, a garden at roof or hanging gardens can be virtually indistinguishable from a garden planted directly in the earth. Hanging garden can be employed to bring nature back to cities.

Well, the above innuendo about hanging gardens gives the reader a simple glimpse, much hidden things are to be noted and considered. For instance, amateurists have to consider the natural growing habit of the plant species, their interaction with the neighbouring plants (allelopathy), insect attraction and repulsion, weight, soil and nutrient requirement etc. All these definitely vary when present in hanging garden environment especially the ubiquitous intensive type. Furthermore as all the plants are well grown majority of pests and diseases incidence are at bay but physiological disorders may come into picture as the plants are forced to grow to maturity on a hard pan structure.

Though the treatment is difficult but can be prevented if the horticulturist and builder exchange their plans and go in synch during the course of establishment.

Though in most of the tropical developing countries, hanging gardens are still a distant dream or a project of fantasies; considering the urbanization and UHIs created hanging gardens and vertical gardens have to be practised in a need basis. Greenery indices should be formulated by urban bodies before certifying the buildings to meet the standards. Government policies and subsidies are formulated by certain nations but have to be practised in letter and spirit.

Further Reading

Baggs S. and Baggs J. (1996). The Healthy House; London; Thames and Hudson.

Bell, Ellen. "Climbing Plants –Five Types of Climbers to Know"2008. http://www.buzzle.com/articles/climbing-plants-five-types-of-climbers-to-know.html

Cantor, S.L. (2008). Green Roofs in Sustainable Landscape Design, W.W. Norton and Company.

Dunnett N. and Kingsbury N. (2004). Planting Green Roofs and Living Walls pp. 1 and 125. Timber Press.

John Dransfield, Natalie W. Uhl, Conny B. Asmussen, William J. Baker, Madeline Harley, and Carl Lewis., 2008. Genera Palmarum, The evolution and classification of palms.

Johnson B. (2001). Green grass and a great view. Retrieved 21 Dec 2001 from http://www.finace-commerce.com

Johnston, J. and Newton, J. (2004). Building Green: A Guide to Using Plants on Roofs, Walls and Pavements, London Ecology Unit.

Mallin, H., (1980). A highrise oasis. *Garden* 105:47-52.

National Parks Board.Sky greenery, Singapore, (2006). Retrieved from the World Wide Web: http://www.nparks.gov.sg/gardencity/skyrise.shtml.

Osmundson, Theodore (1999). Roof Gardens: History, Design, and Construction. New York: W.W. Norton and Company Ltd.

Peck, S., Callaghan, C., Peck and Associates, Bass, B. and Kuhn, M. (1999). Greenbacks from Green Roofs: Forging a New Industry in Canada. Prepared for Canada Mortgage and Housing Corporation. Ottawa: CMHC SCHL Canada.

Scholz-Barth, K. (2001). Green Roofs: Storm water Management from the Top Down. Environmental Design & Construction Home. Retrieved from http://www.edcmag.com/CDA/ArticleInformation

Snodgrass, E.C., Snodgrass, L.L. (2006). Green Roof Plants: A Resource and Planting Guide. Timber Press, Hong Kong, 203 pp.

Stainless Steel Solutions. "Planting advice on Green Walls & Façades"(2009). http://www.s3i.co.uk/facadeplantingadvice.php

Sue Olsen, (2007). Encyclopedia of Garden Ferns,

Wong, N.H., Chen, Y., Ong, C.L. and Sia. A. (2002). Handbook on Skyrise Greening in Singapore. National Parks Board & National University of Singapore, Singapore.

ZinCo GMBH, Green Roofs: Recommended Standards for Designing and Installation on Roofs (6th ed.), Germany: Zinco GmbH, 2000.

380 m-long rooftop landscape and the tall palms with a natural spectacular look

Balcony Garden

Flating Lush greenery of Hanging gardens

Hanging Gardens of Babylon

A lush vertical Green facade has over 20 species of plants and vines such as Argyreia nervosa, Telosma cordata and Hoya growing on it

Extensive Greenroof

Green Wall in verandas are perfect link between house and garden

Greening Pedestrian Overhead Bridges, Bougainvilleas oftens the outlines of Pedestrain bridges across street

Hanging Greenery, Columns of carpark are shrouded by vines

Hanging vertical vegetation on walls

Intensive Greenroof on housing building with semi-mature trees and plants form a seamless canopy

Intensive Greenroof where a grove of trees is planted

Silhouettes of palms growing in the roof of Hanging garden

Sky Bridge, Plants with various forms and colours growing to soften and act as ecological link in urban concrete jungle

Ornamental Horticulture, pp 131-146

New India Publishing Agency, New Delhi - 110034, India

6

Lawn Development and Maintenance

S.S. Sindhu

Lawn is an integral part of any garden and it gives pleasure while sitting during evening in summer and one can enjoy good sunshine during winter. Hence, it is often referred as "heart of the garden". It is covered with perennial grass and provides open landscape to the house and adds aesthetic beauty to the surroundings. Good lawn is essential to have in a good garden. It is permanent feature, which occupies 60% share of total garden area. Here in this chapter lawn is divided into two parts. There are approximately 620 genera and 10,000 species of grasses worldwide. Grasses are used extensively used throughout the world for their beauty and utility. Grasses are found from temperate to tropical to arid semi arid regions and suits in most extreme climatic conditions and have adapted to a worldwide climate and zones. Grasses are used as fodder, food, medicinal and for aesthetics purpose and it belongs to family *Gramineae.*

Uses of grasses in the garden

- **Aesthetic utility**: Grasses are pleasing to the eyes and soothing effect to sole.
- **Sports and activity:** Grasses provide appropriate surface for play and sport. One can enjoy the life more being with natural greens and open spaces far from crunchy city spaces.

Division of Floriculture and Landscaping, IARI, New Delhi

- **Being with family:** lawn provides ultimate platform for family gatherings and enjoyment.
- **Body enrichment:** Being with nature means living in oxygen enriched climate where one can boast aerobics and yoga and simply encourage time spent outside.

Lawn in Front of House

Other uses of lawn

- Lawn adds space to house and surroundings.
- Improve water percolation and water absorption capacity of soil
- Slow down soil erosion and rain-water runoff thus improves soil physical and chemical properties to sustain life.
- Uplift property and add value to the property.

Lawn comprises 60% of total garden features. Morning bare-foot walk on lush green lawn freshen body and enhance blood circulation, at afternoon it gives cooling effect and at evening its nice place where one can stumble on it after hectic day-work.

Soil Preparation and Sterilization

Physical and chemical nature of the soil

Well drained fertile soil with good water holding capacity, free from weeds, rocks and crocks in sub soil is essential to develop an ideal lawn. The optimum pH of soil should be 5.5-7.0. Many warm season grasses grow better at pH 6-7. It is important to realise that the material in which the grass roots grow is not merely a solid mass. The soil consists of solid particles which may be mineral or organic in nature with pore space between them. The pore space, which may form about 50% of the total volume, is occupied by air and/or water. Also, the soil is the home of various forms of life including numerous micro-organisms (bacteria and fungi).

Initial planning

The procedure for making a new lawn depends on the nature of the site on which the lawn is to be made and on the kind of lawn it is intended to produce, taking into account how the lawn is to be used and looked after. A new lawn is sometimes made on an area which has been used as flower beds or vegetable garden and this situation creates few problems.

Selection of site and shape

To develop a good lawn, soil should be preferably sandy loam, which is rich in organic matter, with good drainage facility. Sunny places have been considered good. The shape of lawn may be regular or irregular. However, rectangular shape is most popular. Lawn may follow the graceful curves of slopes to add beauty in the landscape, which could be surrounded by rockery or flowering borders.

Drainage

After selecting the site, the next thing will be to ensure proper drainage. Grasses are shallow-rooted herbs and therefore, no deep drainage is necessary. But at the same time water stagnation should not be there in the rooting zone. In clayey soil, some kind of drainage must be provided. This may be done by drainage pipes or by putting a layer of broken pieces of bricks and rubbish 90 cm below the surface. Ordinary drainage work can be carried out in conjunction with grading or leveling. Generally 2-3 cm slope is necessary per 50 m towards the natural direction of drainage.

Digging

Rough surface leveling by eye estimation should be done prior to digging. If during rough leveling a lot of shifting and filling of soil is necessary, the surface soil should first be taken out and kept separately, which should be laid on the top after final leveling.

After rough leveling is completed the digging work will commence. Thorough preparation of the ground is most essential in the success of a lawn. Very often the digging work is done quite hastily. This is a bad practice and contributes to the failure of a lawn. The digging operation should be done by the trenching method; otherwise the gardener may fail to dig to the desired depth. To ensue this, a trench of 60 cm deep and 45 cm width is dug at one end of the site along its entire length and the soil is kept outside the lawn area. During the process of digging all stones, old masonry, grass roots, etc. should be removed. Special care should be taken to remove the roots of motha grass or nut grass (*Cyperus rotundus*).

Manuring

After digging soil upto 45 cm depth during summer season, the soil is to be manured. If the soil was originally fertile, it is better not to apply any organic manure, as organic manure contains weed seeds which may contaminate the lawn with weeds, which are difficult to remove. But poor soils will need proper dressing with organic manure and well rotten cowdung manure at the rate of 5 kg per m^2 and mix it thoroughly followed by rolling. Apply super phosphate or D.A.P. or bone meal per m^2 in soil mixture and mix well into the top layer of the soil by racking. Night soil manure, F.Y.M. or old stable manure is also used for this purpose. The manure is sieved finely and spread over the surface at the rate of 500 kg per 100 square metres of soil. The amount can be reduced depending on soil fertility level. This is then worked up in the soil to a depth of 15-20 cm.

Moreover, of the many mineral nutrients required by grass for satisfactory growth, only three-nitrogen, phosphate and potash-commonly need supplementing. The special functions of these three major nutrients are:

Nitrogen: It promotes growth of leaf and stem and gives a good colour. It affects root growth since without sufficient top growth there is poor root growth, although too much nitrogen reduces rooting.

Phosphate: It promotes root growth and tillering. Also, help to accelerate the respiration process of plant system through activate the enzymes.

Potash: It necessary to co-ordinate various growths processes and helps disease and drought resistance. It improves the quality of grass.

Selection of Grass

Selection of grass is made based on the climate where lawn is to be developed.

Warmer Climate

Bermuda grass (*Cynodon dactylon*): Commonly referred as doob grass, it is widely used for planting lawn due to its fast growth, hardiness, tolerance to frost, less water requirements, makes an excellent turf and responded well to frequent mowing. The appearance of stolon is very common feature in doob grass if, mowing is not done at week interval especially during summer season.

The recent trend is to use a strain of Bermuda grass known as Selection-1 which is very fine, soft and easy to maintain but, it does grow well under sunny conditions. It forms high quality carpet grass.

Japanese grass (*Zoysia Japonica*): This type of grass is highly suitable for smaller areas and home lawns. It grown slowly and hence frequent mowing is not required. It forms a high quality lawn but it is slow growing turf.

Mexican grass (*Lolium species*): It is very fine grass with shining surface but it requires a lot of maintenance. Grass does not grow equally and uniformly. It grow in bunches and mowing become difficult.

Cooler Climate

Certain grass strains of *Agrostis tenuis, Festuca rubra, Poa* sp.(also called blue grass) are suitable for temperate climate. The common species *Poa pratensis* which is also known as Kentucky blue grass is grown on its own and *Lolium perenne* (rye grass) are used under temperate climate. These species do well under cold climate under low sunshine, shaded place, along wall and underneath the trees. The details description of different cool season grass species are given below.

Methods of Planting

- **Seed Method:** 2.5 gm of seed per m^2 is optimum which is generally mixed with garden soil or sand so as to give even distribution and spread in the area. It is being followed by proper raking of the soil slightly for covering the seeds. Sprinkle the water immediately after sowing and continue till seeds are fully germinated.
- **Dibbling:** This is most widely used practice in India. Grass pieces along with roots are taken from weed free areas and dibbled at a distance of 10-15 cm both ways followed by immediate irrigation.
- **Turfing:** This is the quickest method of making a lawn within a day. The intact pieces of uniform thickness of turf with soil are taken from the place where grass is compact and free from weeds. Then these pieces of carpets are spread on the ground evenly and covered with fine soil followed by watering.

Top dressing

In lawn management top dressing means the application of bulky material like sandy compost, spread as evenly as possible at 1.5 –3.0 kg per m^2 by hand or with a shovel and then worked into the sole of the turf by suitable equipment such as a drag brush, drag mat, lute or even simply the back of a wooden rake.

Top dressing materials

The ideal material for top dressing is a special form of sandy compost produced by making a compost heap with alternating layers of sandy soil and rotted farmyard manure and allowing it to stand for a year or so before breaking down and mixing with suitable sand to produce material with the right sandy texture. Obviously this procedure is not very practicable for the lawn owner who, like many professionals, needs to make up an acceptable substitute. This might well take the form of a mixture of:

- 6 parts suitable sand
- 3 parts sandy loam soil
- 1 part granulated peat

Good, clean (weed-free) sandy loam should be obtained for the mixture if possible but other types of soil can be used with suitable adjustment of the proportion of sand added, *i.e.* less sand for a sandy soil, more sand for a heavy soil. Really heavy or clay soil should be avoided since it is almost impossible to produce a good uniform friable mix from them. Organic matter for the mix is conveniently provided by crumbly granulated sphagnum or sedge peat, but other organic materials such as leaf mould may also be used, again possibly with an adjustment of the proportions shown; for example, the leaf mould may be of a decidedly soily nature so that the proportion of sand may need increasing.

Maintenance

Rolling, Mowing and sweeping

The object of rolling is to help the grass anchor itself securely and also to keep the surface leveled. Rolling should be avoided when the soil is wet. There is difference of opinion regarding the rolling of doob lawns. It is argued that doob grass has a growth pattern completely different from the English lawn grass, in that while the former needs rolling at regular intervals the later does not.

Mowing is another important operation in the proper maintenance of good lawn. Newly planted lawns are not generally moved till it has established well. The first thing is to obtain a good machine which will cut evenly at a correct height. The frequency of mowing is determined by the amount of growth and will vary from season to season. But grass should not be allowed to grow more than 5-6 cm in length during any season.

Sweeping the lawn thoroughly after each mowing is essential to clean the cut grasses which might have fallen from the mower box. Sweeping is also done every morning to clean the fallen leaves and other debris. Sweeping may have to be repeated two or three times in a day during the season when the deciduous trees shed their leaves.

Mowing

Whatever the standard aimed at, regular mowing is essential to give anything like a reasonable lawn. Therefore, a good lawn mower in good condition is very important.

Mowing equipments

For very small patches of lawn in front of small terrace houses, some people make do with garden shears. These perform most satisfactorily if they are cared for properly and kept sharp. At the other end of the scale, small paddocks which need only to be kept reasonably tidy (even if grazed) may be mown conveniently with a machine with a cutter bar (like an agricultural mower), with a flail-type cutter or with a large rotary mower; a large motor mower of the conventional cylinder type might be appropriate if a slightly higher standard was required for the paddock.

Effects of mowing

The quality of a lawn is conditioned mainly by the type of mowing coupled with its frequency and regularity. Mowing affects the growth habits of the grasses, their root and shoot development and the botanical composition of the sward. Within limits mowing encourages tillering and thus a dense sward, but for each of the common lawn species there is a height of cut below which they survive with difficulty.

The weed population is also affected in amount and type by mowing. If a lawn is cut too keenly for the grasses contained in it then weeds can invade the weakened turf. The types of weed in the lawn are those which can survive under the cutting conditions prevailing.

Height of cut

Since the height of mown grass is very difficult to measure, the cutting height is usually described in terms of the bench setting of the mower. When a conventional mower of the roller type, is used the height of cuts the distance between a straight-edge placed so as to pass from front to back roller and the cutting edge of the bottom blade. With side-wheel mowers and ordinary rotary mowers the straight-edge is laid from roller to side wheel.

Return or removal of cuttings

Allowing the cuttings to remain on the lawn as a regular practice has some advantages and some disadvantages. An important advantage to many lawn owners is that it does away with the chore of disposing of boxes of cuttings which can be rather laborious, even though these will rot down to make a good organic compost for general use in the garden. Perhaps more important is the fact that when the cuttings are allowed to stay on the lawn lesser quantities of nutrients are removed from the site.

Lawn edge trimming

When a lawn has been mown its appearance is greatly improved if the grass growing over the edges is trimmed. A pair of garden shears may be satisfactory for the very small lawn but for the larger lawn something less back-breaking is well worth while.

Weeding

One of the main aspects of maintenance is the control of weeds. Without close attention or care a time will come when weeds will overcome the lawn grass, the soil will become sick and the grasses become coarse. Weed is common in both new and old lawn. Therefore, as soon as a lawn is established weeding should start and continue at regular intervals or whenever the weeds come out. The frequency of weeding obviously will be more during the rains than in the cooler months. Motha, also called as nut grass (*Cyperus rotundus)* is very tenacious and common weed of lawn and it is difficult to eradicate. Its root goes as deep as 90 cm. These roots should be removed after deep digging at the time of soil preparation. In a freshly prepared lawn the growth of motha is profuse and unless this growth is checked the weed will smoother and hamper the growth of doob grass. This should be removed with the roots as deep as possible with a long narrow bladed (1-1.5 cm) Khurpa. It is almost impossible to eradicate motha completely. Hand weeding is possible in small areas and it is the only solution to check motha if arises despite using pre-emergence weedicides. Weedicide for dicot weeds can be used e.g. 0.5% solution of 2, 4-D. Alternatively higher concentration of Glyphosate (Round up) or Stomp weedicides may also be used. But, once the lawn is well established with regular mowing the growth of motha is kept in check. A red creeping, dudi (*Euphorbia thymefolia*) often invades the lawn and is not difficult to eradicate if controlled at the initial stages. All weeds should be removed with the roots and these should never be allowed to seed.

After Care

Liming of lawn

A View of Lush Green Grass

Lawns very rarely need lime. Lime is needed to correct excessive soil acidity but excess of lime encourage weeds, earthworm activity and fungal diseases. Even when the soil gets too acid so that the vigour of the grass and its reaction to added fertilizer are reduced the application of lime, though necessary, may have some detrimental effects. When lime is applied to a lawn it falls on the surface and thus makes the surface alkaline. It takes a long time, even years, before leaching of lime to the lower layers or to drainage brings the surface pH down to the desired levels.

Irrigation

Lawn requires plenty of water. During winter months, watering is done at an interval of 7 days whereas during summer season, it should be done on alternate days. Doob grass is shallow-rooted and therefore, frequent light irrigation is better than copious flooding after long intervals. Here again some people prefer flooding at long intervals as this saves labour. The water-hose is left in the soil and shifted occasionally, while the gardener attends to other work in between. Labour as well as water can be saved to a considcrable extent if sprinkler irrigation is used.

Mechanical operations

Much mechanical treatment is involved in efficient maintenance, of the turf on golf greens, bowling greens and cricket tables but such treatment may be of very little importance for the average domestic lawn. Indeed any benefit conferred may not be worth the time and effort put into the work which may even have harmful results.

Scrapping and raking of lawn

Continuous rolling, treading and mowing may result in the formation of a hard crust and the lower part of the lawn may get matted and woody. For such lawns, the grass is scraped at the ground level with the help of a khurpi or with diesel lawn mowcr by setting at zero level during the months of April and May when growth become stagnate and it is followed by a proper mixture (sand + soil + bone meal + fertilizers in equal ratio) and watering. Scraping is followed by raking to break the crust. Where the condition of the lawn is good, hard and thorough raking is done both ways to loosen the old runners and to aerate the soil.

Top Dressing with compost and fertilization

After scraping or raking compost consisting of good garden soil, coarse sand and leaf-mould in the proportion of 1:2:1 (in sandy soil the proportion of sand should be reduced or eliminated altogether) is spread over the lawn to depth of 3-5 cm. To cover such depth compost @ 100 kg per 100 m^2 area will be needed. Bonemeal is also applied at the rate of 1 kg per 10 m^2 area. The same compost is used as top dressing again during September or October. From October to April, ammonium sulphate is applied once every month at the rate of 1 kg per 50 m^2 area followed by watering. This can be applied twice a year (October and May-June). Spraying of urea at the rate of 0.2% is also found useful. Ammonium sulphate 1 kg/100 m^2 every month from October–February gives lush green lawn with a thick growth like a carpet.

Top dressing methods and materials

In lawn management top dressing means the application of bulky material like sandy compost, spread as evenly as possible at 1.5 –3.0 kg per m^2 by hand or with a shovel and then worked into the sole of the turf by suitable equipment such as a drag brush, drag mat, lute or even simply the back of a wooden rake. The ideal material for top dressing is a special form of sandy compost produced by making a compost heap with alternating layers of sandy soil and rotted farmyard manure and allowing it to stand for a year or so before breaking down and mixing with suitable sand to produce material with the right sandy texture. Obviously this procedure is not very practicable for the lawn owner who, like many professionals, needs to make up an acceptable substitute. This might well take the form of a mixture of: 6 parts suitable sand, 3 parts sandy loam soil and 1 part granulated peat.

Good, clean (weed-free) sandy loam should be obtained for the mixture if possible but other types of soil can be used with suitable adjustment of the proportion of sand added, *i.e.* less sand for a sandy soil, more sand for a heavy soil.

The kind of fertilizer to use

Few lawns need more than one dressing a year and this may conveniently take the form of a balanced complete fertilizer (*i.e.* one containing the three main mineral nutrients, nitrogen, phosphate and potash, in suitable proportions) in spring. For those few who wish to make up their own mixture the following formula will suit average conditions:

	Per 100 m^2
Sulphate of ammonia	2.0 kg
Fine hoof and horn meal	0.9 kg
Dried blood	0.9 kg
Powdered superphosphate	2.0 kg
Fine bone meal	0.9 kg
Sulphate of potash	0.9 kg

Liquid fertilizers

In recent years there has been some interest in the use of fertilizers designed to be applied in solution but it would seem that these have limited value in lawn maintenance, partly because the need for solubility and compatibility restricts the kind of fertilizer which can be used and partly because of the difficulty of ensuring that every bit of turf gets the same amount. Uniform application of solutions by sprinkler or watering can is very difficult, Spraying on relatively concentrated solutions can cause damage by scorch and within the lawn there will be some unevenness in the rate at which sprinkled water infiltrates various areas because of compaction or an uneven surface. The stream of water in the hose-pipe draws up concentrated solution from a vessel an dilutes it considerably so that it can be applied through a fine rose as a shower over the area of the lawn which, immediately afterwards, can be watered with the same equipment after the fertilizer solution has been used up.

Scarification

For intensively maintained and used sports turf, particularly fine turf, regular surface scarification with spring-tined wire rakes with mechanical scarifying equipment is necessary to groom the turf and produce good playing surfaces which permit and encourage skilful play. Few lawns need the same amount of intensive treatment. Occasional brushing with a drag brush or stiff broom or light raking, by bringing up horizontal growth, helps to keep a lawn looking trim and to prevent the formation of an excessive layer of fibrous material at the surface.

A very old lawn (usually of fine grasses) which has developed a mat or thatch of fibrous material may need more severe treatment involving vigorous use of a wire rake. For ease and effectiveness, the prudent use of a mechanical scarifier in which rotating knives or rotating wire rakes is power-driven. Such severe treatment should be carried out when good growth conditions are assured so that growth will restore the surface. Usually late summer or early autumn prove the most appropriate times to do this kind of work. Lawn owners who have a severe problem with excess fiber will probably try and borrow or hire a mechanical scarifier.

Plant Protection

Diseases

***Fusarium* patch disease:** *Fusarium* patch disease is caused by the fungus *Fusarium nivale*. It occurs most frequently during mild, moist weather at almost any time of the year but especially in the spring and autumn, particularly the latter. The first signs are usually small irregularly shaped yellow patches (often overlooked) which in the right conditions grow larger and darken in colour through reddish or bright brown (when the disease is at its most active) to dark brown when the grass looks rotten and is probably dead. When the disease is rampant, faintly pinkish white mycelium may be seen round the peri-meter of the light brown patches.

Preventive applications of fungicide are not normally recommended for lawns but if the disease occurs and the weather conditions favour it, then early application of fungicide is necessary, especially on fine lawns, to ensure that unacceptable damage is not caused. Indeed in severe attacks repeated applications may be necessary. There are various types of proprietary lawn fungicides available for use against fusarium disease. They are usually produced for application with added water by means of a suitable sprayer or with a watering can, preferably fitted with a dribble bar. The makers' recommendations should be strictly adhered to and thorough safety precautions observed regarding toxicity hazards.

Corticium disease: Corticium disease (caused by *Corticium fuciforme*) seems to becoming a more common problem on amenity turf than it used to be. In the past the disease seems to have had little more the nuisance value in the summer particularly in relatively weather with dewy nights. But nowadays it is sometimes more serious and seems to attack throughout the year. On domestic lawns it can be said that the situation is pretty much as it has always been. During summer, patch of grass may take on a bleached but brownish look and und close observation, small pink or red branched needles of the fungus may be found on the grass blades and sheaths.

Surface basidiomycetes: A fungal trouble which seems to have become more prevalent in recent years and which has not yet acquired a common name is caused by superficial Basidiomycetes which attack surface fibrous material in which they can be seen as white mycelium. There are believed to be several fungi concerned, one being *Cristella confinis* but the others are as yet unidentified. They cause depressions in the surface. Especially on fine lawns with a fairly thick fibrous layer by reducing the thickness in the patches attacked. The fungi seem difficult to control but, if control measures are desired, it is probably worth while giving the patches repeated applications of organic fungicide.

Fairy rings: Fairy rings are caused by soil-inhabiting fungi belonging to the Basidiomycete family which, starting from a central point, spread out in ever-widening circles, the mycelium of the fungus ramifying in the soil and, in the worst cases, more or less making it waterproof. There are several types of fairy ring. The worst one (so called Grade I fairy ring) is caused by the fungus *Marasmius oreades*. This usually appears as two dark green rings with a brown or bare zone between; they may be quite small (less than a dinner plate) or very large indeed and stretching over whole fields. The soil, impregnated to a depth of 100-300 mm or more with white mycelium, smells strongly of mushrooms and the fruiting bodies which appear on the surface are, in fact, tan-coloured toadstools.

Grade 2 fairy rings show up as dark green ribbons or rings of grass on which puff balls or mushrooms may occasionally be found. One of several different types of fungus may be responsible and though they may dry out the soil there is no brown or dead area and the dense white fungal mycelium does not go as deeply into the soil as it does with Grade I rings. Once again people usually tolerate these since their effect is so small. In some cases they can be controlled by application (after spiking) of organic fungicide well watered-in and this will usually at least prevent the production of fruiting bodies.

Grade 3 fairy rings are also produced by many types of fungus. There is no apparent effect on the grass but rings or groups of toadstools may appear. Control is probably unnecessary since the toadstools can be swept off or mown off but local application of organic mercury fungicide will usually stop the trouble if required.

Control: Drenching of soil with Bordeaux mixture (4:4:50) or bavistin powder or any other suitable fungicide.

Pests

Birds: Starlings, rooks, jackdaws and other birds pay attention to turf mainly when there are grubs or insects of some kind in it. They may therefore serve as a warning that action is needed to deal with the infestation and, once the cause of their interest is removed, they soon go away.

Cats and dogs: On many lawns the chief pests are the cats and dogs, particularly the latter, and there are severe limits as to what can be done to stop their attentions. Once one dog or cat has used the lawn or garden as a lavatory it may well soon become a public lavatory for dogs or cats and continue so for a very long time. Surface excreta should be removed despite the unpleasantness of the task and very often it is a good idea to give a copious application of water. Bitches do most damage in that as their urine may cause severe scorch which is not seen until it is too late to do much about it. If it is seen in time then thorough watering minimises the damage.

Field mice: Occasionally small holes made by field mice can be found and many people prefer to take no action since the field mouse seems to be regarded with some affection. Proud owners of very fine lawns, however, may have other ideas and try to get rid of them by trapping or gassing. Sometimes filling the holes and making life uncomfortable for the little animals succeeds in driving them away.

Moles: Lawns in urban areas are seldom disturbed by moles but elsewhere moles may invade from nearby land where they are outside the lawn owners' control. Moles feed on earthworms for which they have a pressing and voracious appetite so that probably the best way of discouraging moles and their activity in producing mole hills, which can more or less destroy a good lawn, is to kill off the worms. Poisoning moles with baits is dangerous and trapping is best carried out by a professional mole-trapper.

Rabbits: Rarely a pest of domestic lawns, these animals can do a surprising amount of harm if they do occur. Their droppings are most noticeable and seem to foul the land. It is interesting to note that rabbits are selective grazers. On turf trial grounds it has been noted that with various types of grass sward to select from, they show distinct preferences which vary from time to time without a very consistent pattern. They certainly favour succulent grass on newly-sown areas. Where rabbits are a problem, trapping them or wire fencing them out seem to be the most appropriate measures.

Rats: These occasionally cause damage to lawns which are near concentrations of the creatures in farm buildings or river banks. There are special baits obtainable from chemists which can be effective although they must be used with very great care. In some cases gassing as for moles proves a better approach. Trapping is seldom practicable.

Earthworms: The most common natural pest is the earthworm. There are many species of earthworm but only few of them produce casts of earth on the surface of turf. In the garden as a whole, worms do a good job of work in mixing and aerating the soil, but worm casts on the surface of a lawn are unsightly and produce a muddy surface in wet weather, being especially a nuisance when the grass has to be mown. Recently, a earth species known as Jai Gopal has been domesticated at IVRI, Bareilly which survive at high temperature.

Ants: Small heaps of fine soil is deposited on the surface of the lawn mainly in summer. It caused by Ants (*Lasius flavus)* usually yellow brown, which carry soil particles up to the surface and make a underground passage. For control brush it away when soil is dry, water the nest and dust the area with alrdin or thimate or any insecticide in powder form.

Other Problems

Pale or yellow lawn: Yellowing of lawn may be due to several reasons and hence proper care must be taken in checking the problem. This may be due to water logging conditions or due to poor drainage in heavy rainfall areas or lack of water. Deficiency of nitrogen is also responsible for yellowing of the grass, which can be prevented by frequent application of nitrogenous fertilizers.

Frost-injury: In north India, the grass is injured due to frost. This can be avoided to a great extent if the grass is sprayed with water every evening and in the early morning after frost.

References

Bhattacharjee, S.K. (2006). In: Advances in Ornamental Horticulture, Vol. 5. Agarwal Printing Press, Jaipur, India.

Bhattacharjee S.K. (2004). Landscape Gardening and Design with Plants, Aavishkar Publishers, Distributors, Jaipur, India.

Bose, T.K., Dhua, R.S. and Maity, R.C. (1999). Floriculture and Landscaping. Naya Prakash, Kolkata, India.

Randhawa G..S. and Mukhopadhyay, (1986). Floriculture in India. Published by Allied Publishers Pvt. Ltd.

Swarup V. (1997). Ornamental Horticulture. Published by Macmillan India Limited.

Ornamental Horticulture, pp 147-169

New India Publishing Agency, New Delhi - 110034, India

7

CAD for Landscape Gardening

S.Y. Chandrashekar[1] *and B. Hemla Naik*[2]

Computer-Aided Design (**CAD**) is the use of Computer Technology for the design of objects which are real or virtual. CAD involves more than just shapes. As in the manual drafting of technical and engineering drawings, the output of CAD must convey information such as materials, processes, dimensions, and tolerances according to application-specific conventions. CAD may be used to design curves and figures in two-dimensional (2D) space; or curves, surfaces, and solids in three-dimensional (3D) objects.

Utilities of CAD

1. It is an important industrial art extensively used in many applications including
   - Automotive
   - Shipbuilding
   - Aerospace industries
   - Industrial and Architectural Design
   - Prosthetics
   - Landscape Designs
   - And many more

Residence plan

[1&2]Department of Floriculture and Landscape Architecture, College of Horticulture (UHS, Bagalkot), Mudigere, Chikmagalur District, Karnataka

2. CAD is also widely used to produce computer animation for special effects in movies, advertising and technical manuals.
3. Because of its enormous economic importance, CAD is a major driving force for research in computational geometry, computer graphics and discrete differential geometry.
4. It is used in the design of tools and machinery and in the drafting and design of all types of buildings, from small residential types to the largest commercial and industrial structures.
5. It is mainly used for detailed engineering of 3D models and/or 2D drawings of physical components. It can also be used to design objects.
6. CAD enables designers to layout and develops work on screen, print it out and save it for future editing, saving time on their drawings.
7. Occupations which use CAD include designers, landscape architectures, architects and developers.

Current Computer-Aided Design software packages range from 2D vector-based drafting systems to 3D solid and surface modellers. Modern CAD packages can also frequently allow rotations in three dimensions, allowing viewing of a designed object from any desired angle, even from the inside looking out. Some CAD software is capable of dynamic mathematic modelling, in which case it may be marketed as CADD — Computer-Aided Design and Drafting.

The CADD (Computer Aided Design and Drafting)

In the Computer Aided Design and Drafting (CADD), we need not use the essential drawing board tools: paper, pencil, T-square, compass, eraser or scale for designing or make a drawing. Even one of these tools is missing, it will be difficult to draw manually. CADD eliminates use of any one of them! CADD is an electronic tool that enables quick and accurate drawings with the use of a computer.

CADs' image

Advantages of CADD

- With the use of CADD one can sit back in an easy chair and create wonderful drawings just by clicking the buttons of a keyboard and mouse.

- The drawings created with CADD are neat, clean and highly presentable.
- The electronic drawings can be modified quite easily and can be presented in a variety of formats.

There are hundreds of CADD programs available. Some are intended for general drawing work while others are focused on specific engineering applications. There are programs which enable to do 2D drawings, 3D drawings, renderings, shadings, engineering calculations, space planning, structural design, piping layouts, plant design, landscape design, project management, *etc*.

CADD is primarily intended for single-line drafting. It has very limited capabilities to create artistic impressions. However, CADD's 3D and rendering features are fascinating. One can create a 3D model of an object and view it from any angle. With proper shading and rendering, it can be made to look picture perfect!

What can be done from CADD?

The following are some of the important capabilities of CADD

- Presentations
- Flexibility in editing
- Units & accuracy levels
- Storage and access for drawings
- Sharing CADD drawings
- Project reporting
- Engineering analysis
- Computer Aided Manufacturing (CAM)
- Designing
- Add-on programs

3 D Plan

Presentations: The fine drawings can be created with hundreds of colours, line types, hatch patterns, presentation symbols and text styles.

For example, a site planner can add tree symbols, shrubs, pathways, human figures and other landscape elements to create a site plan. Similarly an architect can use ready-made symbols of doors, windows, plants, furniture, *etc*., to make a presentation.

In addition to preparing impressive presentations on paper, CADD can be used to make on-screen presentations. Just plug computer into a projector and present the ideas on-screen. Advanced CADD programs even allow creating animated images. We can illustrate how a building would appear while walking through it, or how a machine would function when different machine assembly parts operate.

Flexibility in editing: CADD provides the flexibility to make quick alterations to drawings. It is easy to erase any portion of a drawing with pinpoint accuracy. It takes only seconds to do a job that could take hours on a drawing board. In many cases, we won't even have to erase the drawing to make the change. We can rearrange the existing components of the drawing to fit new shape. This enables us to analyze design options with minimal effort.

The following are some of the editing capabilities of CADD:

- Move or copy drawing elements
- Enlarge or reduce parts of a drawing
- Add one drawing to another
- Stretch a drawing to fit new dimensions
- Make multiple copies of a drawing element
- Change the size and style of text
- Change units of measure, accuracy and style of dimensions

Units and accuracy levels: CADD allows us to work with great accuracy to create highly accurate geometrical shapes. It avoids time consuming mathematical calculations. These units are represented in standard formats commonly used by professionals.

Example: When working with engineering units, one can specify whether all the dimensions should be represented in inches, feet-inches, centimetres or meters and angular units of measurement such as decimal degrees, minutes, seconds or radians.

We can set an extremely high accuracy for the units of measurement. We can work with as high precision as $1/1000^{th}$ of an inch! However, such accuracy is seldom required.

Storage and Access of Drawings: It is quick and convenient to organize drawings in a computer. We can have thousands of drawings on a computers' hard disk and can open any one of them within seconds. A computers' electronic filing system has the following advantages over the traditional filing system:

- It enables to create a highly organized environment
- It contributes to savings in working space
- An electronic drawing never gets old and faded
- Needs a drawing any time; one can print a new copy from the disks

Sharing of drawings: The electronic drawings can be shared by a number of users, allowing them to coordinate projects and work as a team.

Project reporting: The computer can be used to prepare project reports such as records of areas, quantities and cost estimates. Using the database capabilities of CADD, one can link non-graphic information (such as text or value) with the graphic elements of the drawing.

Example: An architect can link text attributes associated with the symbols of plants, doors and windows in a drawing. The attributes can describe the size of the plants, door, material, hardware, cost, *etc.* Later on the computer can automatically prepare a door schedule listing all the doors and windows in the drawing.

Engineering analysis: There is a separate category of programs called Computer Aided Engineering (CAE) that can use CADD drawings for engineering analysis. The CAE programs have a number of applications in structural design, civil engineering, mechanical engineering and electrical engineering.

Computer Aided Manufacturing (CAM): CADD extends its power to yet another branch of engineering called Computer Aided Manufacturing (CAM). CAM is a common method of manufacturing used by large corporations.

Design: It provides a convenient means to create designs for almost every engineering discipline. It can be used for architectural design, landscape design, interior design, civil and surveying, mechanical design, electrical engineering, plant design, industrial design, duct design, electronic circuit design, plumbing design, textile design and product design.

Add-on programs: There are a number of separate programs available that can enhance the power of CADD. The add-on programs work as an extension of CADD to accomplish specific tasks. An architectural add-on program can allow us to instantly draw symbols of doors, windows, kitchens, bathrooms, staircases, *etc.* Shading and rendering program can be used to enhance the look of 3D images. A plumbing design program includes special functions to draw pipes, drains and plumbing joints.

CADD's Hardware and Software

There are two parts of a CAD computer system, hardware and software. The Computer hardware is the physical components of the computer such as system unit, monitor and plotter. Computer software is the program that determines the application of a system.

There are three main categories of computers with respect to hardware:

- Mainframe
- Minicomputer
- Microcomputers, for example personal computers (PCs)

Mainframe computers have a lot of data processing power and their size is quite big. A single mainframe computer performs all the data processing and is accessed *via* terminals connected to it. Minicomputers are smaller versions of mainframe computers.

Microcomputers (PCs) are the desktop or laptop computers of today and are used for individual computing needs.

There are two main categories of computer software:

- System software
- Application program

The system software manages the internal operations of the computer. The application programs are tools which help us accomplish our work, such as CADD.

Hardware: The following are the main hardware components:

- System unit
    - i Central processing unit
    - ii Memory
    - iii Hard disk, floppy disk, CD-ROM
- External storage devices
- Monitor
- Printers and plotters
- Keyboard
- Digitizer, Puck and Mouse

CAD Mouse

System unit: The system unit is used for all data processing. The main components of the system units are the Central Processing Unit (CPU) and memory. In mainframe and minicomputers, CPU and memory are usually separate compartments which house thousands of devices. In todays' PCs, however, they all fit in a small box commonly known as a desktop computer. The desktop computers are equipped with a hard disk; floppy disk drive and CD ROM. The parts of the system unit are:

- Central Processing Unit
- Memory
- Hard disk, floppy disk, CD-ROM

External storage devices: There are a number of external storage devices available such as magnetic tapes, zip drives and removable hard disks. They are commonly used to keep backup copies of electronic files for safekeeping.

Magnetic tapes are quite common for storing large volumes of data. A magnetic tape that looks like a small videocassette can store thousands of megabytes of data.

Zip drives were introduced recently; they use zip disks to store data. A zip disk is like an upgraded version of floppy disk that can store about 100MB of data. They are particularly helpful if we plan to use CADD or other graphic programs that generate quite big data files. We can store an entire project on one zip disk that could take 50 or more floppy disks.

Another new option for data storage is the removable hard disk. We can remove the entire hard disk and use it on another computer.

The monitor: The monitor is the computer screen and is used to display information. A good monitor is very important to display fine graphics. A color monitor is essential because many CADD drawing techniques are based on colors. The main factor that determines the quality of a monitor is the resolution. The term resolution refers to sharpness of an image displayed on the screen. Resolution is measured by the number of picture elements (pixels) that a screen can display. The more pixels and the closer they

CAD network

are, the sharper the image. The distance between pixels is called the 'dot-pitch". The smaller the dot-pitch, the sharper will be the image.

Printers and plotters: The drawings are printed using fine-quality printers and plotters. Drawings are neat and clean and as accurate as the naked eye can see. We can print drawings at as much as 1200 dpi (dots per inch) accuracy. This means 1200 dots are printed in a one-inch-long line. All the text dimensions and other graphics are printed highly accurate, neat and crisp.

Digitizer, puck and mouse: The digitizer (also known as a graphic tablet) and the puck are the data input devices most commonly used in CADD systems. These devices allow us to enter point locations on the screen and to make selections from the menus. As the puck is moved over the surface of the digitizer, it moves the indicator (cursor) on the screen relatively. To enter a point, we need to position the cursor at the appropriate position on the screen and then press the 'Enter"button on the puck. Digitizers are available in many sizes and styles. A number of commands are printed on the digitizer surface. To enter a command, place the puck over the desired command and press the 'Enter"button. The selected command is instantly entered. The puck buttons are configured to perform many other tasks. For example, one button is used to make selections, another to enter the data, another to return to the previous menu and another to cancel the last command.

A mouse is another pointing device that can be used with CADD. Like the puck, the mouse allows us to control the position of the cursor on the screen by rolling it across a flat surface, but it does not require a digitizer. Some programs support working with a mouse only, while others support both the mouse and the digitizer. A mouse is much cheaper than a digitizer or puck, but provides only limited data entry options.

CADD Software

The program contains hundreds of functions which enable us to accomplish specific drawing tasks. A task may involve drawing an object, editing an existing drawing, displaying a view of the drawing, printing or saving it, or controlling any other operation of the computer. The functions contain a number of commands which enable us to specify exactly what we want to do and how we want to do it. The functions are organized into modules which provide easy access to all the commands. The program is divided into modules such as draw, edit, data

CAD model

output, function control, data storage and management. A program may also have a number of specialized functions such as layers, database and 3D. Lets' have a look at the CADD modules:

- Draw
- Edit
- Data output
- System control
- Data storage and management
- Special features

Draw: The draw module provides access to all the drawing functions. The draw module enables us to draw lines, arcs, circles, ellipses, text, dimensions, symbols, borders and many other drawing components. Draw is the most frequently used module because all drawing work are accomplished using it.

Edit: The edit module lets to change existing drawing elements and manipulate them in a number of ways. One can move, copy or erase drawing components. We can enlarge or reduce the sizes of diagrams or change the color and line type of drawing components. We can also change the size and style of text and dimensions, as well as edit a dimension to show different units of measurement.

Data output: The data output module enables us to display drawings on the screen and then print them on paper. There are two separate sets of functions that help accomplish this:

- View-display functions
- Print/plot functions

The view-display functions allow us to display different views of a drawing on the screen. These functions are used quite often, because every time we need to draw something or edit something, we need to focus on that portion of the drawing. With the help of view-display functions, we can zoom in on a specific portion of the drawing.

Data storage and management: The data storage and management module allow us to store and manage drawing data. Through the use of the functions in this module, we can store drawings as files on the hard disk. We can manage the files in directories and sub-directories and move, copy or delete them as needed.

System control: The system control module allows us to control how CADD works. CADD programs are designed for a broad range of professionals, including architects, landscape designers, engineers and surveyors. With the help of system control functions, we can set the working environment of CADD to suit our needs.

Special features: CADD programs usually offer a number of special features which make working with CADD easier and allow us to automate many drawing tasks.

For example, we can create layers in a drawing that allow us to segregate drawing components. We can develop spreadsheets and databases that can be used to create many types of project reports. We can create three dimensional (3D) drawings, such as isometrics and perspectives, with the help of 3D functions.

3D view of Garden

Components of a CADD Drawing

CADD drawings are created by drawing individual components (also called drawing objects or entities) of the drawing such as lines, arcs, circles, ellipses, symbols, text, dimensions and symbols.

Example: Drawing a square with rounded edges. Using a drawing board, we would quickly draw four lines and round off the edges with a template. Using CADD to draw the similar square, we will have to draw four lines and four arcs separately. We must specify where each line and arc starts, where it ends, the centre point of the arc, the radius, *etc*. There are separate functions to draw text, dimensions, patterns, symbols and other elements of a drawing.

The Coordinate system: The coordinate system is another method of locating points in the drawing area. It enables us to locate points by specifying distances from a fixed reference point. We can locate a point by giving its distance in the horizontal direction, vertical direction, measuring along an angle, etc. The most common coordinate systems are as follows:

- Cartesian coordinates
- Polar coordinates

Cartesian coordinates: Cartesian coordinates is a rectangular system of measurement that enables us to locate points with the help of horizontal and vertical coordinates. The horizontal values, called X-coordinates, are measured along the X-axis. The vertical values, called Y-coordinates, are measured along the Y-axis. The intersection of the X- and Y-axes is called the origin point, which represents the 0, 0 location of the coordinate system.

Polar coordinates: Polar coordinates allow us to define a point by specifying the distance and the direction from a given point. This mode of measurement is quite helpful in working with angles. To draw a line at an angle, we need to specify how long a line we want to draw and specify the angle.

The formats to enter coordinates: Cartesian or polar coordinate values can be entered in two formats:

- Absolute
- Relative

Absolute format is a way of measuring distances from a fixed reference location (origin point), which is the 0, 0 location of the coordinate system. Absolute coordinates are primarily used to adjust the alignment of diagrams in a drawing, to align one drawing with another or to make plotting adjustments. Relative format is a way of measuring distances from the last point entered. All measurements are taken the same way as the absolute coordinates, with the only difference being that the relative coordinates are measured from the last point entered instead of the origin point.

User-Defined Coordinate System: CADD allows us to create a user-defined coordinate system that can help simplify drawing. When we need to work with a complex drawing that has many odd angles this mode of measurement is very useful. Let's say we need to draw or modify an odd-shaped diagram, it is very difficult to use cartesian or polar coordinates because they would involve extensive calculations. In this case, we can create a custom coordinate system that aligns with the odd angles of the diagram. The user-defined coordinate system is especially helpful when we are working with 3D.

CAD drawing top floor

A Prototype CADD Drawing: A CADD program is designed to meet the needs of a vast CAD drawing top floor number of individuals. It comes with a number of options that can be customized. For example, architects, engineers and surveyors use different annotations to measure units and to draw dimensions. We may have a number of preferences regarding a specific style of text, standard sheet sizes and names of layers, symbol styles, borders, or line types to be used in a project.

How to Correct the mistake?

If we make a mistake in issuing a command or in entering any data, we can fix it without many hassles. Most CADD programs have a built-in function (commonly known as UNDO) that instantly reverses the effect of the last command entered.

When to save the drawing or quit

When we feel that sufficient work has been completed, we should save the drawing.

Automatic data saving features: Most CADD programs automatically create a back-up file when we save a drawing. The back-up file contains the original information that was in the file before we saved it. so, if we save a drawing and later realize that we need the earlier version, we can retrieve the original information from the back-up file.

CADD filing system: The computer stores information in blocks of data called files. Each drawing created with CADD is stored as a separate data file.

Getting help: Most of the CADD programs come with built-in help features in the following ways

- We can choose to use a basic tutorial
- We can display an index that lists all the commands available in the program
- We can view topics on any of the commands listed in the index
- We can usually access task-based help in on-line help provided with the program
- We can consult the printed user guides and reference manuals

The Drawing Tools of CADD

The following are the basic drawing tools found in a CADD program:

- Line types
- Flexible curves
- Ellipses and elliptical arcs
- Multiple parallel lines
- Arcs and circles
- Text

- Dimensions
- Polygons
- Hatch patterns
- Arrows

CAD Lines and Curves

Using line types: There are a number of line types available in CADD. They are continuous lines, dotted lines, centre lines, construction lines *etc*. CADD enables us to follow both geometrical and engineering drawing standards. We can use line types to represent different annotations in a drawing.

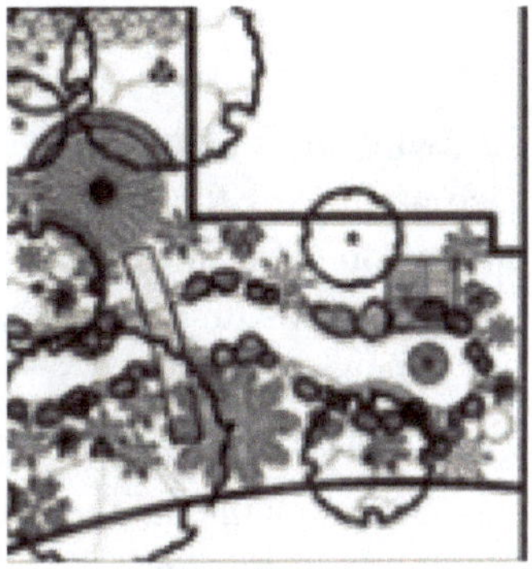
CAD lines and curves

Drawing multiple parallel Lines: CADD allows us to draw parallel lines simultaneously just by indicating a starting point and an end point. These lines can be used to draw something with heavy lines or double lines. For example, they can be used to draw the walls of a building plan, roads of a site map or for any other presentation that requires parallel lines.

A number of add-on programs use multiple lines to represent specific drawing features. For example, an architectural program has a special function called 'wall.' When we use this option, it automatically draws parallel lines representing walls of specified style and thickness.

Drawing flexible curves: CADD allows us to draw flexible curves (often called splines) that can be used to draw almost any shape. They can be used to create the smooth curves of a sculpture, contours of a landscape plan or roads and boundaries of a map.

Drawing arcs and circles: CADD provides many ways to draw arcs and circles. There are a number of advanced techniques available for drawing arcs and circles which can simplify many geometrical drawing problems. We can draw an arc by specifying circumference and radius, radius and rotation angle, chord length and radius, *etc*.

Circles and Lines

Drawing ellipses and elliptical Arcs: Ellipses are much easier to draw with CADD than on a drawing board. On a drawing board, we need to find the right size template or draw a series of arcs individually to draw an ellipse.

Adding text to drawings: CADD allows us to add fine lettering to your drawings. We can use text to write notes, specifications and to describe the components of a drawing. Text created with CADD is neat, stylish and can be easily edited.

Drawing dimensions: CADD's dimensioning functions provide a fast and accurate means for drawing dimensions. CADD automatically calculates the dimension value and draws all the necessary annotations. The annotations that form a dimension are dimension line, dimension text, dimension terminators and extension lines.

Adding hatch patterns to drawings: The look of CADD drawings can be enhanced with the hatch patterns available in CADD. The patterns can be used to emphasize portions of the drawing and to represent various materials, finishes and spaces. Several ready-made patterns are available in CADD that can be instantly added to drawings. Hatch patterns are quite easy to draw. We don't need to draw each element of a pattern one by one. We just need to specify an area where the pattern is to be drawn by selecting all the drawing objects that surround the area. The selected objects must enclose the area completely, like a closed polygon. When the area is enclosed, a list of available patterns is displayed. Select a pattern and the specified area is filled.

Drawing symbols: Symbols provide a convenient way to draw geometrical shapes. We may compare this function with the multi-purpose templates commonly used on a drawing board. To draw a geometrical shape, such as a pentagon or hexagon, select an appropriate symbol from the menu, specify the size of the symbol and it is drawn at the indicated point.

Drawing arrows: Arrows (or pointers) in a drawing are commonly used to indicate which note or specification relates to which portion of the drawing, or to specify a direction for any reason.

The view-display functions: CADD's view-display functions enable us to display various views of the drawing. We need to constantly change views in order to work on different portions of the drawing. For example, to draw a line from one corner of a large drawing to another, we may need to change the view two to three times. First, we need to magnify the corner where the

3D visual display

line is to be started, then shrink the display to see the entire drawing and locate the other corner of the diagram, and then magnify that corner to indicate the end point of the line. Working with the view-display functions is similar to using a telescopic camera. We can zoom out to view larger objects, and zoom in to get a closer view of objects.

The Concept of Layers in CADD

Layers are imaginary transparent surfaces that can be created within a CADD drawing. We can draw on these imaginary surfaces and group drawing objects on different layers. This helps to organize CADD drawings and makes editing much easier. We can turn the layers on or off to view or edit specific objects. When we turn a layer off, the objects on that layer disappear from the screen; when we turn it back on, they are displayed. We may create dozens of layers in a drawing to segregate drawing objects. We may create separate layers to show dimensions, hatch patterns and for any drawing objects that are different in nature. When drawings are complex and a number of drawing objects are overlapping, we can turn off the unwanted layers. This makes it much easier to select the desired objects for editing.

Layers in CAD

Creating layers for an engineers' office: We place walls and partitions on one layer, furniture and equipment on another, and so on for text, symbols, and construction lines, etc. We can also create a separate layer to show the ceiling and lighting plan, floor finish plan, or any other drawing objects that we want to include in the drawing.

The tools to work with layers: CADD provides a number of tools that allow us to work efficiently with layers. Most programs use a layer control window that allows us to manage all aspects of working with layers.

Are layers absolutely necessary?

Layers are not absolutely necessary; however they provide a convenient way to manage information in CADD drawings. With the help of layers, we can create just one composite drawing. The composite drawing can contain all the plans. It makes it easier to correlate information from one plan to another.

3D Drawings

What is the use of 3D?

3D capabilities allow us to draw pictorial views such as isometrics, oblique views and perspectives. The views drawn with CADD have a number of advantages as compared to views drawn on a drawing board. The views drawn with CADD are very accurate and provide a lot of flexibility in terms of editing and display. We can rotate a model on the screen just like an actual model and display views from different angles. Designers often use 3D to visualize designs and to make presentations. It helps them understand how an object will appear from different angles. Using additional rendering programs can further enhance the drawings.

Working with 3D

Pictorial views concept

There are two distinct ways to draw 3D views with CADD: we can draw views using simple 2D functions or using CADDs' special 3D functions. The 2D functions allow us to draw views just like on a drawing board. We can draw a view using lines, arcs or other 2D objects. This is the quickest method to draw simple isometric and oblique views. CADD provides special 3D functions which allow us to create 3D drawings that are true representations of an actual model. These drawings can be viewed from any angle just like a physical model. That is why 3D CADD drawings are called 3D models. The major distinction between a 2D drawing and a 3D model is that a 2D drawing is defined only with two coordinates (X and Y). A 3D model is defined with three coordinates (X, Y and Z). The Z-coordinate determines the height of an object.

3D Garden view

3D Modelling

3D Model Residence

CADD's 3D modelling capabilities allow us to create 3D images which are as realistic as the actual objects. These images are called 3D models because, just like a physical model, they can be rotated on the screen. We can display views from a 3D model, such as isometrics or perspectives, from any angle with a few simple steps.

Working with 3D Coordinates: 3D coordinates are measured with the help of three axes: X, Y, and Z. The axes meet at a point in the shape of a tripod. This point is called the origin point, which is the 0, 0, 0 locations of all coordinates. All distances can be measured using this point as a reference .The three axes form three imaginary planes: XY plane, XZ plane and YZ plane. The XY plane is the horizontal plane and the XZ and YZ are the two vertical planes. When we need to draw something horizontal, such as the plan of a building, we draw it in the XY plane using X and Y coordinates. This generates a plan view. When we need to draw something vertical, such as an elevation of a building, we draw it using the XZ or YZ planes.

The 3D coordinates can be entered using the following formats:

- Cartesian coordinates
- Spherical coordinates
- Cylindrical coordinates

Coordinates in CAD

Cartesian coordinates: Cartesian coordinates are based on a rectangular system of measurement. Positive Z coordinate values are used when we need to measure distances above the XY plane; negative values are used for the distances below the XY plane. Coordinate values are entered separated by commas (X, Y, and Z). The coordinates can be measured from the origin point (absolute coordinates) or from the last reference location of the cursor (relative coordinates).

Spherical coordinates: Spherical coordinates are based on the longitude and latitude system of measurement. Consider the origin point of the coordinate system at the center of the earth or a transparent globe. Then consider a horizontal plane (XY plane) passing through the center of the globe. To locate a point in 3D, first locate a point in the XY plane by specifying a radius and an angle (polar coordinates). To specify the height, enter an angle up or down from the XY plane (latitude).

Cylindrical coordinates: Cylindrical coordinates are commonly used to draw cylindrical shapes. They are based on a cylindrical system of measurement. Consider a cylinder placed vertically and the origin point at the center of the cylinder. Cylindrical coordinates are quite similar to spherical coordinates, the difference being that the Z-coordinate is specified by height and not angle

View of Residence Garden

Displaying views: We can rotate a 3D model on the screen and display different views. The viewpoint represents the position of the camera from where a picture of the view is to be taken. We can define a viewpoint with the help of any of the coordinate methods.

There are two main protocols used to display views:

- View coordinate geometry
- Object coordinate geometry

View coordinate geometry: View coordinate geometry assumes that the camera (viewpoint) remains stationary and the 3D model is rotated to display a desired view. The model can be rotated around the X, Y, or Z axis.

Object coordinate geometry: Object coordinate geometry assumes that the model remains stationary and the camera (viewpoint) is moved to a display a desired view. We can use any of the coordinate methods to specify an exact viewpoint. Spherical coordinates are particularly helpful to indicate a viewpoint.

Comparison: View coordinate geometry can be compared to holding a small model in your hand and rotating it on its sides to get a desired view. Object coordinate geometry can be compared to viewing a building from the sky. The building remains stationary, while the camera is moved to get a desired view.

Displaying isometric views: To display an isometric, we need to specify the direction from which the view is to be taken. The most appropriate method to indicate direction is with the help of spherical coordinates.

Displaying plans and elevations: We can display standard 2D views such as plans and elevations by specifying the direction of the view. To display a plan view, we need to view the model from the top, that is, enter the angle from the XY plane (latitude) as 90^0.

Displaying perspective views: CADD allows us to display perspective views from any angle of the model. We can display a true perspective by specifying an exact distance between the viewpoint and the model.

3D Drawing-aid functions: The common 3D drawing-aid functions of CADD are described as follows:

- 3D ready-made shapes
- Linear extrusion
- Radial extrusion
- Shading and rendering

3D Ready-made shapes: CADD allows us to draw a number of 3D ready-made shapes in a few simple steps. To draw a cube, we don't need to draw all the lines or 3D faces for each of its sides. We can instantly draw a cube by specifying its dimensions.

Extruding objects in the linear direction: CADD allows us to extrude 3D shapes from 2D profiles. We can extrude a square to form a cube, a circle to form a cylinder, or a triangle to form a prism. When we use the linear extrusion function, we are prompted to select the objects to be extruded and specify the direction of extrusion (axis of extrusion). We can select any profile made of lines, arcs, poly-lines, or other objects and extrude it in any direction by specifying the axis of extrusion.

Extruding objects in the circular direction: CADD allows us to extrude drawing objects along a circular path (called radial extrusion). For example, we can draw a section of a cylinder and extrude it to form the complete cylinder.

Shading and rendering: There are a number of shading and rendering programs available that can be used to make 3D drawings very realistic. These programs allow us to create colors, shades and shadows exactly as they would appear in a picture. With the help of rendering programs, we can specify a number of shading and rendering parameters and create a 3D scene. We can assign colors and textures to different surfaces of a model. We can specify light sources in a scene and specify what kind of light is used and how it is directed. We can create a special setting for the model, such as a landscape or interior.

Printing and Plotting

Printing with accuracy

CADD drawings are printed using a printer or a plotter. The process of printing is as simple as selecting the print or plot function from the menu. This action sends data from the computer to a printer or plotter, which produces the final drawing. The drawings are neat, clean and, depending on the quality of the printer, highly accurate. We can specify a number of parameters to control the size and the quality of a plot. We can plot a drawing to any size by applying an appropriate scale factor. We can specify line thickness and colors for different drawing objects.

The following are the important considerations for plotting:

- Selecting a scale for drawings
- Composing a drawing layout
- Selecting text and dimension heights
- Choosing pens colors and line weights

Selecting a scale for drawings: When working on a drawing board, we use a specific scale to draw diagrams.

For example, when we need to draw a plan of a building or a township, we reduce the size of the diagrams to 1/100 or 1/1000 of its actual size, that is, we use a 1:100 or 1:1000 scale.

Composing a drawing layout: CADD provides a number of special functions to compose a drawing layout. We can arrange diagrams on a sheet as we like and apply any scale factor.

The following are some of the standard sheet sizes (in inches)

Mark	Size	Mark	Size	Mark	Size
A	8.5x11	A4	8.3x11.7	A	9x12
B	11x17	A3	11.7x16.5	B	12x18
C	17x22	A2	16.5x23.4	C	18x24
D	22x34	A1	23.4x33.1	D	24x36
E	34x44	A0	33.1x46.8	E	36x48

Selecting text and dimension heights: As diagrams are enlarged or reduced by applying a scale factor, the size of text, dimensions, patterns and symbols are also changed. When we place different scale diagrams on the same sheet, we may get different sized text for each diagram. This is generally not acceptable for professional drawings.

Choosing pens, colors and line weights: CADD allows us to work with a variety of colors and line weights depending upon the plotter. In most CADD programs, the colors we use on-screen are configured with a specific line weight in the plotter. For example, the objects drawn with red color on-screen may be printed with 0.5 mm line weight; the objects drawn with blue color may be printed with 0.2 mm line weight. These are called pen assignments.

Plotting area: We can plot a part of the drawing or the entire drawing. We will be able to indicate the plotting area by indicating a window (an imaginary rectangle formed by two diagonal points) or by selecting a specific view for plotting.

Plotting scale factor: Enter a scale factor based on how big or small we want to print the drawing and the sheet size used.

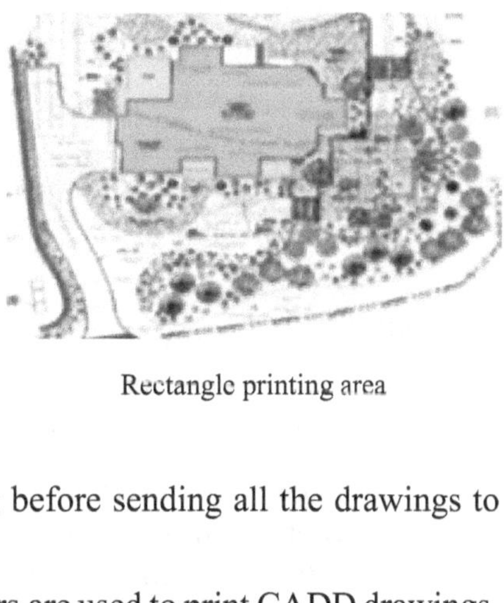

Rectangle printing area

Plotting origin: The plotting origin is a point that allows us to align the drawing shown on the screen with the paper in the plotter. We can place the diagrams on the paper anywhere by entering the exact coordinates of the plotting origin.

Important tips: It's always better to do a test plot of a small area of the drawing before sending all the drawings to plotter or printer.

Printers and plotters: Printers and plotters are used to print CADD drawings. The drawings are generally printed at about 300-600 dpi (dots per inch) accuracy, which is considered quite high precision for engineering drawings. There are machines available that can print at 1200 dpi or higher accuracy as well.

The following are some examples of printers and plotters.

- Pen Plotters
- Ink-Jet Printers/Plotters

- Laser Printers
- Electrostatic Printers/Plotters
- Auto CAD, Micro Station and Cadkey Terms

Tips for selection of CADD

The following criteria may help us make a CADD program selection:

- Is CADD intended for personal use or professional? A low-end product may be accepted for personal use, but for professional use, we may want to consider a high-end product.
- Select a program compatible with the programs used by your clients and consultants. It makes sharing drawings much easier. Although we will often hear from the CADD vendors that their programs are 100% compatible with other programs, it may not be the case. Ask for a trial version of the program and confirm it.
- Review how many customization features a CADD program offers. Customization plays a major role in making CADD more productive.
- Decide whether we need to buy a brand name product or any workable CADD program. Both may provide the same capabilities, but a brand name may cost twice as much. Selecting a brand name program may give us certain advantages such as finding add-on programs and experienced users.
- Find out if the dealer offers technical support. In the beginning we may have a lot of questions about the program. Without technical support, we may have to pay hourly charges to a consultant.
- Ask if the dealer offer free or low-cost updates of the program. Most CADD programs get revised every year. We may have to pay double in four or five years to get all the updates.
- New technology advancements are taking place every day. CADD programs are getting highly specialized to meet the needs of specific professionals. With some research we may be able to find exactly what we need and we may not have to spend a bundle.

Leading CADD Products

- General-purpose
- Mechanical CAD

- Solid modeling
- Visualization and rendering
- Facility management
- CAD/CAM/CAE

References

Ambrosius, L. and David, B. (2006). Auto CAD & Auto CAD LT ® a l l-in-onedesk reference: For DUMmIES, Wiley Publishing, Inc.: pp1-819.

Carl Stelnitz, (2009). Landscape Planning: A History of Influential Deas: Landscaping Architecture, 99 (2):74-85.

David, F. and Jon McFarland, (2008). Autocad® 2008 and autocdLT® 2008, Wiley Publishing, Inc. pp:1- 723.

Farin G.E., Josef, H. and Myung-Soo K. (2002). Handbook of computer aided geometric design. Elsevier Science B V, Netherlands. pp:1-735.

Kamakoti, V. and Shankar Balachandran, (2001). CAD for VLSI Design, pp :1-11.

www.caddprimer.com

www.Wikipedia.com

Ornamental Horticulture, pp 171-185

New India Publishing Agency, New Delhi - 110034, India

8

LED Lighting in Landscape Gardening

Mam C. Singh

The landscape gardening for aesthetic appreciation is an integral part of the progress achieved through the growing world floriculture business that has reached a $ 40 billion mark and valueless aesthetic emotional gains. The industry is continuously rising through transitional changes to have more accepted and unique applications as a wide and economic choice to the consumers with regards to the energy saving in the system. The use of energy for sustenance of the landscape plants before and after planting can be a benchmark in time to come as no one can believe that the plants instead can grow with still better traits either be it the foliage colour, variegation, shape, size and or form beside the fact that the consumption of such altered/compact plants has increased manifold in house, gardens, parks and indoors with a large number of products worldwide in a eco-friendly manner.

At the same time there is great environmental consciousness towards transforming landscape gardening with more natural and unspoiled system of growing plants and maintaining them in the landscape as perennial ones. The system being independent of distance of use or application offers an ease of transport into a mass for large-scale uniform planting. This is the fact that in the recent times, the commercial growers do not rely on the nature to introduce the genetic mutation that will make them millionaires but depends on the art and scientific knowhow to create a novelty in its own. This has resulted in the

Centre for Protected Cultivation Technology (CPCT), IARI, New Delhi seriousness and painstaking effort of the growers to produce the plants round the year in new forms and shapes to put forward a new plant with extra-ordinary marketable characteristics every year, e.g. compact chrysanthemum, kalanchoe, bromeliads, orchids, *etc* are some of the common products with an unusual and eye-catching hue in the auction markets. Similarly, a large number of annual and perennial plants perfectly suited for various kind of landscaping, slope, soil types and moisture regimes under the specific and artificial environment created for a specific growth of the plant.

Compact landscape plants produced under LEDs lighting exposure

The Principle how LEDs are Used in Plants

Since plants only use a tiny percentage of the full light spectrum and chlorophyll being green acts as a 'Sun-block' to remove 95% of the unwanted and damaging light wavelengths due to higher irradiance at high temperature (Hughes and Cockshull, 1972) resulting in poor growth. On the other hand red and blue (LED) lighting have such narrow bandwidths the plants without green chlorophyll can still survive and will be safe against the harmful light wavelengths. Its is further to note that nearly 100% of the light an LED grow light emits is completely absorbed by the

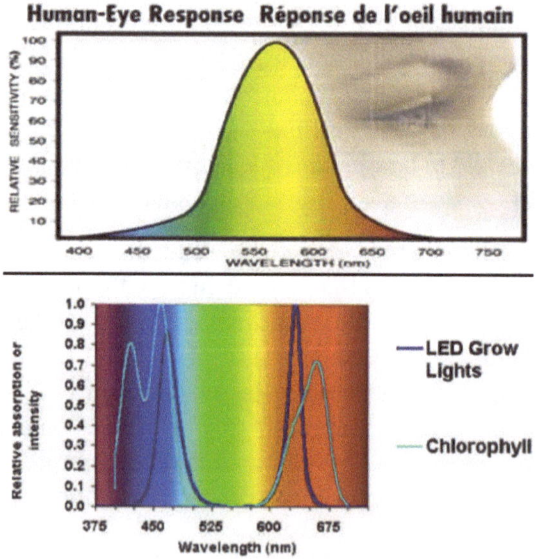

The usage spectrum by plants for photosynthesis

plants. LEDs' can also be used for 24 hours a day without stressing plants since the chlorophyll does not have to do battle with unwanted light-waves as they produce a very small amount of heart over the plants canopy. If we measure the quantum of LED grow lights in Einsteins' it means that how many photons of light strike an area per second which sometimes is just vague in measuring the quantum absorbed in the plants as tiny fraction. However, the irradiance power factor or amount of light that actually gets used in plant growth or chlorophyll production as phytochrome response can be measured as $\mu mol\ m^{-1}s^{-1}$ in the conditions when the lights are hardly 95% growth efficient. Also in that case the 95% light is wasted since plants only use very narrow bandwidths of the total light energy.

What are LEDs

A Light Emitting Diode (LED) is a semiconductor diode that emits incoherent narrow-spectrum of light when electrically biased in the forward direction in the form of an electroluminescence. This light source is usually a small area, often with extra optics added to the chip that shapes its radiation pattern and flow over the plant canopy in different colours. The color of the emitted light depends on the composition and condition of the semiconducting material used which can be infrared, visible or near-ultraviolet.

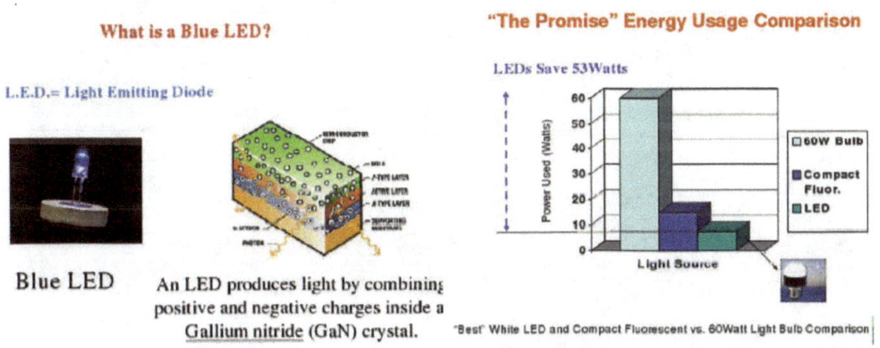

Useful LEDs source and energy usage

Infact, the light emitting diodes are rectifying semiconductor device which converts electrical energy into electromagnetic radiation. The wavelength of the emitted radiation ranges from the near-ultraviolet to the near-infrared, that is, from about 400 to over 1500 nanometers. This light can be used as a regular household light source and most commercial light emitting diodes, both visible and infrared, are fabricated from type III-V semiconductors. Wherein, these compounds contain elements such as gallium, indium, and aluminum from column-III (or group 13) of the periodic table, as well as arsenic, phosphorus and nitrogen from column-V (or group 15) of the periodic table. There are also LED products made of II-VI (or group 12-16) semiconductors, for example Zn, Se and related compounds. If we combine them together, these semiconductors possess the proper band-gap energies to produce radiation at all required wavelengths. Most of these compounds have direct band gaps and therefore, these are efficient in the conversion of electrical energy into radiation

Need of Lighting

Alternatively, in the current scene there is an ardent need to explore the creation of diverse product forms in commercial floriculture using an artificial lighting for plant growth in many ways to provide sufficient light that a plant needs to grow without the shift to a suitable location and its climatic demands. The supplementing sunlight and photoperiodic manipulation, especially in winter months when daylight hours are short and by increasing the length of the 'day"in order to trigger specific growth and flowering is just perfect for producing a desired product in least possible time frame. Therefore, its obvious

Use of Garden LED Lights

that if we use normal lighting in the growing system it's a high energy usage and to exploit the potential of saving electricity by using semiconductor such as Light Emitting Diodes (LED), which at the same time is advantageous in its applications and under such conditions the plant grows faster without disturbing their response to the days and night cycle.

It is universally known that the plant require light as source of energy in photosynthesis, attain growth and later to sustain it for maintenance. Basically there are two main requirements dominate the utilization of artificial light sources in floriculture in a commercial greenhouse, the first being efficiency as seen with high pressure sodium lamps, generally adopted to offer the highest efficiency in terms of plant growth rate and economy. However, to remain within current standards of farm products (such as leaf greenness and coloration, internodes length, stem diameter, and leaf thickness), a combination with metal halide lamps is recommended. The second requirement is related to the aesthetic improvement not for its further growth but the maintenance in its natural appearance. The efficiency of light is not a prerequisite, but the balance of quality light becomes important in order to bring out the essential color characteristics of plants and flowers as well as maintaining plant health. Therefore, to meet the requirement for high color rendering lights using LED and LD are being recommended in commercial greenhouses and growth chambers.

How LED Lighting Works

Among all the light intensities intercepted by the plants red and blue light is essential for plant growth as its fact that the chlorophyll, leaves perform the process of photosynthesis and make red and blue spectrum or wavelengths most essential for photosynthesis. The blue light is absorbed to its maximum in the spectrum region of 430-450 nm while the red in the spectrum with 650-670 nm. It has been estimated that the ideal balance for photosynthesis is 92% red LEDs and 8% blue LEDs if it is practiced artificially. In this case the blue lights have a smaller influence than those of the red lights and it makes easy to select anything between 1% and 20% of blue light, depending on requirements of the plants for their growth and developments. The commercial LED lights are available as per colour and the frequency of light used by plants for efficient and healthy growth so as to provide an efficient photosynthetic energy than the use of conventional grow lights. In general, red light being in the vicinity of a plant in the first peak of light absorption spectrum i.e. 660 nm it contributes maximum to the plant photosynthesis and total leaf dry mass. However, the red light, when combined with blue light, encourages flowering and most often used in higher doses to stimulate flowering, seeding and fruiting in certain

How LED exposed landscape plants grows faster

plants of economic importance. Whereas, the blue light is responsible primarily for vegetative growth (leaf) as infect the chlorophyll has its second distinct absorption peak in the vicinity of blue light wavelengths (450 nm). The blue light is indispensable to the morphologically healthy growth plant and enriched more sunlight energy in the blue-green wavelengths.

It is evident that every plant contains photo sensors that respond to the specific part of the spectrum from the ultra violet radiation (UV) to 740 nm spectrum which is known to control various aspects of plant growth and flowering. In last decade, several researchers worldwide have studied the effects of light on plants to explore that which part of the daylight provides most suitable spectrum for photosynthesis. Its further more critical that how low intensities of LED lights (on the order of 1 W/m2 or 100µmoles m-2sec-1in total) can be used to control the desired characteristics in the plants for final consumer as a finished product whether be it size, shape and colour of the foliage or uniform compactness and architecture of the plant as whole. Its been already standardized that the plants use light from 400 to 700 nm for photosynthesis, with some wavelengths having the greatest effect at certain times of the day and others at various stages of a plants' growth.

Commercial Use of LED Lighting

In majority of the flower growing countries, conditions are perfect and the days are always 12 hours long; sunlight is intense, and nights are cool. However, the most commercial sectors are now moving to the still better climatic conditions, into the developing countries, where they take over scarce resources, like water and arable farmland, leaving less for native agriculture or industries and very slowly the host economies become dependent on the growers for example, in Ecuador, flowers are now the third most important product, after oil and bananas. Despite the hard economics, the flower business is thriving well and there are several flowering plants those respond to the natural short day (light) conditions. To cut short their vegetative phase into reproductive with sufficiently elongated stems and quality flower buds we may have provision of providing the plants an extra photoperiodic as well as photo-synthetic light to make them accessible to enough exposure, especially in winter. The blue LED spectrum not only avoid photo-damage in plants but also can affect properties such as compactness of shape and time to flowering. Not only that in such plants grown for food or medicinal use, it can change the plants' chemical content, which can affect flavor, color, and nutritional and/or pharmaceutical value.

Commercial Production of Landscape Plants under LEDs

What is more important is to optimize the lighting requirement for a given effect on the plants. For example, the various scientists in UK and Europe have revealed that a particular red wavelength causes a fuchsia plant to flower four weeks earlier than it does under the same conditions without the additional LED light-ing using daylight as the main source of radiant energy with an additional LED light-ing. It has also been reported that the system could reduce

LEDs Application has Increased 20% Stem Yields

the cost to an economically viable level and is affordable with present-day LEDs commercially. LEDs application has high application efficiency in terms of stem elongation and higher biomass and the availability of narrow emission spectra across the relevant wavelength range over the plant canopy. This is now being used in horticulture for a quite a long time, but they provide the photosynthetic light as well as the control light, which adds to cost. Their use in special transmittance spectra under greenhouses as a recent introduction placed over benches for the pot plants and as visible and near-IR lights is commonly being used on large scale in the roses and chrysanthemum to induce long day response without disturbing dark period under greenhouses in France to improve grape quality.

Landscape Gardening and Aesthetics in Horticulture

LED lights can be successfully incorporated into a landscape in a variety of ways. Specifically, LED strips are effective for lighting long paths, stairs or pools, thanks to their lengthy design. Throw down a long strip, and a large amount of real estate can be covered and properly lit. Mounted LED fixtures that are strategically placed can illuminate a specific spot on a lawn. LEDs can also be installed directly into other devices, such as lanterns, for enhanced lighting, Also, thanks to their immense flexibility; LEDs can be brighter or softer depending on your security needs or the mood that needs to be set. Voltages and amps for the LED are highly configurable. In terms of a landscape job, this means that non-intrusive, low-light bulbs can be used for illuminating footpaths and sidewalks, while brighter lights can be housed just outside the home, or on a stoop. Additionally, lights of wildly different colors can be used to spice up a landscape and further enhance a desired mood. Above all, if we see the aesthetics in landscape lighting even a nighttime lighting is becoming common in and outdoor, almost at no cost is elegance beyond imagination. The following are kind of styles recently being used in gardening.

Key Night Time LED Lighting

LED Garden Stick Lights

LED garden stick lights are being used mainly for energy saving, long working life and easy 'stick and play' installation along with weather-proof and super bright LED light properties. These garden lights look like star at night or fireflies shuttling in your garden. These lights serve as great supplement for garden or home decoration.

Solar Powered LED Garden Stick Lights

These LED garden stick lights can be solar powered as super bright LED color changing light having almost no electric cord wiring, long working hours in an aluminum stick body and durable weather-proof super bright LED light bulbs. These lights normally are built in two solar panels and with one AAA Ni-Nh rechargeable battery and can light up 10 hours after each full recharge. These are best fit for yard decoration and perfect gifts for Christmas and holiday seasons. The most valuable advantage of this kid of lights in the garden is that they can be put simply in water as plug and play, easy to relocate and need no hardware installation. These make a great ease for display in pond, party event or it can be sit on top of the regular fountain to make a beautiful lighted water spray.

Submersible Water Garden LED Lights

These lights in the garden are basically LED light rings and named as ocean mist. These are used as an innovative light source for shinning up running water at fountain water outlet, waterfalls, water fountain spray nozzle or water taps. Each LED ring has a low voltage transformer, a long submersible cord and is

Submersible Water Garden LED Lights

visible underwater. Submersible LED lights can be connected with fountain mounted on pre-filter 1.5 HP pump in the form of a light ring (LR-A120) and a nozzle (FT-05). Often the fountain shooting about 20 cm high and 15 cm wide with 3 tier nozzle head, it has two other spraying patterns with the other two nozzle heads, spraying high/wide could be adjusted. LED light ring (LR-A120) are available in automatically changing light colours. These lights are atleast 15 cm deep, easy in installation and offer a great display in pond or lake. Normally, floating water fountain with 48 LED light ring and a water pump are available commercially and are adjustable to higher or lower by reducing and increasing the spraying jet hole. These LED lights are super bright and available in changing color lights or all white lights and it can be made sit on top of the regular fountain to make a beautiful lighted water spray. Nabesei exhibited a completely watertight LED light in a tank. The lights are also available in a bendable type, which is suitable for interior lighting and plant cultivation requiring water sprinkling.

Full Spectrum LED (red & blue) Grow Lights

These garden lights are high power (90 watt) bulbs and has no ballasts to burn out like other plant lights. These exert a warm temperature rather than very high heat which is common with most other inefficient High Pressure Sodium (HPS) plant lights and reduces the need to water so often and keeps rooms with plant lights from getting uncomfortably hot in the summer months requiring additional air conditioning. This state of the art LED grow light is extremely energy efficient and saves electricity (80-90%) consumption compared to fluorescent tubes and HPS. Over 95% lights of LEDs can be absorbed by plants against 10% by HPS and reduce grow room temperature by 15 degrees compare to HPS and last 50,000 hours. A wide angle projection of these garden grow lights insures uniform leaf

Full Spectrum LED (red & blue grow lights)

canopy coverage. The major advantage of these lights being a design all in one that offers full spectrum grow light that allows maximum level photosynthesis food production, growth and flowering all in one. These garden lights are full spectrum lighting system providing maximum plant health, beauty and productivity.

Solar Garden LED Lights

The solar power LED lights are large variants of solar garden lights e.g., solar fairy lights with changing colours and having almost 100 LED bulbs. These bulbs are also fixed as solar powered illuminated rock lights to provide an ambient lighting for the garden or patio. These garden lights save energy while the lighting unit illuminates automatically. In general, solar garden lights, as with many things, new solar LED technology allows light discharge of upto 12 hours per night. The individual advantage is that these solar energy LED garden lights require no wiring as they charge through the daytime from the suns' rays (they require a sunny position, although will gain some charge on an overcast day). The solar based lED lights come on automatically at dusk, although have a manual over-ride switch. The Solar LEDs panels are also highly efficient, mono-crystal or multi-crystal silicon solar based panels and have rechargeable battery (Ni-Cd/Ni-MH), preferable and adjustable LED color e.g., super bright white, yellow, red, green, blue etc. and the all lights in panel are in automatically turned on mode.

Garden Solar LED Lights

LED Lit Sea Urchins and Lanterns

The new fancy LED lights have been developed in a matching shape of sea objects to give an impression of sea-life by illumination objects in a open garden corners. These lights look like tiny sea urchin shells and often are available at beach shops in US and Europe. These water garden landscape lights are thin and lighten as eggshells. However the sea urchin lamps were available before, but they've always been made with large (i.e., sturdy) and colorful ones. In contrast, these tiny ones might be better to hang around

Christmas lights like little paper lanterns. These when lightens look like living urchins and one will always be surprised by how elegant and mobile they are. These lights very well fit in the 10 mm LEDs holders and required a slightly smaller battery than lamps (CR2032) to fit in here, maybe a low input bulb (CR1632) can be suitable one for tiny illuminated sites.

Temporary and inexpensive garden LED lanterns

While using solar power based LED lights in the garden ultra-low cost temporary garden lights using LEDs, lithium coin cells, and mason jars can be made to start with. For which we need LEDs and CR2032 lithium coin cells one each per jar. The best kind of LED for this design is an ultra bright LED with a diffused lens available in commercial stores and they are effective in casting the light by LED chip every direction. One may also opt for an easy access, with 10 mm diffused white LEDs and there are still similar LEDs available elsewhere as well.

Usefulness and Limitations of LED Lighting

Advantages

- The LED lighting is more cool and aesthetic component of landscape especially in the evening and early morning for the dwellers.
- The LED lights are more efficient to produce light per watt than incandescent bulbs. These can emit light of an intended color without using the color filters that is required in traditional lighting methods and thus can lower the initial costs.
- The size of LEDs can vary from very small (>2 mm^2) to a large size and can be easily populated onto the circuit board. Frequency of their lighting on and off make them convenient to use as these can achieve full brightness in microseconds.
- The LEDs are ideal for use in applications that are subject to frequent on-off cycling, unlike fluorescent lamps that burn out more quickly when cycled frequently, or HID lamps that require a long time before restarting.

- The LEDs can very easily be dimmed either by pulse -width modulation or lowering the forward electric current.
- In contrast to most light sources, LEDs generate light and waste heat by different mechanisms and a relatively cool light stream can be produced.
- The LED lights are mostly failing by dimming over time, rather than the abrupt burn-out of incandescent bulbs.
- The LEDs can have a relatively long useful life and estimated to have up to 35,000 to 50,000 hours if we compare with the fluorescent tubes, which are typically rated at about 10,000 to 15,000 hours, depending on the conditions of use and incandescent light bulbs can be rated only at 1,000–2,000 hours.
- The solid package of the LED can be designed to focus its light but the incandescent and fluorescent sources often require an external reflector to collect light and direct it in a usable manner.
- The LEDs, being solid state components are difficult to damage with external shock, unlike fluorescent and incandescent bulbs which are fragile. The LEDs do not contain mercury unlike the fluorescent lamps and avoid the environmental pollution.

Disadvantages

- LEDs are currently more expensive on an initial cost, than most conventional lighting technologies.
- The performance of LEDs are largely depends on the ambient temperature of the operating environment as over-driving in high ambient temperatures may result in overheating of the LED package, eventually leading to device failure.
- These are voltage sensitive and must be supplied with the voltage above the threshold, a current below the rating for the series resistors or current-regulated power supplies
- The whitest lights (LEDs) have spectra different significantly from a black body radiator like the sun or an incandescent light. However, the color rendering properties of common fluorescent lamps are often inferior to that of available white LED lights.
- LEDs do not approximate a "point source" of light and cannot be used in applications required a spherical light field and are not capable in divergence below a few degrees.

- There is an increasing concern that the blue and white LED lights are now capable of exceeding safe limits but still the eye safety specifications are still needed.
- These are a cause of concerns as light pollution (scatter 2.7 times more than the white light) by the Earths' atmosphere.

References

Anonymous (2010). http://www.articlesnatch.com/Article/Lighting-Your-Landscape-With-Led-Lights/1063664.

Bakker, M., Van Dijk, J.M., Van der Knaap, J., Van Marrewijk, P., Mul, M., Rijgersberg, A. Sonneveld, J. and Zuidgeest, P. (1990) *Fides Mum Manual*: for all year round chrysanthemums. De Lier: Fides Holland BV.

Bertram, L. and Karlsen, P. (1994). Patterns in stem elongation rate in chrysanthemum and tomato plants in relation to irradiance and day/night temperature. *Scientia Horticulturae,* 58:139-150

Bula, R.J., Morrow, R.C., Tibbetts, T.W., Ignatius, R.W., Martin, T.S. and Barta, D.J. (1991). Light Emitting Diodes as a light radiation source for plants.*Hort Science* 26 : 203-205

Cathey, H.M. (1974). Participation of phytochrome in regulating internode elongation of *chrysanthemum moriflolium* (Ramat) Hemsl. *Journal of the American Society of Horticultural Science* 99: 17-23.

Cockshull, K.E. and Kofrank, A.M. (1985). Long day flower initiation by chrysanthemum. *Hortscience* 20:296-298.

Dougher, T.A. and Bugbee, B.G. (1988). Is blue light good or bad for plants? *Life Support Biosph. Sci.,* 5(2): 129-136.

Furuta, T. (1954). Photoperiod and flowering of *chrysanthemum morifolium. Proceedings of the American society of Horticultural Science* 63: 457-461.

Gisleord, H.R. and Nelson, P.V. (1997). Effect of relative air humidity and irradiance on growth of *Dandranthema x grandiflorum* (Ramat) Kitamura. *Gartenbau wissenschaft* 62: 214-218.

Hughes, A.P. and Cockshull, K.E. (1972). Further effects of light intensity, carbon dioxide concentration, and day temperature on growth of *chrysanthemum morifolium* cv. Bright Golden Anne in controlled environments. *Annals of Botany* 36: 333-350.

Ieperen, Wim van; Singh, M.C. and Heuvelink, Ep (2009). Flowering Preserved in Short-day Chrysanthemum plants grown under Long-Day conditions using PAR with a variable spectrum over the day. In: *Proceedings of 6th International Symposium on Light in Horticulture.* November 15-19, 2009, Tsukuba, Japan. p92.

Keeler, B. and Nicole, S. (2010). PhlatLight is a registered trademark of Luminus Devices Schwartz Communications, *Luminus Media*.781-684-0770. www.luminus.com

Kher, M.A. (1975). *Chrysanthemum. Bulletin of the National Botanical Garden,* 1:1-41 (New Series).

Kofrank, A.M. (1992). Cut Chrysanthemums: In: Larson RA, ed. *Introduction to Floriculture.* 2nd edition. London Academic Press Inc. 3-42.

Mortensen, L.M. and Moe, R. (1983). Growth responses of some greenhouse plants to environment. VI. Effect of CO_2 and artificial light on growth and flowering of *Chrysanthemum x morifolium* Ramat. *Scientia Horticulturae* 19: 141-147.

Singh, M.C., Wim van Ieperen and Ep Heuvelink. (2009). Effect of day length extension on growth and flowering in *Chrysanthemum morifolium* cvs Zembla and LeMans'. In: *Proceedings of 6[Th] International Symposium on Light in Horticulture.* November 15-19, Tsukoba, Japan. p145.

Singh, M.C., Wim van Ieperen and Heuvelink, Ep (2009) Flower induction in *Chrysanthemum morifolium* under long days using LEDs. (M.Sabu Award Paper). In: *International Symposium on "Angiosperm Systematics & Phylogeny: Retrospects' and Prospects* ". November 12-14, 2009. NBRI (CSIR), Lucknow, UP

Yoshikawa, M. (2009). Is Pink Best LED Color to Grow Plants? Apr 21, 2009 17:13; Nikkei Microdevices, Nabesei Co Ltd.

Ornamental Horticulture, pp 187-223

New India Publishing Agency, New Delhi - 110034, India

9

Herbaceous Perennials of Kashmir Valley for Landscape Use

Sumina Ramzan[1], T. M. Pal[2] and Sajad, A. Bhat[3]

The state of Jammu and Kashmir is highly mountainous except for a short belt joining the Punjab plains. Geographically, the state of Jammu and Kashmir is divided into three regions -alpine desert of Ladakh, subtropical region of Jammu and predominantly temperate region of Kashmir valley. The valley is rich in herbaceous flora with huge potential for landscape use, however, many of which are still not exploited much for the purpose. Plants which live and bloom year after year are known as perennials which have an importance of their own and are backbone of any well-planned garden. The stems of herbaceous perennials die at the ground level each winter. They comprise a vast group of plants widely differing in size, leaf texture, flower color and are found in all the gardening countries of the world. The monotonous spread of concrete and glass in cities can have a dehumanizing effect. Growing colourful flowers in the house compound and in the parks will greatly enhance the beauty of surroundings in the countryside, the towns and cities. Flowers also bring happiness to life and boast the sagging spirits. The assemblage and display of colourful and attractive herbaceous perennials can achieve very significant and telling effects on the environment particularly in the landscaping of the house compounds, gardens,

[1,2&3] Division of Floriculture and Landscaping, SKUAST-K, Shalimar, Kashmir Jammu and Kashmir

parks, health resorts and hotels. Now a day herbaceous plants are also used therapeutically to cure several problems related to humans. Judicious selection of native and introduced herbaceous perennials should camouflage with the type of habitation, building pattern, colour of the countryside etc. These plants should occupy only well thought of position in the landscape. There are always a number of shady spots in our lawns/gardens especially towards north-east side of a structure where lawn grass fails to grow or show poor growth. These spots can be converted into perennial herbaceous border of foliage plants.

Trees and shrubs are often the most dominant plants in landscapes but ground covers and vines play an important role too. The low spreading form of ground covers can connect and unify major planting areas and even serve as a focal point when the plants are flowering. In contrast to annuals, perennials provide savings in time, labour and expenses each year because where adapted to a specific climate zone, they can successfully over come winter and re-establish for more than a single growing season. Perennials are probably the most diverse plant group, providing a huge variety of shape, form, colour, texture and scent with which to design a planting. Since there is an increasing trend and awareness of planned landscaping in the public and in view of its value in preserving the environment, there is an urgent need to identify and select the appropriate native, naturalized and exotic planting material for specific sites /locations. Information regarding herbaceous perennials surveyed and identified in Kashmir Valley with prospective use in landscaping are given as under:

Achillea filipendulina **Lam**

Clump forming, evergreen perennial with rosettes of oblong, 1 or 2 pinnate, mid- to grey-green leaves. Strong leafy stems bear golden yellow flower heads in flat corymbs 12 cm across from early summer to early autumn. Fruit an achene, without any pappus.

Achillea milifolium **L.**

Mat forming and invasive perennial with 2 to 3 pinnate, lanceolate leaves the lower being petiolate. Stems erect, simple, somewhat woolly. Inflorescence composed of numerous heads arranged in a corymb. Heads 5 mm in diameter. Bracts of the involucre with brown or blackish papery margins. Ray florets female, disk florets hermaphrodite. Achenes without pappus 2 mm long not ribbed. Plant strongly scented.

Achillea ptarmica **L.**

Strong, rhizomatous perennial with simple linear lanceolate, sessile, finely toothed dark green leaves 1.5 to 7.5 cm long. Stem erect, angular. Inflorescence a lax corymb of heads. Heads heterogamous, 1.2 to 1.7 cm in diameter. Ray florets female, ligulate white. Disk florets hermaphrodite, tubular creamy or greenish white. Achenes without pappus, 1.5 mm long.

Aconitum heterophyllum **Wall**

Perennial, erect, leafy, lower leaves petioled, upper ovate cordate, incised, amplexicaule; racemes panicled, follicles 5 and flowers blue in colour. Fruit is a follicle. Follicles-five in number.

Aconitum napellus **L.**

Variable erect perennials with rounded, deeply 5-7 lobed, dark green leaves, the lobes toothed or further divided. Bears dense racemes of indigo blue flowers in mid and late summer. Fruit is of 3-5 hairy follicles, cylindrical in shape.

Agapanthus praecox **Hoffmg**

Clump forming evergreen perennials with broad, strap-shaped dark green leaves 30 to 70 cm long. Bears large, rounded umbels, 15 to 30 cm across of trumpet shaped rich mid blue flowers 4 to 6 cm long in late summer and early autumn. Fruit is loculicidal capsule.

Ajuga reptans **L.**

Creeping, evergreen rhizomatous perennial with decussate, leaves 7.5 cm long, obovate to oblong margins usually crenate; the lower forming the basal rosette. Flowers in several usually distant whorls, supported by leaf like bluish bracts, zygomorphic hermaphrodite. Calyx-campanulate, Corolla-bilabiate. Fruit of 4 nutlets, obovoid, surface reticulate.

Althea rosea **Cav.**

Vigorous, upright perennial producing rounded roughly hairy, light green leaves to 3.5 cm long, cut in to 3 to 7 shallow lobes. Long terminal racemes of single, purple, pink, white or yellow flowers 5 to 10 cm across are borne in early and mid summer. Consists of compressed beakless. Friut is one seeded carpel.

Amaryllis belladonna L.

Bulbous perennial producing, stout, purple or purple-green stems with umbels of 6 or more funnel shaped, scented pink flowers, 6 to 10 cm long, in autumn strap shaped, fleshy leaves, 22 to 40 cm long are produced after flowering. Fruit a capsule, with several seeds, opening in 3 valves.

Anemone biflora D.C

A small tuberous perennial, with radical leaves, 2-5 cm in diameter, long petioled, 3-partite or 3 foliolate, the leaf segments rounded or cuneate, deeply cut into narrow lobes or toothed. Flowers appearing at the same time as the leaves; when freshly opened pure white or bright carmine or white diffused with carmine, turning brownish or dull red when over mature; 2-3 cm in diameter on a leafless scape bearing 1 to 3 pedicelled flowers with 3 involucral bracts at the base of pedicles. Achenes with short styles imbedded in dense wool.

Anemone x hybrida Paxt.

Vigorous, erect woody based perennial with suckering shoots and oval, usually 3-palmate, toothed mid green leaves, softly hairy beneath. Basal leaves are 10 to 20 cm long. Stem leaves are 5 to 12 cm long. Branched stems bear umbels of 12 to 18 cm, single white flowers, pink tinged on the reverse with golden yellow stamens in late summer and mid-autumn. Fruit a group of achenes, oblong in shape.

Aquilegia canadensis L.

Airy, perennial with 2 ternate, fern like, dark green leaves divided into leaflets, 1 to 2 cm long. Bears terminal, leafy racemes of up to 20 nodding flowers from mid-spring to mid-summer. The sepals are scarlet and forward pointing, lemon yellow petals taper into straight, erect red spurs, 1.5 cm long; the stamens and styles extend outwards and downwards. Fruit is a follicle.

Aquilegia vulgaris L.

These aquilegias of rare beauty are numbered amongst the tall mountain perennials. Upright perennial with finely divided ternate or 2 ternate bluish green leaves divided into leaflets, 3 cm long. Bears terminal leafy racemes of 2 or 3 nodding blue flowers, sometimes with white petal tips and with straight and curving spurs, to 2.5 cm long in late spring. Fruit is a follicle.

Aster alpines L.

Spreading clump forming perennial with short-stalked, spoon shaped to narrowly lance shaped mid green leaves, 9 cm long. Bears solitary violet flower heads, to 5 cm across, with deep yellow disc florets, on erect stems in early and mid summer. Fruit is an achene.

Begonia evansiana Andr.

Tuberous, winter dormant perennial with pointed glossy bright to dark green leaves and reddish on the lower surface. The flowers are borne in small clusters consisting of two small female flowers and one showy, frequently double male flower. The stem is branched and succulent. Fruit is a capsule.

Begonia x semperflorens Link & Otto

These hybrids have glossy foliage which accounts for the name wax begonias by which the group is sometimes known. Flowers white, slightly tinged pink flowers and pale green shiny foliage. The colour range too, has become quite wide, both in leaves and flowers. There are some sorts with almost variegated leaves and atleast one with yellowish leaves. Others have bronze-red or dark red foliage, while there is a few which have white leaves of a shape which has caused them to be described as the calla leaved begonias. Stem glabrous and leaves glossy green with reddish ribs. Flowers 2.5 cm, pink or white. Fruit-green, with prominent red ringed wings.

Begonia x tuberhybrida Voss.

The group comprises numerous hybrid tuberous begonias of complex derivation, and are variable in habit, size and colour. They are very popular flowering plants upto 60 cm tall. The stems are short, reddish green and the leaves alternate, irregular, rough and fleshy. The flowers come in a wide range of colours except blue and mauve. Separate male and female flowers appear on same plant. Fruit is a capsule.

Belamcanda chinensis DC

Tuberous rooted herbaceous perennial, with sword-shaped mid to deep green leaves, 20 cm long closely resembling those of Iris leaves. In summer, bears a succession of upto 12 loosely arranged clusters of wide opening flowers, 4 cm across, bright yellow to orange red with maroon spots, followed by large black seeds. Fruit is a capsule.

Bellis perennis **L.**

Stoloniferous perennial with narrowly obovate or obovate-spathulate leaves in a basal rosette, bright-green 1 to 6 cm long. Inflorescence a solitary head, 15 to 25 cm in diameter. Ray florets white or rarely pink in a single row, female; disc florets yellow, hermaphrodite. Achenes compressed without ribs, 1.5 to 2 mm long, without pappus. From late winter to late summer produce flower heads.

Bergenia ciliata **(How) Sternb**

Clump forming perennial with very broadly obovate, hairy mid green leaves, to 35 cm long. In early spring, flower stems bear more or less erect flowers which are pink or white fading to pinkish white, with rose pink calyces usually 2 cm across on short branched often red or purple flower stems. In frost-prone areas, foliage dies back in winter. Bears fruit stalks.

Caltha palustris **L.**

A robust aquatic perennial, with kidney shaped leaves, glabrous, with usually finely toothed margins and long petiolated. Flowers white, 2-5 cm diameter, terminal with usually 5 or 6, petiolated sepals, many stamens, carpels free with no petals. Fruit is a head of narrow flattened follicles.

Campanula latifolia **L.**

Vigorous, upright perennial with stout unbranched stems arising from basal clumps of ovate oblong, long-stalked, rough-textured toothed mid green leaves, 7 to 12 cm long. Stem leaves are similar but stalkless and sharply pointed, decreasing in size towards tops of stems. In summer, flowers are produced single or in twos or threes from axils of upper leaves, forming narrow leafy spike-like racemes. The flowers are broadly tubular-bell shaped, 4 to 6 cm long, pale to deep violet and have wide spreading corolla lobes. Fruit a capsule ½"glabrous or nearly so.

Campanula medium **L.**

Slow growing, clump-forming and downy biennial producing lance-shaped to elliptic, toothed mid green basal leaves 12 to 15 cm long and smaller stem leaves. Lax racemes of single or double, bell shaped white pink or blue flowers 2.5 to 4 cm long, open from spring to summer. Fruit is an ovoid capsule.

Centaurea montana L.

Rhizomatous, mat forming perennial with ovate to broadly lance shaped, entire to pinnatified, some times slightly toothed, mid green leaves, 6 cm long, wooly beneath with densely wooly stems. Blue flower heads 5 cm across, with reddish violet florets. Fruit is a compressed achene.

Cerastium tomentosum L.

Rampant, carpet or mat forming perennial producing linear or linear lance shaped, white or silver woolly leaves and 1 to 3 cm long. Profuse cymes of star-shaped white flowers to 2.5 cm across are borne in late spring and summer. Fruit is a capsule opening at the top.

Cheiranthus cheiri L.

A perennial of longer duration and more woody than the common stock more branched and less hoary, the hairs forked at the base and closely pressed on the surface, or often quite green and nearly glabrous. Leaves narrow, pointed quite entire. Flowers rather large, generally sweet scented, pods 2.5 inches long, the valves with a slightly prominent midrib pod siliquose.

Chrysanthemum leucanthemum L.

Extremely variable rhizomatous perennial with the lowest leaves in a rosette, sub circular to obovate –spathulate, pedicelled; the upper ones oblong, sessile. Inflorescence a solitary head, 2.5 –4 cm in diameter. Involucral bracts with papery, narrow, purplish edges. Ray florets white, female. Disk floriets yellow tubular, hermaphrodite. Fruit an achene. Ribbed 3 mm long, cylindrical without pappus.

Clivia miniata Regel

Evergreen perennial with strap shaped leaves, to 60 cm long. Bears large umbels of up to 20 semipendents, open tubular to funnel shaped, yellow, red, or orange flowers 5 - 7 cm long, from spring to summer. Fruit is a globose red berry.

Colchicum luteum Baker

A small perennial with the base of the stem below ground thickened in to a solid thick, fleshy ovoid body termed as corm. Leaves are radical appearing with the flowers and becoming much elongated after the flowering season is passed, linear and obtuse. Flowers 2.5-4 cm in diameter, erect, golden yellow with very short flowering stem or scape, bearing 1 or 2 flowers. Fruit a capsule 3-4 cm in length, the valves with long recurved beaks.

Convallaria majalis L.

Rhizomatous herbaceous perennial producing tough, slender and creeping branching rhizomes bearing pairs of ovate-lance-shaped to elliptic, stalked, hairless basal leaves, 4 to 20 cm long. Arching racemes of pendent, spherical-bell-shaped, strongly scented, waxy white flowers 1 cm across are produced on leafless stems in late spring. Fruit a berry, red, globose but poisonous.

Coreopsis grandiflora Hogg.

Clump forming almost hairless, perennial, often growth as an annual, with simple lance shaped or palmately lobed lower leaves to 10 cm long. Flowering stems produce 3 –5 pinnate leaves with linear leaflets. From late spring to late summer, produces solitary flower heads to 6 cm across, consisting of golden yellow ray florets, with evently cut outer margins and darker yellow disc florets. Fruits is an achene.

Crocosmia masonorum N.E.Br

Robust cormous perennial bearing pleated mid-green leaves 60 to 100 cm long. In mid summer, it bears upward-facing, orange red flowers 5 cm long, in arching, usually unbranched spikes. Fruit is a capsule.

Crocus chrysanthus Herb.

Late winter and early spring flowering, crocus, with dull green leaves and upto 4 rounded and scented flowers, 1.5 to 4 cm long, cream to deep golden yellow, often suffused or veined bronze-maroon outside has fruit capsule buried among leaves.

Crocus sativus L.

Autumn flowering crocus produces 1 to 5 widely open rich liliac flowers, to 5 cm in long, with dark purple veins. Dull green leaves are borne with or shortly after the flowers. Saffron is obtained from long, conspicuous, 3-branched deep red style. Sterile plants are increasing only by division. Fruit is a capsule buried among the leaves.

Cynara cardunculus L.

The plant is clump forming perennial with pinnatified or 2 –pinnatified, usually spiny with silvery grey leaves to 50 cm long, with ovate to linear –

lance shaped segments and deep basal leaves. From early summer to early autumn, purple flower heads 4-8 cm across, are produced on grey-woody stems. Branched leaf stalks and midribs are edible. Fruit is an achene.

Delphinium elatum L.

Clump forming perennial with fleshy crowns producing flowering spikes. The spikes bear numerous double flowers with 5 large outer sepals and an éye'formed by 8 inner sepals. The flowers atleast 6 cm across are usually larger at base of spike. Each stem also bears lateral shoots flowering after the main spike with slightly smaller flowers.

Digitalis lanata Ehrh

Clump forming perennial or biennial with oblong, lance –shaped or inversely lance –shaped, mid green leaves, to12 cm long, hairless beneath or with toothed margins. Dense, racemes of pale cream or fawn flowers 3 cm long, each with brown or violet - brown veins and a lighter cream lower lip, are produced from mid to late summer. Fruit is a capsule.

Digitalis purpurea L.

Hairy perennial with leaves ovate to lanceolate or lanceolate –oblong, 22 cm long, greyish –tomentose beneath; petiole winfed. Stem erect, stout simple or rarely branched. Inflorescence a terminal, many –flowered raceme. Flowers zygomorphic, hypogynous, hermaphrodite, rather showy. Calyx deeply 5-lobed, green; the lobes ovate, overlapping. Fruit is an ovoid sepeticidal capsule, exceeding the calyx.

Echinacea purpurea Moench

Erect perennial with smooth, sometimes rough –hairy, red –tinted green stems, ovate, toothed, rough-hairy basal leaves, 15 cm long, and ovate –lance shaped, toothed stem leaves. From mid summer to early autumn, bears flower heads to 12 cm across, with golden brown, cone shaped discs and partly reflexed, purplish red ray –florets, 3 –8 cm long. Fruit is an achene.

Eranthis hyemalis Salisb

Clump-forming, tuberous perennial producing bright yellow flowers, 3 cm across, each above a ruff of dissected, bright green leaves, in late winter and early spring. Quickly forms large colonies particularly in alkaline soil. Fruit of 5-8 stalked follicles.

Eryngium biebersteinianum Nevski ex. Bobrov

Stiff, hard herbs, usually perennial and with prickly leaves and involucres. Flowers in a compact spike or head, with a scale or bract on the common receptacle under each flower. Petals erect, with a long inflected pointed or prickly teeth of the calyx. Fruit-ovoid, without vitas, crowned by the pointed or prickly teeth of calyx.

Euphorbia stracheyi Bioss

Rhizomatous, sometimes invasive herbaceous perennial with vigorous, erect stems, reddish green when young. Lance shaped to linear oblong, dark green leaves 12 cm long, with red midribs, turn red and yellow in autumn. In early summer, bears terminal cymes, 10 cm across, of yellow cyathia cupped within orange-red to red involvucres. Fruit of three carpels, each with a single seed.

Fritillaria imperials Linn

A tall handsome, robust perennial with lanceolate, sessile leaves, the lower ones opposite and the upper ones longer and in whorls. Flowers are about 2-3 cm in length, brownish orange colour in a terminal umbel of about 5 to 8 pedicelled and pendulous flowers. Fruit, a capsule 5 cm long, obovoid almost 6 winged.

Gagea elegans Wallich-ex.D.Don

Leaves up to 40 cm, long 8 mm broad, with parallel veination, hooded at the apex. Inflorescence subtended by large leaf like bracts, umbel like, 1-5 flowered, loose. Flowers actinomorphic, erect. Perianth of 6 segments in two whorls; the segments spreading 1.5 cm long, lanceolate-oblong, yellow with a green stripe underneath. Another 6 stem leafless (Except for the large leaf like bracts). Fruit is a capsule, nearly globose.

Galanthus nivalis L.

Bulbous perennial with narrow, glaucous leaves having parallel veination, linear, bluish –green 20cm long. Flowers solitary, actinomorphc, always nooding, enclosed in a papery spathe before flowering free, spreading (white); the inner 3 only half as long, deeply emarginated white with a green spot at the middle near the apex. Fruit is a capsule, ovoid in shape.

Geranium wallichianum, D. Don

Herbaceous perennials with forked, stems often swollen at the nodes, opposite palmately divided leaves and purplish flowers, solitary on axillary peduncles. Stamens-10, of which five shorter, but generally with anthers. Ovary 5 –lobed, terminating in long beak with five stigmas on the top, the lobes being all whorled –beaked receptacles. Fruit is a capsule, separating into 5-1 seeded carpels which curve upwards, with a long elastic awn, detached from the beak and glabrous inside.

Gladiolus byzanthus Mill.

Vigorous perennial producing linear leaves, 60 cm long. From late spring to early summer it bears deep magenta flowers, 5 cm across, with paler marks on the lip tepals. Fruit is a capsule.

Gloriasa rothschildiana OBrein

Climbing perennial with ovate-lance-shaped to oblong glossy, bright green leaves, 6 cm long which narrow to form terminal tendrils, 4 cm long. Nooding flowers, 8 cm across are borne from the upper leaf axils. Flowers have 6 reflexed, wavy –margined, sometimes entirely yellow, with long and protruding stamens.

Gyposphila paniculata L.

Tap rooted herbaceous perennial with branching stems forming an airy mound of linear –lance –shaped, usually hairless, glaucous leaves, 6 cm loose, many flowered panicles of shallowly trumpet shaped white flowers to 8 mm across, forming mounds to 45 cm or more across. Fruit is a capsule.

Hedychium gardnerianum Wall.

Upright, rhizomatous perennial with lance-shaped, greyish green leaves, 35 cm long, lemon–yellow flowers, with bright red stamens, borne in dense, broadly cylindrical, terminal racemes, 30 cm long. Fruit is a capsule.

Helenium autumnale L.

Upright clump forming perennial-shaped toothed leaves 15 cm long. From late summer to mid autumn, bears flower heads to 5 cm long with yellow ray florets which reflex as the brown disc florets open. Fruit is a 4-angled achene.

Helianthus autumnale Nutt

Tall, coarse plant with large simple bristly, alternate leaves, the showy daisy –like flower heads with sterile ray florets are borne in loose corymbs and have yellow disc florets. Fruit is an obovate achene.

Hemerocallis fulva **L.**

Semi evergreen, rhizomatous perennial with wide; dark bluish green leaves, 75 cm long. Branched scapes bear trumpet-shaped, orange-brown flowers 7 cm across, with recurved tepals, in mid and late summer. Fruit is a capsule.

Hesperis matronalis **L.**

Leaves shortly stalked, or tapering at the base, ovate-lanceolate or lanceolate 9 cm long, or the upper ones smaller. Flowers usually fragrant in the evening stems one meter high, usually slightly branched nearly cylindrical, but much contracted between the seeds. Pods 9 cm long.

Hippeastrum puniceum **Urban**

Bulbous perennial with bright green leaves emerging just after the flowers. Bears umbel of upto 3-funnel shaped scarlet flowers. Fruit is a globose capsule.

Hosta lancifolia **Engler**

Perennial forming a dense mound of arching, narrowly lanceolate, thin glossy, dark green leaves 15 cm long. In late summer, slender very leafy, red dotted scapes 60 cm long, bear narrowly funnel shaped, deep purple flowers. Fruit is a capsule.

Hosta plantaginea **Aschers**

Clump –forming perennial producing ovate to heart shaped, slightly wavy, glossy, light green leaves, 20 cm long with prominent, widely spaced, raised veins. Trumpet-shaped long tubed, very fragrant white flowers 10 cm long borne on leafy, bright green scapes 70 cm long, in late summer and early autumn. Fruit is a capsule.

Hyacinthus orientalis **L.**

Bulbous perennial with linear to lance –shaped, channelled, bright green leaves, 30 cm long. In early spring bears erect racemes of up to 40 tubular – bell shaped, waxy, very fragrant, single flowers 30 cm long, pale violet –blue at the bases and almost white above, with spreading, then recurved lobes. Fruit is a 3-angled capsule.

Iris ensata *Thunb*

Rhizomatous, beardless laevigatae irises with single occassionally branched stems. Leaves are with prominent mid ribs. Each stem bears 3-4 blue flowers. Standards are erect and smaller than the falls. Fruit is a capsule.

Iris germanica L.

The most commonly found garden iris is the bearded *Iris germanica*, the rhizomatous species with 3-10 basal sword shaped leaves. The inflorescences are fan shaped and contain one or more symmetrical six-lobed flowers. The fruit is a capsule containing numerous seeds within.

Iris japonica Thunb

Rhizomatous crested irises spread by surface rhizomes with fans of strap-shaped evergreen glossy dark green leaves to 45 cm long. In spring, each branched stem bears 3 or 4 flattened and frilly white or pale lavender-blue flowers. Falls have purple patches and orange crests. Fruit is a capsule.

Iris kashmiriana Baker

Flowers large, usually mauvish-blue; falls 5.5-9 cm and 4-6 cm broad with yellow beard; standards broadly elliptic, 4.5-6 cm broad. Spathes longwith a green base and a papery apex. Leaves 2-3.5 cm broad. Stems 40-90 cm. Fruit is a capsule.

Iris reticulata M. Bieb

Bulbous reticulate irises are with square sectioned leaves. It bears a solitary fragrant flower varying from pale to deep violet-blue or reddish purple with a yellow central ridge on each fall. Bulbs often split after blooming and may take some years to reach maturity. Fruit is a capsule.

Iris tectorum Maxim

Rhizomatous crested iris with fans of broad, ribbed, glossy, dark green leaves. Each branched stem bears 2 to 3 lilac flowers with darker veins and a white crest on each fall. Fruit is a capsule.

Iris unguicularis Poir

Rhizomatous crested iris with fans of strap shaped evergreen, dark green grass-like leaves. Pale lavender to deep violet with contrasting veins and a central band of yellow on the fall and with perianth tube.They are born singly on very short stem arising from a branching rhizome.

Ixiolirion karateginum Lipsky

Linear, lance shaped leaves in basal rosettes with racemes or umbels of funnel shaped usually deep blue or violet flower. Fruit is a capsule.

Kniphofia uvaria Hook.

Evergreen perennial with lax, linear, keeled finely toothed, coarse leaves, to 60 cm long. Flowers 3 –4 cm long, red in bud, opening to orange, and fading to yellow, are borne in slender, oblong ovoid racemes in early autumn.

Lathyrus latifolius L.

Herbaceous perennials with weak stems, sometimes climbing with sagittate stipules; the leaves usually pinnate mid green in colour usually consisting of one pair of ovate to elliptic leaflets to 5 cm long. Flowers pink purple and red flowers. Petals usually broad, especially the standard upper stamen free style flattened below the stigma, quite glabrous on the outer side. Fruit is a pod is cylindrical with several seeds.

Leucojum aestivum L.

Robust, bulbous perennial with erect strap –shaped, glossy, dark green leaves, to 40 cm long. In spring, leafless stems bear up to 8 bell shaped, faintly chocolate scented white flowers, 2 cm long, with green tips. Fruit is a capsule.

Liatris spicata Willd

Perennial with hairless stems and linear-lance shaped basal leaves, stem leaves are smaller and linear. Long lasting, purple flower heads, to 1 cm across, are borne in dense spikes 60 cm long. Fruit is a capsule.

Limonium latifolium Kuntze

Rosette forming perennial with elliptic to spoon shaped, mid to dark green leaves, usually to 30 cm long, occasionally to 60 cm. in late summer, branched, wiry stems bear panicles of spikelets which consist of shortly tubular deep lavender blue flowers 6mm long, with white calyces. Fruit is a capsule.

Lunaria annua L.

Annual or biennial with ovate or heart shaped, coarsely toothed, light to mid-green leaves to 15 cm long. Cross shaped light purple flowers to 1 cm across are borne in broad leafy racemes to 18 cm long. Flat seed pods 6 cm long are rounded and silvery.

Lupinus polyphyllus Lindl.

The stout erect to 1-1.5 m ft., glabrate, or sparsely hairy-pubescent above. Leaves very long-stalked, the petioles 15-30 cm long or the upper ones shorter, leaflets 9-16, lanceolate to oblanceolate, 6-15 cm inch long, acute, glabrous above and sparsely hairy beneath; stipules pointed, adnate to petiole for one-third or one-half their length. Flowers with purplish standard, blue wings and mostly dark-tipped point to keel 1.5 cm long, many and alternate on slender pedicels in a rather dense terminal raceme 30-60 cm long. Pod-3-4.5 cm long, wooly hairy, seed about 6; oblong, brown-black and glossy.

Lycoris radiata Herb

Bulbous perennial with wavy-margined rose-red or deep red flowers, 4 cm long with strength reflexed tepals and conspicuous, protruding stamens, borne in umbels, leaves semi-erect strap shaped and dark green appear after the flowers. Fruit is a capsule.

Lythrum salicaria L.

Clump forming perennial, usually in damp situations with erect stiff, branched stems bearing lance-shaped downy leaves. Spike like raceme bears star shaped bright purple-pink flowers. Fruit is a capsule.

Matthiola incana Br.

Erect, usually perennial and more or less woody at the base, but not of long duration; leaves oblong-linear, obtuse, quite, entire, soft and hairy on both sides with short crisped hairs. Flowers-purple or reddish, rather large, the petals obovate, fruit a pod, pod 12 cm long, crowned by short stigmas, which are rather thickened at the base.

Mimulus nepalensis Benth

Flowers yellow, tubular, weaky two-liped, the upper lip erect and two lobed. The lower spreading, three lobed. Flowers solitary, axillary, borne on stalks usually longer than the leaf. Erect leafy stems 15-20 cm opposite, elliptic, coarsely toothed, stalked leaves, mostly 2-2.5 cm. Calyx-inflated in fruit. Fruit is a capsule.

Mirabilis jalapa L.

Bushy perennial with ovate leaves 8 cm long. Fragrant, red pink, magenta, yellow or white flowers to 5 cm long, some striped, and often with several colours present on the same plant, are borne. Individual plants open in late afternoon and die by morning. Fruit is a dry capsule.

Muscari armeniacum **Leicht**

Bulbous perennial producing semi-erect narrowly linear to linear inversely lanceolate, green leaves, 30 cm long. Tubular bright blue flowers with distinct, constricted white mouths are borne in dense racemes 2 to 8 cm long, in spring. Fruit is a 3-angled capsule.

Narcissus pseudonarcissus **L.**

Single flowered, wild daffodils with leaves linear, erect having parallel veining, bullish green. Flowers actinomorphic, solitary, drooping or horizontal, showy 4.5 cm in diameter. Corolla a perianth, with a corona (paracorolla) inserted between segments and stamens. Fruit is a capsule, globose, often obtusely-trigonous.

Narcissus tazetta **L.**

Very variable wild species of daffodil with erect, broad, twisted, keeled, glaucous and mid-green leaves 40 cm long. Bears upto 20 sweetly scented flowers 4 cm across, with white perianth segments and yellow cups. Fruit is a globose capsule.

Nelumbo nucifera **Gaertn.**

An aquatic plant, with a whitish creeping rhizome from which sprout long and upright stalks which emerge from the water. The leaves are downy (50 cm) in diameter with their blades borne slightly above the water surface and the stalk sometimes one cm long, in the center. The flowers, on peduncles of 1 m are large, pinkish-white and reversed cone. Fruit is a spongy nut.

Nepeta nervosa **Royle ex-benth**

Bushy perennial bearing erect, unbranched stems with narrowly lance shaped, entire to slightly toothed, conspicuously veined, hairly faintly aromatic, mid to gray green leaves up to 10 cm long. Dense, cylindrical and spike like whorled cymes of purplish blue flowers to 12 mm long, are borne. Fruit is a nutlet.

Nerine bowdenii **Watson**

Robust bulbous perennial with broad, strap –shaped leaves, up to 30 cm long. bears open umbels of upto 7 or more funnel –shaped faintly scented pink flowers, up to 8 cm across, with recurved, wavy-marginated tepals. Fruit is a loculicidal capsule.

Nymphae stellata Willd

A perennial aquatic plant with very beautiful star like flowers with rounded, cordate and glossy leaves floating on the surface of water at the end of long, stout flexible stems. Flowers solitary. Fruit is a spongy berry ripening under water, seeds buried in the pulp and each covered with a fleshy aril.

Oenothera biennis L.

A biennial with stems almost simple and more or less hairy; ovate lanceolate or lanceolate slightly toothed hoary or downy. Flowers yellow large and fragrant, in a long terminal spike often leafy at the base. Ovary sessile about 20 cm long, the tube of the calyx at least 3 cm longer, the petals broad and spreading. Fruit is an oblong capsule.

Ornithogalum umbellatum L.

Bulbous perennial with linear leaves having parallel veination 30 cm long each with a central silver stripe above; the leaves wither as the flowers open in early summer. Inflorescence a raceme, corymbose 8-10 flowered, flower actinomorphic, erect with a perigon, petals white with a green band at the back. Fruit is a hexagonal capsule 1-1.5 cm in diameter.

Paeonia emodi Wall. ex Royle

A stout, shrubby perennial, 3 to 7.5 cm in height with exstipulate, alternate, glabrous, 15 cm leaves, petiolated, divided in to distinct leaflets or deeply cut into segments. Flowers 9 to 12 cm diameter, white, usually solitary (or sometimes 2 or 3) flower bud globose, bracts 2 or 3 leafy, lying close against the calyx, sepals 5, persistant, outer ones with a leafy point, petals 5 to 10, stamens many, carpels 1 to 5, surrounded below by a fleshy disk, fruit a single follicle (rarely 2) about 3 cm long.

Paeonia lactiflora Pall.

Herbaceous perennial with erect, red-motted stems and dark green leaves each with an elliptic or lance –shaped rough margined leaflets, paler and slight hairy beneath. Bears bowl shaped fragrant and white to pale pink flowers. Fruit is an oblong follicle.

Penstemon barbatus (Cav) Roth

Erect perennial with semi –evergreen basal rosettes and deciduous stems bearing lance-shaped to linear, entire, sometimes glaucous, green leaves to 20 cm long. Bears long panicles of pendent, tubular red flowers tinged pink to carmine red 4 cm long, the reflexed lower lips with yellow beards the upper ones projecting over them. Fruit is a capsule.

Phlox paniculata L.

Erect herbaceous perennial with ovate or lance-shaped to elliptic, toothed, thin leaves 11 cm long. Panicle like cymes of salver-form, fragrant, dark lilac 2.5 cm across are borne from summer to early autumn. Fruit is an ovoid capsule.

Physalis alkengi L.

Vigorous,spreading,rhizomatous perennial with traingular-ovate to diamond shaped leaves to 12 cm long Nooding, bell shaped cream flowers 2 cm long, with star-shaped mouths,re-produced from the leaf axils,flowed by large, bright orange-scarlet berries and closed in a papery red calyses to 5 cm across.

Platycodon grandiflorum A. DC.

Compact-clump forming perennial with ovate to ovate-lanceolate, toothed, blushed green leaves, to 5 cm long, borne in whorls on the lower stem alternately higher up. Large baloon like buds open to shallow bell shaped, 5-petalled purple-blue flower to 5 cm across, with darker blue veins and pointed tips to the petals. Fruit is a capsule.

Primula denticulata Smith

The commonest of the primulas and one of the earliest flowers to bloom at elevations above 2200 m manual. It is a perennial with radical, simple, ob-ovate, spathulate, sessile or the base narrowed into a short winged petiole. The lower surface is sparingly covered with a yellowish powder. At the time of flowering the leaves are small, but subsequently become much elongated. Flowers, 2.5-3.5 cm in diameter, yellow, in compact round clusters of shortly pedicelled flowers, the clusters about 3-6 cm in diameter. Fruit is a capsule.

Primula rosea Royle

A beautiful primula with simple, radical, elliptic or ob-ovate, spathulate leaves with the base narrowed into a winged petiole, green and glabrous on both surfaces. Flowers 2.5 cm in diameter, bright or pale rose-pink with a yellow centre, in lax umbels of 3 to 10 pedicelled flowers on a scape of 8-20 cm in length. Fruit, a capsule, globose sunk in the calyx.

Primula vulgaris Huds.

Leaves oblong or obovate-oblong about 2 cm long, obtuse, narrowed into petiole shorter than blades, crenulate, more or less pilose or glabrescent, wrinkled. Flowers solitary on pubescent pedicels to 15 cm long, pale yellow, purple or blue, becoming green in drying, 3-4.5 cm across, calyx pubescent, the narrow lanceolate-acuminate lobes several times shorter than tube; throat of corolla slightly contracted and with scale-like folds.

Ranunculus falcatus L.

An erect perennial covered with long, closely appressed hairs, stem 30-60 cm without runners, usually much branched many flowered, hairs white. Radical leaves 6-12 cm in and more across, long stalked, deeply 3 –lobed, segments deeply cut, nearly equal in length. Flowers long stalked bright yellow. Fruit a globose head of many smooth, flattened, not dotted achenes, margins of achenes thickened.

Rudbeckia fulgida Ait.

Hairy per. 30-60 cm high. Leaves mostly entire, narrowly to oblong-lanceolate, the lowest and basal spatulate-lanceolate. Petiolate, 6-12 cm long. Foliaceous bracts of involucre often wide, 1.5-3.0 cm long; rays 12-14 golden-yellow or with orange base, as long as the bracts; disk-black-purple, over 1.5 cm diameter; style-branches short and obtuse. Pappus a minute crown.

Rudbeckia laciniata L.

Branching per. 60-360 cm high, smooth and glabrous or more or less roughned with short stiff hairs. Leaves broad, the basal pinnately divided into 5-7 after 2-3 cleft segms; straight leaves 3-5 parted or cleft those of the branches few-toothed or entire. Heads-solitary or few, long-peduncled, involucral brackets and irregular, rays yellow, few or several, oblanceolate, 3-4.5 cm long, disk greenish yellow, style-branches short and obtuse; pappus a short crown.

Salvia sclarea L.

Erect, many branched, glandular hairy perennial or biennial with ovate to oblong, notched to irregularly toothed, wrinkled, mid green leaves to 23 cm long with heart-shaped or perfoliate bases. Bears many flowered, terminal panicles or racemes of blue flowers to 3 cm with prominent liliac bracts. Fruit is of 4 round nutlets.

Solidago canadensis L.

A tall, graceful, perennial herb, with a few branches, with angular edges, rather rough. Stems erect, stout. Leaves lance - shaped, basal ones short-stalked, elliptic, coarsely toothed, sharp –pointed or blunt, upper leaves smaller, narrower entire. Head, 1-4 on short axillary stalks, gathered on a long leafy panicle. Ray-flowers (outer flowers) 10-12, central flowers 12 –20, all golden yellow. Fruit (achene) hairless or slightly hairy and ribbed.

Sprekelia formosissima Herb.

Bulbous perennial with linear leaves 30 cm long and one (rarely two) pink flower-scape bearing a single funnel shaped flower 10 cm long, bright red with the three upper petals broader. Fruit is a loculicidal capsule.

Sternbergia lutea Roem and Schult

Bulbous perennial with upright trigonal stem, grooved linear leaves, and single flower, yellow with perianth composed of a 2 cm tube and spatulate segments. Fruit is an achene.

Thymus serphyllum L.

Mat forming perennial with finely hairy, trailing stems and linear to elliptic or elliptic ovate, mid green leaves. Purple flowers are brown in congested whorls. Bears ovoid nutlets.

Tulipa clusiana DC.

Tulip with 2-5 linear, hairless, glaucous, grey green leaves, to 15 cm long, bowl shaped, later star shaped flowers, to 10 cm across, with rounded bases, is produced singly or in pairs. Flowers are white, stripped pink outside with purple marks and purple stamens. Fruit is a capsule oblong in shape.

Tulipa lanata Rgl

Tulips with 15-60 cm stems, downy. Leaves 4 or 5, glaucous, downy, margin waxy, red, ciliate; flower solitary; perianth-segments red, somewhat silvery on the back and downy on the upper half, basal blotch olive green with a yellow border.

Typha angustifolia Bory and Chaub.

Aquatic perennial with linear leaves, to 1.5 m long, sheathed at the bases. Brown flower spikes, 20 cm long are borne in mid-summer. Male and female

Tulip Garden

General View

Tulips in Full Bloom

Bed Planting of Tulips

flowers are 7 cm apart, the females with dark raddish brown scales. Fruit is a nutlet.

Verbascum thapsus L.

A stout erect biennial, simple or branched clothed with soft woolly hairs, leaves oblong, pointed, slightly toothed, narrowed at the base into 2 wings running a long way down the stem; the lower ones often stalked. Inflorescence a terminal spike like raceme, many flowered, dense. Flowers zygomorphic, hypogynous showy, 2.5 cm in diameter. Fruit a capsule, septicidal and exceeding the calyx.

Viola odorata L.

Perennial stock short, but sometimes branched, knotted with the remains of the old leaf stalks and stipules and usually emitting creeping runners or scions. Leaves in radical (or rather terminal) tufts, broadly cordate, rounded at the top and crenate, downy or shortly hairy with longer stalks. Stipules narrow lanceolate or linear and entire. Peduncles about as long as the leaf stalks with a pair of small bracts. Stigma pointed horizontal or turned downwards. Fruit, a capsule.

Zantedeschia aethiopica Spreng.

Clump forming rhizomatous perennial, ever green in mild areas, with semi-erect, arrow shaped, glossy, bright green leaves, up to 40 cm long. Bears a succession of large, pure white spathes up to 25 long with creamy yellow spadices. Bears baccate, red tinged wings.

Zantedeshia elliottiana Engler

Rhizomatous perennial with erect, heart shaped, dark green, basal leaves up to 45 cm long, covered with translucent white spots. In summer, bears golden yellow spadices. Bears baccate and red tinged wings.

Herbaceous perennials for landscape use

Beds

Herbaceous perennials such as *Aquilegia canadensis*, *Achillea ptarmica* and *Aster alpinus* are suitable for growing in beds.

Borders

Herbaceous perennials suitable for border are *Aconitum napellus*, *Agapanthus praecox*, *Achillea millefolium*, *Althea rosea*, *Caltha palustrus etc.*

Woodlands

Herbaceous perennials like *Aconitum heterophyllum*, *Campanula medium*, *Digitalis lanata etc.* are suitable for woodlands.

Rock garden

Herbaceous perennials generally suitable for rock garden are *Achillea filipendulina*, *Centaurea montana*, *Coreopsis lanceolata*, *Chrysanthemum leucanthemum etc.*

Companion plant

Species *of Paeonia, emodi* and *lactiflora* are used as companion plants.

Specimen plant

Agapanthus praecox, *Arundo donax*, *Cynara cardunculus*, *Digitalis lanata* and *D. purpurea* are suitable for growing as specimen plants.

Back drop

Aquilegia canadensis, *A. vulgaris*, *Althea rosea* and *Arundo donax* can be used as back drop in landscape designs.

Pool

Herbaceous perennials like *Arundo donax*, *Adiantum capillus* Veneris, *Caltha palustrus*, *Lythrum salicaria*, *Nelumbo nucifera*, *Nymphae stellata* and *Typha angustifolia* are adaptable to moist conditions and are suitable for pool.

Bulbous plants for landscape use

Beds

Anemone biflora, *A × hybrida*, *Begonia evansiana*, *Canna indica*, and *Clivia miniata* are suitable for growing in beds.

Borders

Alstroemeria aurantiaca, *Belamcanda chinensis* and *Crocosmia masoniorum* are the best suited for borders.

Pathways

Bulbous plants likes *Nerine bowdenii* and *Lycoris radiata* are suitable for growing in pathways.

Rock garden

Bulbous plants like *Crocus chrysanthus, C. sativus, Fritillaria imperialis* and *Galanthus nivalis* are the best suited for rock gardens.

Specimen plant

Dahlia variablis was observed to be suitable for growing as **specimen** plants.

Pots

Flowering pots add charm and beauty to landscape design. *Iris reticulate, Iris japonica* etc. are suitable for growing in pots.

Ground covers

Bulbous plants like *Colchicum luteum, Muscari armeniacum* etc. are suitable for ground cover.

Ground covers for landscape use

Beds

Species namely *Ajuga reptans, Bellis perennis, Bergenia ciliata* and *Eranthis hyemalis* can be grown in beds.

Borders

Species most ideal for growing in borders are *Hosta palantaginea, H. lancifolia, Caltha palustrus, Lycoris radiata, Primula denticulata, P. rosea, P. vulgaris, Saxifraga stonilfera* and *Sedum spectabile.*

Woodlands

The most attractive species which can be grown in woodlands are *Ajuga reptans, Aspidistra elatior, Aptenia cordifolia, Bellis perennis, Bergenia ciliate, Convallaria majalis, Geranium wallichianum, Hosta plantaginea, H. lancifolia, Saxifraga stolonifera* and *Viola odorata.*

Lightly shaded areas

Species for lightly shaded areas are *Eranthis hyemalis, Hosta lancifolia, H. plntaginea, Leucojum aestivum, Muscari armeniacum, Thymus serphyllum, Tradescantia sillamontana* and *Viola odorata.*

Pathways

Lycoris radiata and *Nerine bowdenii* are suitable for planting along pathways.

Sunny areas

Suitable species are *Bellis perennis, Cerastium tomentosum, Geranium rectum, G. wallichianum* and *Santolina chamaecyparissus*.

Pool

Ajuga reptens, *Bergenia ciliata* and *Caltha palustris* are suitable for pool situations.

Rock gardens

Species are suitable for plantine in rock gardens are (*Aubrieta deltoidea* and *Bergenia ciliata*).

Plants are the major elements used in landscape design. Their constantly changing form enhance a design concept in a number of ways. They contribute not only to the visual quality of mans' environment but also to the physical quality, and have relevant design features like form, colour and texture that can be used by designers to generate character of a landscape. They provide contrast of colour, texture and form in a built environment, introducing the shapes, colours and feelings for nature into man-made geometric patterns of roads and buildings. The horticulture notes incorporating habit, habitat, shape, size, texture, fragrance, flower colour, leaf, and fruit characters of landscape value and uses are pre-requisites for preparing landscape planting plan.

Table 1: Herbaceous perennials of Kashmir Valley

S. No.	Scientific name	English name	Family	Habitat preference	Mode of propagation	Planting Time	Blooming time	Flower colour
1.	*Achillea filipendulina* Lam	Fern leaf yarrow	Asteraceae	FS/FWDS	Seed, Division	Early March	June-Aug	Golden yellow
2.	*Achillea millefolium* L.	Milfoil	Asteraceae	FS/FWDS	Seed, Division	Mid March	June-Aug	Naples Yellow
3.	*Achillea ptarmica* L.	Sneezewort	Asteraceae	FS/FMWDS	Seed, Division	Early April	July-Aug	Greenish white
4.	*Aconitum heterophyllum* Wall.	Aconite	Ranunculaceae	SS/FMWDS	Seed, Division	Early April	July-Sep	Blue
5.	*Aconitum napellus* L.	Aconite, Monkshood	Ranunculaceae	SS/FMWDS	Seed, Division	Early March	July-Aug	Indigo blue
6.	*Agapanthus africanus* Hoffmgg	Blue African lily, African lily	Liliaceae	FS/FWDS	Seed, Division	Early April	July-Sep	Mid blue
7.	*Ajuga reptans* L.	Bugle	Labiatae	SS/FMWDS	Seed, sucker	Early October	April-May	Mellow purple
8.	*Althea rosea* Cav.	Hollyhock	Malvaceae	SS/FMWDS	Seed, Division	Early March	June-July	Purple, Pink,Yellow
9.	*Amaryllis belladonna* L.	Belladonna lily	Amaryllidaceae	SS/FWDS	Seed, bulb	Early March	July-Aug	Persian rose
10.	*Anemone biflora* D.C.	Windflower	Ranunculaceae	SS/FMWDS	Seed, Division	Early October	Feb-March	Pure white With magenta
11.	*Anemone x hybrida* Paxt	Japanese anemone	Ranunculaceae	FS/FMWDS	Seed, Division	Mid Sep	Late Oct	White with Pinkish tinge
12.	*Aquilegia canadensis* L.	Columbine	Ranunculaceae	FS/SS/FMWDS	Seed, Division	Mid Sep	April -May	Mimosa yellow and garnet brown
13.	*Aquilegia vulgaris* L.	Alpine columbine, Granny's bonnet	Ranunculaceae	FS/SS/FMWDS	Seed	Late Sep	April -June	Blue
14.	*Aster alpinus* L.	Michaelmas, Easter Daisy	Asteraceae	SS/FWDS	Seed,cutting, Division	Mid Sep	June-july	Sea lavender

(Contd.)

S. No.	Scientific name	English name	Family	Habitat preference	Mode of propagation	Planting Time	Blooming time	Flower colour
15.	*Begonia evansiana* Andr.	Begonia	Begoniaceae	SS/FWDS	Seed,cutting, Division	Late Oct	April -May	Magenta
16.	*Begonia* x *semperflorens* Link & Otto	Fibrous rooted Begonia	Begoniaceae	FS/FWDS	Seed,cutting, Division	Late Oct	April-June	Blood red
17.	*Begonia tuberhybrida* Voss.	Multiflora and pendula begonias	Begoniaceae	SS/FWDS	Seed,cutting, Division, bulb	Early March	May-June	Magenta
18.	*Belamcanda chinensis* D.C	Blackberry-lily, Leopard lily	Iridaceae	FS/FMWDS	Seed, Division	Early March	June-July	Mandarian red
19.	*Bellis perennis* L.	True or English Daisy, Common Daisy, Shasta Daisy	Asteraceae	SS/FWDS	Seed, Division	Early Oct	April-May	Rose madder, fuchia purple
20.	*Bergenia ciliata* (How.) sternb	Elephant eared Saxifrage, Elephant's ears	Saxifragaceae	SS/FMWDS	Seed, Division	Late Sep	March-April	White and Magenta
21.	*Caltha palustris* L.	King cup, Marsh marigold, Bigwatercup, Blob	Ranunculaceae	SS/FMWDS	Seed, Division	Early March	April-May	Aureolin
22.	*Campanula latifolia* L.	Giant bell flower	Campanulaceae	FS/FMWDS	Seed,cutting, Division	Early Oct	June-July	Veronica violet
23.	*Campanula medium* L.	Canterbury bells	Campanulaceae	FS/FMWDS	Seed, Division	Early Oct	May-June	Mineral Violet
24.	*Centaurea montana* L.	Mountain Bluet, Hardheads, Knapweed	Asteraceae	FS/FWDS	Seed, Cutting, Division	Late Sep	May-June	Phlox Purple

(Contd.)

S. No.	Scientific name	English name	Family	Habitat preference	Mode of propagation	Planting Time	Blooming time	Flower colour
25.	*Cerastium tomentosum* L.	Snow in Summer	Caryophyllacea	FS/FWDS	Seed,cutting, Division	Early April	June-July	Pure white
26.	*Cheiranthus cheiri* L.	Wall flower,Bee flower	Cruciferaceae	FS/FWDS	Seed,cutting, division	Mid March	April-May	Mineral violet
27.	*Chrysanthemum leucanthemum* L.	Marguerite daisy, Ox-eye Daisy, White weed	Asteraceae	FS/SS/FWDS	Seed,cutting, division	Late Oct	June-Aug	Pure white
28.	*Clivia miniata* Regel.	Kaffir lily	Amaryllidaceae	SS/FWDS	Seed, Division	Mid Oct	April-May	Nasturtium red
29.	*Colchicum luteum Baker*	Autumcrocus, Naked ladies	Ranunculaceae	FS/FWDS	Seed, bulb	Mid Sep	Jan-Feb	Buttercup yellow
30.	*Convallaria majalis* L.	Lily of the valley, May flower	Convallariaceae	FS/SS/FMWDS	Seed, Sucker	Early Oct	April-May	Pure white
31.	*Coreopsis grandiflora* Hogg.	Tickseed	Asteraceae	FS/FWDS	Seed,cutting, division	Mid Oct	May-July	Lemon yellow
32.	*Crocosmia masoniorum* L.	Montbretia	Iridaceae	FS/SS/FMWDS	Seed, Division	Early Oct	June-July	Delft rose
33.	*Crocus chrysanthus* Herb	Winter crocus	Iridaceae	FS/FWDS	Seed,Bulb	Mid Oct	March-April	Buttercup yellow
34.	*Crocus sativus* L.	Saffron crocus	Iridaceae	FS/FWDS	Seed,Bulb	Late April	Oct-Nov	Campanula violet
35.	*Cynara cardunculus* L.	Cardoon	Asteraceae	FS/FWDS	Seed, Division	Early March	June-Sep	Bluebird blue
36.	*Delphinium elatum* L.	Larkspur, Candle larkspur	Ranunculaceae	FS/FWDS	Seed, Cutting	Early March	June-Aug	Cornflower blue
37.	*Digitalis lanata* Ehrh.	Foxglove, Grecian Foxglove	Scrophuliaria-ceae	SS/FWDS	Seed	Mid April	July Sep	Palp cream

(Contd.)

S. No.	Scientific name	English name	Family	Habitat preference	Mode of propagation	Planting Time	Blooming time	Flower colour
38.	*Digitalis purpurea* L.	Common Foxglove	Scrophulariaceae	SS/FWDS	Seed	Mid Oct	May-July	Mallow purple
39.	*Echinacea purpurea* Moench	Cone flower	Asteraceae	FS/FWDS	Seed,Cutting, Division	Early Oct	May-July	Roseine Purple
40.	*Eranthis hyemalis* Salisb	Winter aconite	Ranunculaceae	FS/SS/FMWDS	Seed, Division	Mid Sep	Feb-March	Lemon yellow
41.	*Eryngium biebersteinianum* Nevski ex Bobrov	Eryngo	Umbelliferae	FS/FMWDS	Seed,Cutting, Division	Early March	June-Aug	Greyish green
42.	*Euphorbia stracheyi* Bioss	Milkweed, Spurge	Euphorbiaceae	FS/SS/FMWDS	Seed,Cutting, Division	Mid oct	May-July	Orange red
43.	*Fritillaria imperialis* L.	Crown imperial, Fritillary	Liliaceae	FS/FWDS	Seed, Division, Bulb	EarlyOct	March-April	Orpiment orange
44.	*Gagea elegans* Wallich ex. D.Don	Yellow-star-of-Bethelem	Liliaceae	FS/FWDS	Seed, Bulb	Mid Sep	Feb-March	Orpiment orange
45.	*Galanthus nivalis* L.	Common snowdrop	Amaryllidaceae	SS/FMWDS	Seed, Division	Early Oct	Feb-March	Purple white with inner tepals nickel green
46.	*Geranium wallichianum* D.Don	Wallich's Cranesbill	Geraniaceae	SS/FS/FWDS	Seed,Cutting, Division	Late March	May-Aug	Aster violet
47.	*Gladiolus byzanthus*	Gladioli	Iridaceae	FS/FWDS	Seed,Corm	Late Oct	April-May	Rhodaning purple
48.	*Gloriosa rothschildiana* O'Brien	Glory or Climbing lily	Colchicaceae	FS/FWDS	Seed	Early April	July-Aug	Cherry with lemon yellow streaks

(Contd.)

S. No.	Scientific name	English name	Family	Habitat preference	Mode of propagation	Planting Time	Blooming time	Flower colour
49.	*Gypsophila paniculata* L.	Baby's-breath	Caryophyllaceae	FS/FWDS	Seed,Cutting, Division	Early Oct	May-July	Purple white
50.	*Hedychium gardnerianum* Wall.Wall.	Kahili ginger, False ginger lily	Zingiberaceae	SS/FS/FMWDS	Seed, Division	Late April	Aug-Oct	Purple white
51.	*Helenium antumnale* L.	Sneeze weed, Helen's flower	Asteraceae	FS/FMWDS	Seed,Cutting	Early March	June-Aug	Sulphur yellow
52.	*Helianthus autumnale*	False sunflower	Compositae	FS/FMWDS	Seed,Cutting, Division	Late April	July-Sep	Lemon yellow
53.	*Hemerocallis fulva* L.	Common yellow-Day lily	Hemerocalli daceae	FS/FMWDS	Seed, Division	Early March		Lemon yellow
54.	*Hesperis matronalis* L.	Dame's violet, Sweet rocket	Brassicaceae	SS/FS/FMWDS	Seed, Cutting	Mid Sep	April-May	Dauphins Violet
55.	*Hippeastrum puniceum* Urban	Giant Amaryllis	Amaryllidaceae	SS/FS/FMWDS	Seed	Early March	May-July	Dutch vermillion
56.	*Hosta lancifolia* Engler	Narrow-leaved plantain-lily	Hostaceae	SS/FS/FMWDS	Seed, Division	Early March	Aug-Oct	Wisteria blue
57.	*Hosta plantaginea* Aschers	Fragrant-plantain lily	Hostaceae	SS/FS/FMWDS	Seed, Division	Early May	Aug-Oct	Purple white
58.	*Hyacinthus orientalis* L.	Common hyacinth	Hycinthaceae	SS/FS/FMWDS	Bulb	Early Oct	March-April	Spectrum blue
59.	*Iris ensata* Thunb.	Blue flag, Japanese iris	Iridaceae	SS/FS/FWDS	Seed,Division,	Mid Oct Rhizome	April-May	Wisteria blue
60.	*Iris germanica* L.	German iris,	Iridaceae	SS/FS/FWDS	Seed, Division, Bulb	Mid Oct	March-April	Hyacinth blue
61.	*Iris japonica* Thunb	Iris	Iridaceae	FS/SS/FMWDS	Seed,Division, Rhizome	Mid Oct	Apr-May	White and bluebird blue margins

(Contd.)

S. No.	Scientific name	English name	Family	Habitat preference	Mode of propagation	Planting Time	Blooming time	Flower colour
62.	*Iris kashmeriana* Baker	Iris	Iridaceae	FS/FWDS	Seed, Division, bulb	Mid Oct	Apr-May	Campanula violet
63.	*Iris reticulata* M. Bieb	Spanish iris	Iridaceae	FS/SS/FMWDS	Seed, Division, bulb	Mid Oct	Mar-April	Hyacinth blue
64.	*Iris tectorum* Maxim	Roof iris	Iridaceae	FS//FWDS	Seed,Division, Rhizome	Mid Oct	April-May	Campanula violet
65.	*Iris unguicularis* Poir	Iris	Iridaceae	FS/SS/FMWDS	Seed,Division, Rhizome	Early Oct	Mar-April	Veronica violet
66.	*Ixiolirion karateginum* Lipsky	Ixiolirion	Amaryllidaceae	FS//FWDS	Seed	Mid Sep	Mar-April	Lobelia blue
67.	*Kniphofia uvaria* Hook	Red hot poker, Torch lily	Asphodelaceae	SS/FMWDS	Seed, Cutting, Division	Early Oct	April-June	Persimon orange
68.	*Lathyrus latifolius* L.	Everlasting pea	Leguminosae	FS/SS/FWDS	Seed	Early March	June-July	Rose madder
69.	*Leucojum aestivum* L.	Snowflake	Amaryllidaceae	FS/FMWDS	Seed, bulb	Mid Sep	Feb-March	Pure white
70.	*Liatris spicata* Willd.	Blazing star Gay-feather	Asteraceae	FS/FMWDS	Seed, division	Mid March	June-Aug	Victoria violet
71.	*Limonium latifolium* Kuntze	Sea-lavender	Plumbag inaceae	FS/FWDS	Seed, Division	Late Sep	April-May	Campanula violet
72.	*Lunaria annua* L.	Moonwort, Satin-flower, Honesty	Cruciferae	SS/FS/FMWDS	Seed, Division	Mid Sep	April-may	Cyclamen purple
73.	*Lupinus polyphyllus* Lindl.	Lupine	Caryophyllaceae	SS/FWDS	Seed, cutting	Early oct	May-june	Gentian blue
74.	*Lycoris radiata* Herb.	Red spider lily	Amaryllidaceae	FS/FWDS	Seed, Bulb	Mid March	May-July	Scarlet
75.	*Lythrum salicaria* L.	Spiked or Purple loose strife	Lythraceae	FS/FWDS	Seed, Cutting, Division	Early April	July-Sep	Petunia purple
76.	*Matthiola incana* R. Br.	Gilly flower, Stock	Brassicaceae	FS/FMWDS	Seed	Late Sep	April-June	Rhodamine purple

(Contd.)

S. No.	Scientific name	English name	Family	Habitat preference	Mode of propagation	Planting Time	Blooming time	Flower colour
77.	*Mimulus nepalensis* Benth.	Monkey flower, Mush flower	Scrophula riaceae	FS/SS/FMWDS	Seed,Cutting, Division	Mid Sep	April-May	Yellow
78.	*Mirabilis jalapa* L.	Four O'clock flower, Marvel-of-Peru	Nyctaginaceae	FS/FWDS	Seed, Division	Mid April	July-Sep	Phlox purple
79.	*Muscari armeniacum* Leicht.	Grape hyacinth	Hyacinthaceae	FS/FMWDS	Seed, bulb	Early Oct	March-April	Lovelia blue
80.	*Narcissus pseudonarcissus* L.	Trumpet narcissus	Amaryllidaceae	FS/FMWDS	Seed, bulb	Late Sep	Feb-April	Yellowish cream
81.	*Narcissus tazetta* L.	Polyanthus narcissus, white narcissus	Amaryllidaceae	FS/FMWDS	Seed, bulb	Early Oct	Feb-April	Cream
82.	*Nelumbo nucifera* Gaertn.	Lotus, Chinese water lily	Nelumbonaceae	FS/FMWDS	Seed, bulb	Mid March	June-July	Pinkish white
83.	*Nepeta nervosa Royle ex* Benth.	Catmint	Labiatae	SS/FS/FWDS	Seed, Cutting, Division	Early March	June-August	Purplish blue
84.	*Nerine bowdenii*	Spider lily	Amaryllidaceae	FS/FWDS	Seed, Division	Mid March	July-August	Pink
85.	*Nymphae stellata* Willd.	Indian blue water lily	Nymphaeaceae	FS/UW	Seed, Division	Late Oct	May-July	Pink
86.	*Oenothera biennis* L.	Evening primrose	Onagraceae	FS/FWDS	Seed, Cutting, Division	Early April	July-Sep	Sulphur yellow
87.	*Ornithogalum umbellatun*	Star of Bethlehem	Hyacinthaceae	FS/FWDS	Seed, bulb	Early Oct	April-May	Pure white with agathia green base
88.	*Paeonia emodi* Wall. ex Royle	Himalayan peony	Paeoniaceae	SS/FMWDS	Seed, Cutting, Division	Mid Sep	April -May	Pure white
89.	*Paeonia lactiflora* Pall	Peony	Paeoniaceae	SS/FMWDS	Seed,Cutting, Division	Early Oct	May-June	Persian rose
90.	*Penstemon barbatus*	Bearlip penstemon	Scrophula riaceae	SS/FS/FWDS	Seed, Division	Early Oct	May-Aug	Caramine

(Contd.)

S. No.	Scientific name	English name	Family	Habitat preference	Mode of propagation	Planting Time	Blooming time	Flower colour
91.	*Phlox paniculata* L.	Summer perennial phlox	Polemoniaceae	SS/FS/FWDS	Seed, Cutting, Division	Early March	June-Aug	Phlox pink
92.	*Physalis alkekengi* L.	Winter cherry Chinese lantern plant	Solanaceae	SS/FS/FWDS	Seed, Cutting, Division	Mid Oct	May-July	Cream
93.	*Platycodon grandiflorum* A.DC.	Balloon flower	Campanulaceae	SS/FS/FWDS	Seed,Division	Mid April	July-Sep	Campanula violet
94.	*Primula denticulata* Smith	Drum stick primula	Primulaceae	SS/FS/FMWDS	Seed,Division	Mid Sep	March-April	Bishops violet
95.	*Primula rosea* Royle	Primrose	Primulaceae	SS/FS/FWDS	Seed, Cutting, Division	Mid Sep	March-May	Magenta
96.	*Primula vulgaris* Huds	Primrose	Primulaceae	SS/FS/FWDS	Seed, Cutting Division,	Early April	July-Aug	Mauve
97.	*Ranunculus falcatus* L.	Turban flower	Ranunculaceae	SS/FS/FWDS	Seed,Division	Late Sep	April-June	Yellow
98.	*Rudbeckia laciniata* L.	Cone flower	Asteraceae	SS/FS/FWDS	Seed,Division	Early April	July-Aug	Roseine purple
99.	*Salvia sclarea* L.	Biennial clary	Labiatae	SS/FS/FWDS	Seed,Cutting, Division	Late Sep	April-june	Plum purple
100.	*Solidago canadensis* L.	Golden rod	Asteraceae	FS/FWDS	Seed,Division	Late April	July-Oct	Canary yellow
101.	*Sprekelia formosissima* Herb.	Jacobean lily Aztec lily	Amaryllidaceae	FS/FWDS	–	Early March	May-June	Scarlet
102.	*Sternbergia lutea* Roem and Schult	Autumn daffodil	Amaryllidaceae	FS/FWDS	Seed,Division	Mid Sep	Jan-Feb	Lemon yellow
103.	*Thymus serphyllum* L.	Thyme	Labiatae	FS/FWDS	Seed,Division	Late April	May-July	Tyrian rose

(Contd.)

S. No.	Scientific name	English name	Family	Habitat preference	Mode of propagation	Planting Time	Blooming time	Flower colour
104.	*Tulipa clusiana* D.C.	Lady Tulip	Liliaceae	FS/FWDS	Seed, Bulb Division	Mid Oct	March-April	White stripped with tyrianrose
105.	*Tulipa lanata* Rgl	Mughul Tulips	Liliaceae	FS/FWDS	Seed, Bulb	Mid Oct	March-April	Jasper red
106.	*Typha angustifolia* Bory and Chaub	Lesser bulrush	Typhaceae	FS/DMS	Division	Early Oct	May-April	Brown spikes
107.	*Verbascum thapsus* L.	Aaron's rod, great mullein	Scrophula riaceae	FS/FWDS	Seed,Cutting, Division	Late April	Aug-Oct	Primrose yellow
108.	*Viola odorata* L.	English violet, garden violet, sweet violet	Violaceae	SS/FS/FMWDS	Cutting,division	Late Sep	March-April	Cobalt violet
109.	*Zantedeschia aethiopica* Spreng	Arum lily, calla of gardeners	Araceae	FS/FMWDS	Seed, Division	Early march	July-Aug	Pure white
110.	*Zantedeschia elliottiana* Engler	Golden arum, golden calla	Araceae	FS/FMWDS	Seed, Division	Late Oct	April-July	Aureolin

Abbreviations used
SS – semishade, FS – full sun,
FWDS – fertile well drained soil,
FMWDS – Fertile moist drained soil,
UW – undisturbed water, DMS – dry moist soil

Herbaceous Perennials of Kashmir Valley

Aquilegia canadensis

Achillea filipendulina

Bellis perennis

Bergenia ciliate

Caltha palustris

Colchicum luteum

Convallaria majalis

Crocus chrysanthus

Cynara carduncules

Eranthis hyemalis

Fritillaria imperialis

Gagea elegans

Galanthus nivalis

Gypsophila paniculata

Hyacinthus orientalis

Herbaceous Perennials of Kashmir Valley

Iris kashmeriana

Iris japonica

Liatris spicata

Matthiola incana

Muscari armeniacum

Primula rosea

Santolina chamaecyparissus

Senecio cineraria

Solidago canadensis

Sternbergia lutea

Thymus serphyllum

Tradescantina sillamontana

Typha angustifolia

Zantedeschia aethiopica

Zantedeschia elliottiana

References

Bailey, L.H. (1963). The standard encyclopaedia of horticulture. The Macmillan Company, New York. pp. 1-1200.

Bhattacharjee, S.K. (2003). Bulbous plants in landscaping. *Advances in Ornamental Horticulture* 5: 17-19.

Brickell, C. (1996). The Royal Horticultural Society (A-Z) Encyclopaedia of Garden Plants. Dorling Kindersley Limited, London. pp. 1-1080.

Corley, W.L. (1989). Going wild in Dixie. *American Nurserymen* 172(9): 2.

Hartmann, H.T., Kestar, D.E. and Davies, F.T. (1994). Plant Propagation Principles and Practices. Prentice Hall of India Pvt. Ltd., New Delhi, pp. 1-647.

Kanth, T.A. (1985). Stratigraphic profile of Jammu and Kashmir State. *Geography of Jammu and Kashmir*. Ariana Publishing House, New Delhi, pp. 1-320.

Macoboy, S. (1986). Perennials for Beds and Borders. South China Printing Co., Hongkong, pp. 1-160.

Miles, B. (1963). The Wonderful World of Bulbs. D.Van Nostrand Company Limited, Canada, pp. 1-314.

Randhawa, G.S. and Mukopadhyay, A. (2004). Floriculture in India. Allied Publishers, Private Limited., New Delhi, pp. 1-630.

Robin, J. and Carr, G. (1984). Local flora, finding it, growing it, selling it. *Australian Horticulture* 82(3): 40-48.

Shuler, C. (1991). Landscaping for Colour and Fragrance. *Proceedings of a Conference on Landscaping with Wild Flowers and Native Plants.*USA 13-15 March. *Acta Horticulture* 43: 82-84.

Singh, G. and Kachroo, P. (1976). Forest flora of Srinagar and plants of neighbourhood. Bishen Singh Mahendra Pal Singh, Dehradun (India). pp. 1-278.

Ornamental Horticulture, pp 225-240
© Editor, S.S. Sindhu, 2016
New India Publishing Agency, New Delhi - 110034, India

10

Xeriscaping[1]: An Approach of Landscaping for Dry Areas

Balaji S. Kulkarni

Whether you are an experienced gardener or a newcomer to the high desert, learning how to successfully garden the desert can be an exciting challenge. With minimal annual precipitation, a short growing season and fluctuating temperatures, maintaining an attractive, healthy landscape can keep you on your toes. Most gardeners value the natural beauty of our native landscapes, but are interested in enhancing it with additional plants to add more color and texture. Can a more beautiful landscape be created and still be water efficient and sustainable? The answer is YES.

The Meaning of Xeriscape

The word xeriscaping (pronounced zer-iskap- ing) is commonly referred to as water-wise or water-smart gardening. Xeros is derived from the Greek word meaning"dry."However, do not let the Greek derivative conjure up images of a dry, desert-like, "blandscape."Xeriscaping and Xerogardening refer to landscape gardening in ways that reduce or eliminate the need for supplemental irrigation. It is promoted in areas that do not have easily accessible supplies of fresh water and is gaining acceptance in other areas as climate pattern changes. Xeriscape is a proactive approach to curbing wasteful use of a valuable natural resource. Xeriscaping, is a seven step method that combines innovative water conservation techniques with sound horticultural practices. Not only it is possible to have a colorful and beautiful landscape while conserving water, but it can become more sustainable while saving you time, energy and money.

[1]I/c. Department of Floriculture and Landscape Architecture, Kittur Rani Channamma College of Horticulture, (U.H.S. Bagalkot), Arabhavi, Karnataka

Often it is thought that a xeriscape means the use of all native or low water use plants, resulting in limited plant selection. However xeriscapes can incorporate hundreds of plant choices including trees, shrubs, evergreens, perennials and grasses. Plants whose natural requirements are appropriate to the local climate are emphasized, and care is taken to avoid losing water through evaporation and run off. The specific plants used in xeriscaping depend upon the climate. The concept originated in Denver, Colorado, in early 1980s. Because of severe drought conditions, Denver rationed water and prohibited irrigation of lawns and yards. As a result, vegetation in yards withered, and Denver landscapers began promoting what they called xeriscape, a landscaping approach that uses small amounts of water but maintains a traditional look. In some areas, terms likes water-conserving landscapes, drought tolerant landscaping, zeroscaping and smart scaping are used instead.

What are the Benefits of Xeriscape?

1. It saves water, up to 50-60 percent

Reducing peak water demand could extend the long-term capacity of our reservoir and water treatment facility. Because water providers must plan their water storage and water treatment facilities to meet peak demand. If we reduce peak demand we can delay costly expansion and the same time we accommodate new residents and businesses in our state. In the long term, delaying expansion of water facilities will produce significant savings for all consumers. In addition, if we cut our individual water use, we will save on our water bills, although the monthly savings will be fairly small.

Sovacrol and Morgan (2005) studied the monthly water bill for a typical Las Vegas area home for traditional and xeriscape type of garden. They found that in summer, that is in June to August, water bill of a home which follows traditional landscape reaches high. But in case of xeriscape type of garden there is less fluctuation in water bill, because plants have the capacity to manage with less water when water is scarce.

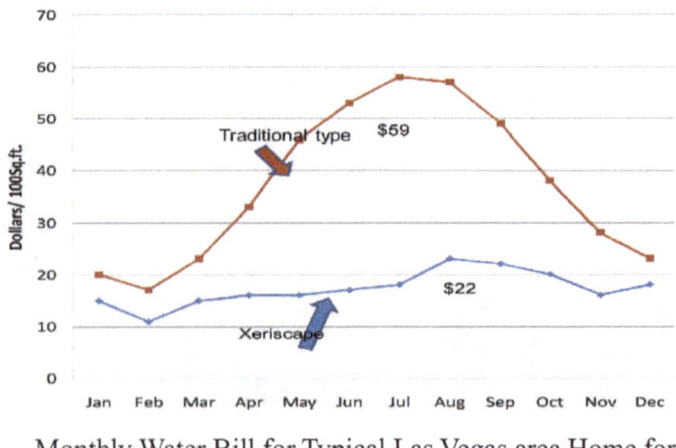

Monthly Water Bill for Typical Las Vegas area Home for Traditional and Xeriscape Type of Garden

But Does Your Yard Really Make a Difference?

Suppose you replace your traditional yard (6000 square-feet average) with a mix of drought tolerant turf and water-wise plants.

You Save:

- 48,000 gallons of water each year
- 2.4 million gallons over a lifetime
- About $75 in water bills per year
- $8,400+ over a lifetime

Suppose you replace your traditional yard with moderately drought tolerant plants.

You Save:

- 90,000 gallons of water each year
- 4.5 million gallons over a lifetime
- About $135 in water bills per year
- $15,800+ over a lifetime

Suppose you replace your traditional yard with highly drought tolerant plants.

You Save:

- 108,000 gallons of water each year
- 5.4 million gallons over a lifetime
- About $160 in water bills per year
- $19,000+ over a lifetime

YES! Your yard does make a difference (Welsh and Welch, 1994)

2. It makes yards less vulnerable to ravages of drought and water shortages.

If a prolonged dry spell or drought should produce water shortage in urban areas, our door use of water –including lawn watering –would have to be restricted. Xeriscaped yards would be less affected or perhaps not affected at all, by water use restrictions because they make much more effective and efficient use of water available.

3. It allows us to develop a high-quality; beautiful landscape using ecologically sound landscaping practices

A well-planned, established landscape that fully incorporates the seven principles of xeriscape requires little or no synthetic fertilizer or pesticide use. This not only creates a more compatible environment for outdoor activities but also eliminates the possibility that runoff from our yards and other landscaped areas might take toxic materials or nutrients to streams and rivers. In addition, if we are able to utilize clippings and leaves from our trees and shrubs for mulch, we can help reduce the volume of solid waste going to the city landfill.

4. Wild life conservation

Providing habitat for native species by attracting the butterflies, birds *etc.*

5. It helps in reducing time, energy and money

Sovacool and Rosales (2001) compared the Xeriscape garden with traditional garden w.r.t. cost and time required for maintenance and opined that xeriscape garden maintenance requires less time and money

Table 1: Maintenance cost and time requirement for traditional and xeriscaping type of garden

Types	Monthly maintenance cost ($)	Annual maintenance cost ($)	Time (hr/ month/ 500 sq. ft.)
Xeriscape garden	1.80	473.93	5.95
Traditional garden	11.60	683.03	8.17

6. It will take full advantages of rainfall

7. When water restrictions are implemented, xeriscape plants tend to survive, while more traditional plants may be unable to adapt.

Basic principles of Xeriscaping

Xeriscaping incorporates seven basic principles which lead to saving water.

1. Planning and design
2. Improve soil
3. Appropriate plant selection
4. Efficient irrigation
5. Creating practical turf
6. Mulching
7. Maintenance

1. Planning and Design

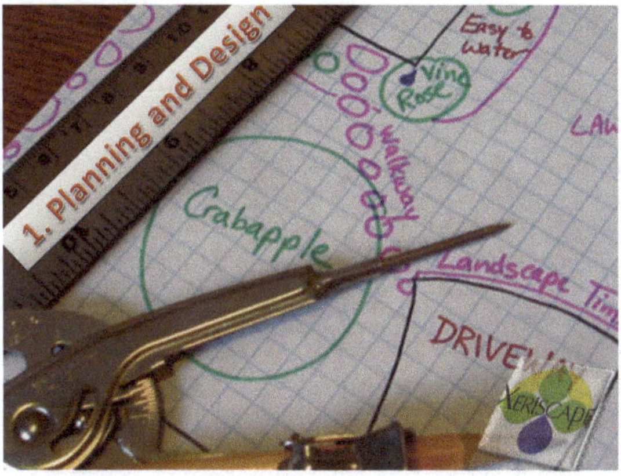

Fig. 1: Plan of Landscape

Whether you are developing a new landscape or renovating an existing landscape, proper planning and design are important. You may wish to start thinking immediately about what plants to use, but first you must solve any environmental and physical problems in an attractive and practical manner. Think about various areas of your landscape in terms of how they should be developed for different uses and how much space you should allot to each area.

Taking the time to carefully plan your landscape will result in a beautiful landscape for many years. A plan provides direction and guidance and will ensure that water saving techniques are implemented in the landscape. The first step is to look at the existing landscape and create a "base plan".

• Begin with a base map

A base map is a plan of the property drawn to scale on graph paper showing location of the house, its orientation to the sun, other structures on the site, unusual features such as stone outcroppings and existing vegetation.

• Plan for different use

To begin your plan, overlay the base map and site analysis sheet with another piece of tracing paper. Consider how these areas will be developed based on space requirements for each activity.

- The ***public*** **area** is the highly visible area that most visitors see, such as the entry to the home. In a traditional landscape, this area typically receives the most care, including the most water. Therefore, the careful design of this area is important for water conservation.
- The ***private*** **area** of the landscape, usually the backyard, is where most outdoor activity occurs. It is generally the family gathering area. It may also include a vegetable garden or fruit orchard.
- The ***service*** **area** is the working or utility area of the landscape, an area usually screened from view and containing such items as garbage cans, outdoor equipment, air-conditioning units or a doghouse.

• Incorporate shade into the design

Shade from trees or structures in the landscape keeps the landscape cooler and reduce water loss while creating a comfortable living environment. A shaded landscape can be as much as 20F cooler than one in the full sun.

A person standing in an open parking lot is bombarded with 1,000 heat units from the sun and another 1,000 heat units reflected from the paved surface (Fig. 2). Walking beneath a shade tree provides immediate relief from the sun because the tree acts like an umbrella, blocking light and heat (passive shade). If the tree is growing in moist soil, it will not only block heat but will also dissipate heat by evaporative loss from the leaves (active shade). A moist soil surface also porates heat and reduces heat load further. That's what shade acts in our landscape

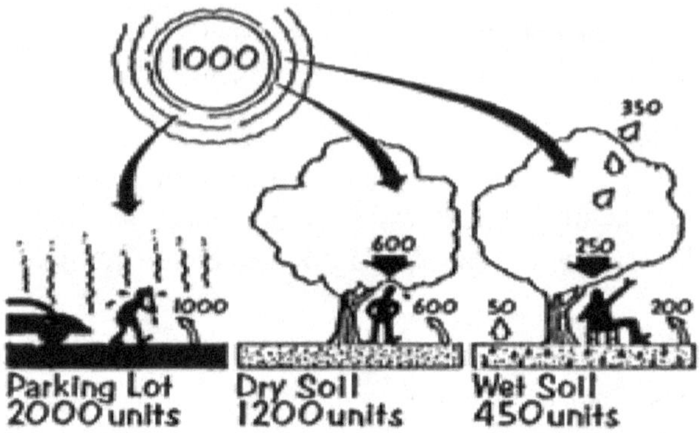

Fig. 2: Effects of Shade on Water Loss

• Renovation of an existing landscape for improved water conservation

The redesign of the residence shows expansion of the shrub beds in the public and private areas of the landscape to provide seasonal interest, variety and reduced maintenance. Shade-tolerant ground covers are used under the existing trees on the left side of the front and right rear of the property.

2. Improve Soil

A thorough analysis of both the physical and chemical characteristics of the soil is important when developing a water-wise landscape. Soil analysis determines whether soil improvement is needed for better water absorption and improve water holding capacity. Each soil has its own unique structure and texture, drainage pattern, pH, nutrient content and need for amendments and fertilizer. To complicate matters, an individual landscape may have several different soil types, or the soil may consist of fill dirt brought onto the site.

In the areas where the soil may be compacted, it will often be necessary to physically improve your soil before you can grow deep-rooted plants. Physical improvement of soil involves tilling to break up compaction and provide aeration and adding organic matter to keep soil porous.

Add organic matter

Xeriscaping must address the need for garden soil amendments. Sandy soil readily absorbs water but drains quickly, whereas clay soils hold moisture but have a low absorption rate. The key, in either case, is to incorporate lots of organic matter in the soil to maximize water absorption and retention.

Do not add organic matter to the planting hole, because matter added to individual planting holes in clay soils acts like a sponge in a bathtub, holding excess moisture around plant roots after irrigation or rainfall. Wet and waterlogged soils that suffocate plant roots are a leading cause of plant death. Even in well drained soils, organic matter encourages the roots of plants to stay within the hole instead of growing out to explore the native soil. The general rule of thumb is to amend your native soil with 1/3 organic matter.

In addition, it may be necessary to chemically improve the soil with nutrients or other materials. Landscape architects emphasize that both kinds of soil improvements (physical and chemical) are important to developing healthy, deep roots, and that heavy fertilizing will not compensate for sufficient physical soil preparation.

Use hydrogels

Water-absorbing polymers, commonly called hydrogels, are popular new products. These man-made crystals, sold under various trade names, absorb several hundred times their weight in water and gradually release it to plant roots. One pound of crystals applied to 100 square feet of bed area will absorb 20 to 25 gallons. Hydrogel of water absorbs about 50 times as much moisture as peat moss. They last from six months to several years in the soil, depending on product.

3. Appropriate Plant Selection

Appropriate plant selection means selecting plants that not only are compatible with the design but also are well suited to the planting site and local environment. Appropriate plant selection keeps the landscape more in tune with the natural environment. Both native and exotic plants make up the huge variety of plants available for xeriscaping. There is a misconception among people that only plants allowed in xeriscape are cactus and other desert perennials. There are a large number of plants which do well in semi-arid climate with little or no supplemental irrigation. The trick to water wise gardening is to put the right plants in the right place. Different plants have different requirements when it comes to light, soil and water.(Table.2) Plants have to be grown in the microclimate that most closely matches their needs. For instance, plants that like sun and dry conditions would do the best in a southern or western exposure, in areas that get watered infrequently. Plants that prefer cooler sites would grow well in an eastern or northern exposure, or in areas that are shaded by other plants. Higher water-using plants would do the best in low-lying drainage sites. To make it easier to irrigate essentially; plants should also be grouped according to their water requirements. It involves selecting plants according to the soil type and light level of the site. Drought tolerance is important in a xeriscape-type landscape. However, it should not be the only criteria used to select plants. Junipers, for instance, are extremely drought tolerant, but they cannot tolerate wet soils or heavy shade.

Other Important Criteria to Consider Include

Mature size and form. Will the plant remain in scale with the rest of the landscape as it matures, or will it likely overgrow the site and compete with other plants for space, nutrients and water?

Growth rate. Slow-growing dwarf shrubs and ground covers used around the base of the home require little routine pruning.

Texture. Is the leaf texture fine, medium or coarse, and does it combine well with the adjacent plants?

Color. Is the flower or foliage color compatible with other plants or the background color of the building?

Functional use. Is the plant suitable for the location and intended purpose; i.e. under low windows, along the perimeter of the property as screening hedge, or as a ground cover?

Select healthy and vigorous plants. Examine the root system for well-developed roots throughout the root ball and an abundance of small white roots (absorptive roots) along the exterior of the root ball. Examine the leaves and stems for insects or diseases and avoid plants weak or appear unhealthy.

Table 2. Some of the important plants suitable for xeriscape

Turf grasses

Common name	Best adapted for
Bermuda grass	Good drought tolerance; produces dense turf; poor shade tolerance; plant seed or sod
Buffalo grass	Excellent drought tolerance; produces thin turf; poor shade tolerance; plant seed or sod
Carpet grass	Adapted to moist sites; tolerates partial shade; many seed heads; thin turf; plant seed
Centipede grass	Low maintenance; tolerates partial shade; tolerates drought; plant seed or sod
St. Augustine grass	Produces dense turf; good shade tolerance; poor drought tolerance; plant sod
Tall fescue	Under irrigation, remains green year-round; good shade tolerance; poor drought tolerance; plant seed
Zoysia grass	Produces dense turf; good shade tolerance; good drought tolerance; plant sod

Ground covers

Common name	Scientific name	Exposure
Ajuga	*Ajuga reptans*	Shade
Asiatic jasmine	*Trachelospermum asiaticum*	Sun/shade
Confederate jasmine	*Trachelospermum jasminoides*	Sun/shade
English ivy	*Hedera helix*	Shade
Liriope	*Liriope muscari*	Sun/shade
Monkey grass	*Ophiopogon japonicus*	Sun/shade
Prostrate rosemary	*Rosmarinus officinalis*	Sun
Light blue flowers	*Santolina santolina* sp.	Sun
Trailing juniper	*Juniperus* sp.	Sun
Perennials		
Autumn sage	*Salvia greggii*	Sun
Baby sun/sunray coreopsis	*Coreopsis grandiflora*	Sun
Bearded iris	*Iris xiphioides*	Sun/shade
Blue plumbago	*Plumbago auriculata*	Sun/shade
Bouncing bet	*Sapinara officinalis*	Sun/shade
Cigar plant	*Cuphea micropetala*	Sun/shade
Daylily	*Hemerocallis* sp	Sun
Garden canna	*Canna x generalis*	Sun/shade
Indian blanket.	*Gaillardia* sp	Sun
Rosemary	*Rosmarinus officinalis*	Sun/shade
Society garlic	*Tulbaghia violacea*	Sun/shade
Shrubs		
Red yucca	*Hesperaloe parvifolia*	Sun
Barberry	*Berberis thunbergii*	Sun

China rose	*Rosa chinensis*	Sun
Indian hawthorne	*Raphiolepis indica*	Sun/shade
Juniper	*Juniperus* sp.	Sun
Tea rose	*Rosa odorata*	Sun
Pomegranate	*Punica granatum*	Sun
Lilac	*Syringa vulgaris*	Sun
Oleander	*Nerium oleander*	Sun
Russian olive	*Eleagnus angustifolia*	Sun/shade

Trees

Common name	Scientific name
American plum	*Prunus americana*
Panicled golden raintree	*Koelreuteria paniculata*
Soapberry	*Sapindus drummondii*
Wild olive	*Corida boissieri*
Live oak	*Quercus virginiana*
Hackberry	*Celtis occidentalis*
Redbud	*Cercis* sp.
Nandina	*Nandina domestica*
Slash pine	*Pinus elliottii*
Winter honeysuckle	*Lonicera fragrantissima*

Irrigate Efficiently

Water-wise landscape requires a minimal amount of supplemental water from irrigation. When irrigation is used, water is applied efficiently and effectively to make every drop count, just as we zone plants in the landscape according to their different water needs. Usage of irrigation water efficiently also requires selection of the appropriate type of irrigation for the plants and for each area of the landscape. Trees and shrubs in the low water-use zone would need supplemental water only during establishment (first 8 to 10 weeks after transplanting); plants in moderate water-use zones require water only during periods of limited rainfall when they show signs of stress. For these plants, a temporary system like a soaker hose or hand watering may be all required.

- In moderate water use zones a turfgrass would be irrigated only when it shows sign of moisture stress.
- In low water zone it requires 1- 2 inches / week and divide these amount into 2 applications / week
- Never apply more than 1 inch at a time

The plants develop a shallow root system, due to daily irrigation with small amount and reduces drought tolerance. Whereas the plants develop deep root because of less availability of water on surface area and those plants will have more drought tolerance. (Fig.3)

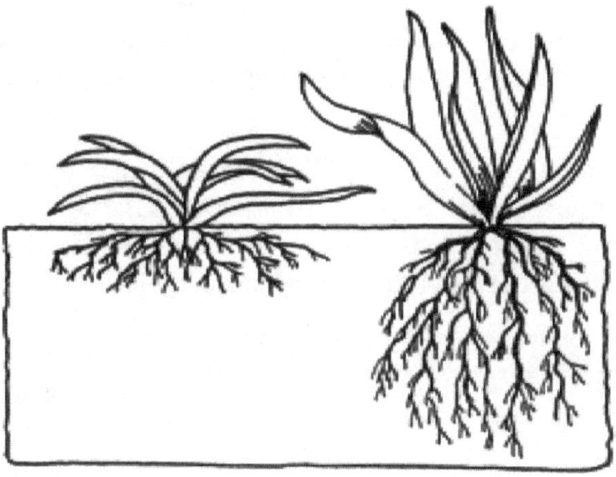

Daily irrigated Twice a week

Fig. 3: The Effect of Irrigation on Root Distribution

The best time to water is during morning or evening hours. Watering in middle of the day when wind and evaporation rates are be highest is not very efficient. Most of the water applied between 10 am and 6 pm barely reach the plant due to evaporation and wind. Stationary sprinkler heads (spray heads) which keep the water fairly low or sprinkles that move horizontally from side to side (rotary heads) are less prone to evaporation and wind drift. Drip, micro-sprays, bubblers or emitters are the most efficient for watering flowers, shrubs and trees. Care has to be taken to design the sprinkler system or plan the base placement so that a given zone or hose placement waters plants with similar watering requirements at the same time. For automatic sprinkler system, different irrigation methods on the same zone should never be mixed. All methods, fixed spray head, rotary head and drip emitters apply water at different rates. Trying to irrigate plants with different water requirements at the same time or mixing irrigation methods result in either over-watering some plants or under watering others. The irrigation system has to be regularly checked for leaks, broken or misaligned heads and the repairs promptly. The irrigation schedule should be changed to reflect seasonal changes in water requirements.

5. Create Practical Turf Areas

Turf grass is one of the most versatile and functional plants in the landscape. It provides one of the best recreational surfaces for outdoor activities. From a water management standpoint, turf is recognized as one of the most effective plant covers to reduce runoff and erosion while recharging the ground water, which results in more efficient use of rainfall. Turf has a tremendous mitigating effect on the environment. For example, research documents that a turf area can be as much as 30F cooler than a concrete or asphalt surface and 10F to 14F cooler than bare soil. This cooling effect from the average lawn is equal to more than eight tons of air conditioning; the average home central-air unit produces three to four tons.

Maximum water conservation with turf is obtained through proper selection, establishment and maintenance. In addition to differing in appearance, turf grasses differ in their tolerance to environmental factors like shade, temperature, soil fertility, water use and drought resistance Table.3.

Table 3: Average water use and drought resistance of selected turf grass

Common name	Water use	Drought resistance
Tifway Bermuda	Very low	Very high
Common Bermuda	Very low	High
Raleeigh St. Agustine	Very low	Very high
Rebel 11 tall fescue	Very low	Medium
Centipedi grass	Low	Medium- high
Meyer Zoysia	Low	Low
K 31 Tall Fescue	Low	Low- medium

Native and drought tolerant turfs area has to become available proposition. When designing turf grass areas in the landscape, the ease or difficulty in watering the proposed area has to be considered. Long narrow and small odd-shaped areas are difficult for any irrigation equipment to efficiently use water. It is appropriate to eliminate long, narrow areas and maintain square areas. (Scott., 2007).

6. Use Mulch

Mulches conserve moisture by preventing evaporative water loss from the soil surface and reducing the need for supplemental irrigation during periods of limited rainfall. A two- to four-inch blanket of mulch conserves water by retaining soil moisture. Mulch can reduce or eliminate weeds, which compete for moisture, nutrients and sunlight. It can also prevent erosion and organic natural mulches break down and decompose to improve soil texture and drainage.

Mulches also prevent crusting of the soil surface and allow water to penetrate readily to plant roots. They insulate the roots of plants from summer heat and winter cold and help control weeds that compete with plants for moisture. By serving as a barrier between the plant and soil, mulches discourage soil borne diseases that stress plants and cause them to have a higher demand for water.

Be careful when choosing your mulch not to get material, such as hay, that contains viable seeds or you will have a multitude of unwanted sprouts throughout your landscape. The same caution applies to using on composted animal manures.

Types of mulches

- Organic–Straw, pine needles bark nuggets, compost, wood chips, sawdust and other wood products.
- Inorganic- Lava rock, rock, plastic, fabric

Adoption of mulching in landscaping with compost yard waste (C: N ratio of 17:1) and ground wood mulch (C: N ratio of 51:1) dramatically increased the soil organic matter content (Fig. 4) and there by increased the water holding capacity of the soil, ultimately reduced frequency of irrigation. (Daniel *el al.*, 1999)

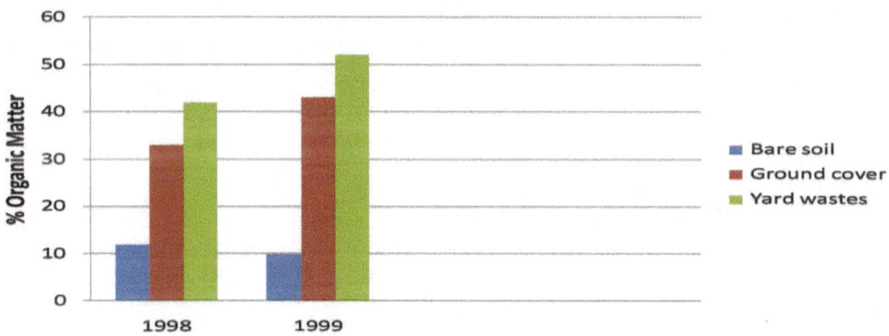

Fig. 4: Effect of Compost Yard Mulch, Ground Wood Mulch, and Bare Soil on Soil Organic Matter

7. Maintain Landscape Properly

Appropriate maintenance preserves the beauty of xeriscape plus saves water. No garden is completely maintenance free. Like more traditional landscape, xeriscapes require regular maintenance to look their best. This includes weeding, pruning shrubs and trees, dead heading perennials, aerating, mowing and fertilizing lawns, irrigating and controlling pests and diseases *etc.*

• Don't let weeds compete with plants for water

Scout the landscape regularly and don't let weeds take over. Hand weeding, chemical herbicides and mulches will keep weeds in check.

• Scout for pests before you spray

Control pests when they begin affecting appearance and overall health of a plant. Target your control measures to affected plants and avoid spraying the entire landscape if the pest problem is confined only to a small area. The most effective way to control garden pests and diseases is through integrated pest and disease management. A variety of techniques including using resistant plants, spraying with water, using botanicals or predictors to manage and using less chemical pesticides or fungicides can be practiced.

• Make every drop count

Where irrigation systems are used, check nozzles and emitters regularly to see if they are operating efficiently and are delivering the right amount of water in the right locations.

Let your plants tell you when they need water

Learn to identify the symptoms shown by plants under water stress. Shrubs under moisture stress will turn a gray-green color and wilt. Trees show premature fall color and shed leaves. Turf grass will turn a dull gray-green color, and the blades wilt and roll inward.

Survival watering during drought or watering restrictions

During drought or watering restrictions, consider replacement cost of plants in the landscape and do what you can to save the most valuable plants. Annual flowers can be replaced more readily than trees and shrubs.

• A pruning

Regular pruning keeps plants looking good and promotes plant vigor. Also, many shrubs bloom on new wood, so pruning necessary to have flowers. The best time to prune most shrubs and trees is in the winter to early spring. Prune dead or unwanted branches slightly away from the trunk, on outer edge of branch.

• Deadheading

Dead heading means removing spent flowers before they mature into seed heads. Dead heading makes the garden look more attractive, prevent unwanted, self-sown seedlings and encourages the plant to produce more flowers. Many plants like salvia will bloom repeatedly during the summer if deadheaded.

• Aeratlng

Regular core aeration will reduce thatch buildup and loosen the soil. This promotes better root growth, and allows water and fertilizer to move into the root zone. Core aeration can be done with a manual aeration device. Core aeration is the best done twice year in spring and late summer.

• Mowing and Fertilizing

Lawn should be mowed to a height of 2-3" and should be mowed frequently enough so that no more than 1/3 of the grass height is removed at any time. Fertilization schedules vary according to the type of grass involved.

How Gardens look before and after xeriscape

Before **After**

Conclusion

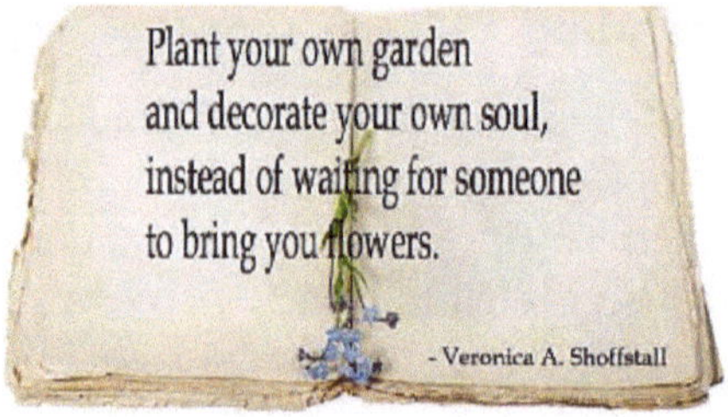

'Xeriscape is one of the important landscape practice most suitable for dry region of India but in our country most of the people shifted towards traditional landscape that is, high water using gardens so that, we are wasting a lot of water in maintaining the gardens, hence there is a lot of scope in adopting the xeriscape technique"

References

Daniel, E.J., Bejamin, R. and Harry, A.J. (1999). Effect of mulches and fertilization on soil organic matter in ornamental plants. Bull. No. 173, Ohio State Univ., Ohio (USA).

Scott, A.J. (2007). Xeriscape - A guide to developing a water wise. *Website:http://www.xeriscape .org*

Sovocool, K.A. and Morgan, M. (2005). Xeriscape for Water Conservation: An In-Depth Investigation. *J. Ameri. Water Works Assoc.,* 84: 260- 272.

Sovocool, K.A. and Rosales, J. (2001). Study finding out the demonstrate water and economic savings from converting traditional garden to xeriscaping, *J. Ameri. Water Works Assoc.,* 52: 136-138.

Welsh, F.D. and Welch, C.W. (1994). Xeriscape - Landscape Water Conservation, *Websitet: http://texaserc.tamu.edu*

Ornamental Horticulture, pp 241-268

New India Publishing Agency, New Delhi - 110034, India

11

Bonsai

Alka Singh

Bonsai is an art of growing and training of a woody plant in a miniature form in a shallow container imitating natural look of an old aged tree. It differs from a pot plant where either the foliage or the flower is the main criterion, but here the artistic look of the tree in a miniature and natural form maintained for many years is desired.

The word Bonsai is a combination of two Japanese words, *Bon* (meaning shallow pan) and *Sai* (meaning plant), which can be translated as "tray planting". By growing the trees in shallow containers, the growth is checked or dwarfed but the plant is not starved. A properly maintained bonsai should outlive a full size tree of the same species. Bonsai is a small tree in comparison to its huge life-sized tree of the same species. Thus, Bonsai is specially trained ordinary tree or plant, not special hybrid dwarfs. Bonsai culture is viewed as a hobby that allows greater understanding and association with nature. This art of bonsai is linked with Zen Buddhism - where this art originated as pastime that merged with natural elements and developed into a unique method of meditation and expression. The bonsai is even expressed as "heaven and earth in one container"

Bonsai is developed from seeds, cuttings, layering or grafting from young trees or from naturally occurring stunted trees transplanted into shallow containers. Most bonsai range in height from 5 centimeter (2 inch) to 1.2 meter (4 ft). Bonsai is kept small and trained by pruning branches and roots, periodic repotting, pinching off new growth and wiring the branches and trunk for giving

Department of Floriculture and Landscaping, Navsari Agricultural University, Navsari, Gujarat, India

desired shape. Tradition holds that three basic virtues are necessary to create a bonsai: *shin-zen-bi* standing for truth, goodness and beauty which are parallel to Hindu concept of *Satyam, Shivam, Sundaram*. The bonsai must look natural and should never show the intervention of human hands (with the exception of Chinese bonsai which in many cases depicts images of dragons and other influential symbols of the culture at the time of origination). Bonsai is primarily kept outdoors in partial shade with exception of some plants suited, trained and indoor grown. In traditional Japanese rooms, there is a *tokonoma*, the alcove, specially designed for the display of bonsai on special occasions. Bonsai can live for hundreds of years with utmost care and the prized specimens are being passed from generation to generation, admired for their age and valued as a reminder of those who have cared for them over the centuries.

History of Development of Bonsai Culture

Bonsai was introduced in Japan during the Kamakura period (1185-1333) by means of Zen Buddhism - which at this time was rapidly spreading around Asia. The art was refined in Japan to an extent not yet approached in China. These simple trees were not just confined to the Buddhist monks and their monasteries, but were introduced to be representative of the aristocracy as a symbol of prestige and honour with passage of time. The ideals and philosophy of bonsai were also greatly changed over the years. Bonsai represented a fusion of strong Eastern philosophies of the harmony amont man, the soul and nature for the Japanese.

It was by the fourteenth century, that the bonsai was viewed as a highly refined form of art. This art became an important part of Japanese life by being displayed on specially designed shelves. The practices of training and pruning in bonsai developed later. In the 17th and 18th century, the Japanese arts reached their peak and were regarded highly prestigious. Bonsai again evolved to a much higher understanding and refinement of nature - although the containers used seemed to be slightly deeper than those used today. At around this time, bonsai also became common place to the general Japanese public - which greatly increased demand for the small trees collected from the wild and firmly established the art form within the culture and traditions of the country.

Bonsai began to take on different styles, which varied immensely from one another with passage of time. Bonsai artists gradually looked into introducing other culturally important elements in their bonsai plantings such as rocks, supplementary and accent plants and even small buildings and people which itself is known as the art of *bon-kei*. They also looked at reproducing miniature landscapes in nature - known as *sai-kei,* which further investigated the diverse range of artistic possibilities for bonsai.

Finally, in the mid-19th century, Japan opened itself up to the rest of the world. The travelers who visited Japan spread the bonsai art. Further, bonsai exhibitions were held in London, Vienna and Paris in the latter part of the century. The Paris World Exhibition of Bonsai in 1900 A.D. opened the worlds' eyes up to bonsai. The growing demand and popularity of bonsai, evolved it as an industry. Several basic styles were adopted, and artists made use of wire, bamboo skewers and growing techniques to do this - allowing the art to evolve even further. The Japanese opened up nurseries dedicated solely to grow, train and export bonsai trees. Bonsai techniques such as raising trees from seed or cuttings, grafting, budding, pruning, training for styling and grafting of unusual material onto hardy root stock were further developed. Bonsai is now a joy shared by executives and factory workers alike. Bonsai has now evolved to reflect changing tastes and times - with a great variety of countries, cultures and conditions in which it is now practiced. The evolution of bonsai is now a well known and respected horticultural art form that has spread throughout the world and is constantly changing and reaching even greater heights by merging the tastes and cultures of different countries within and representing unity of this world.

Criterion for Selection of Plants for Bonsai

In Japan, the art of bonsai has been developed, explored and suitability of various plants for the different types or shapes of bonsai has been tested and established, of which pine, azalea, camellia, bamboo and plum are most often used. However, the plants suitable for bonsai in our region differ. A list of plants for bonsai culture suitable to Indian conditions is given below:

Trees: *Adansonia digitata, Adenanthera pavonina, Bombax malabaricum, Brassaia actinophylla*, Bamboos (e.g. *Phyllostachys nigra* Syn. *Bambusa nigra* commonly called Black Bamboo. *B. bambusoides, Arundinaria simonii, etc.*) *Berberis thunbergii, Butea monosperma, Callistemon lanceolatus, Chorisia speciosa, Erythrina crista-galli, E. parcellii, Ficus religiosa, F. benghalensis, F. infectoria, F. mysorensis, F. retusa, Jacaranda mimosifolia. Kigellia pinnata, Mangifera indica, Melia azedarach, M. azadirachta, Milletia ovalifolia, Punica granatum var. Nana, Putranjiva roxburghi* and *Thespesia populnea.*

Conifers : *Juniperus chinensis, Juniperus prostata, Pinus khasiana, etc.*

Shrubs: *Adenium obesum, Brya ebenus, Berberis asiatica, Hamelia patens, Jatropha podagrica, Ixora, Phyllanthus* and *Murraya exotica*

Climbers: *Bougainvillea, Combretum, Derris scandens, Hiptage madablota, Wisteria chinensis, etc.*

Styles of Bonsai

The two basic styles of bonsai are the classic (formal) or *koten* and the informal or *unjin*. The classic style has the trunk of the tree wider at the base and tapers off towards the top and the opposite is the '*bunjin*'style which is more difficult. Over the years, bonsai enthusiasts have frequently reclassified the styles, and have placed many sub-divisions into which plants can be trained. There are basically five main styles of bonsai based on the trunk, rock styles and some other styles like twin trunk group planting, broom style *etc*. Besides, there is *jin* style representing dead wood effect in bonsai.

I. *The Normal Styles of Bonsai*

The five basic bonsai styles are formal upright, informal upright, slanting or windswept, cascade and semi-cascade (Figure 1). All have their own individual beauty and serenity.

a) Formal Upright

A tree that grows in the open under perfect conditions develops with a style such as formal upright. The most important requirement for this style is that the trunk should be perfectly straight, tapering naturally and evenly from base to apex. The branches are symmetrically spaced and balanced.

Species suitable for the formal upright style: *Ficus, Syzygium cumini, Casuarina equisetifolia, Terminalia, Grevillea robusta, Jacaranda mimosifolia. Kigellia pinnata, Manilkara achras, Mangifera indica, Melia azedarach, M. azadirachta, Milletia ovalifolia, Punica granatum var. Nana, Putranjiva roxburghi, Ixora etc* are all suitable species.

Considerations for this style

i) To achieve an effective formal upright, about one third of the trunk should be visible from the front, either from the base to the first branch or cumulatively, as seen through the tracery of its branches.

ii) The placement of branches follows a pattern. The first branch up from the bottom is the longest, heaviest and in proportion is trained to grow to an equivalent to a third of the total height of the tree. This is the'heaviest' branch almost making a right angle to the trunk. The second branch directly opposes the first branch and is higher on the trunk. As the branch structure ascends, they taper assuming a somewhat cone-like form. The top of the bonsai is usually very thick with foliage - so full and tightly ramified that it is difficult to see its internal structure through the mass of leaves or needles.

iii) The tip of this style of bonsai also has a slight curve, to lean forward and effectively 'look at the viewer.'

iv) To achieve distinctive taper of the branches and trunk of a formal upright bonsai the growing tip of the trunk or branch with each new year is cut off and a new branch is wired into position to form the apex. This is difficult and time consuming process but produces a good effect when the trunk matures and the taper becomes prominent.

b) Informal Upright

The informal upright style develops in nature, when a tree bends or alters its direction away from wind or shade or towards light. In an informal upright bonsai, the trunk should slightly bend to the right or left - but never towards the viewer.

Species suitable for the informal upright style: Most species of plants are suitable for this style, mainly the Tamarind, Ficus species, Bougainvillea, Hibiscus, Largestroemia, *Melia azedarach, Putranjiva roxburghi, Thespesia populnea* and Pomegranate on sub-tropical region. For temperate region Japanese Maple (*Acer palmatum*), Trident Maple (*Acer buergerianum*), Beech, practically all Conifers and other ornamental trees such as the Crab Apple and Cotoneaster can be taken.

An informal upright bonsai basically uses the same principle of the formal upright bonsai only that it is informal.

Considerations for this style

i) This style also requires a tapered trunk but the trunk direction and branch positioning is more informal. The trunk takes on an unexpected curve or series of twists and the branches are thus positioned to balance this effect.

ii) The crown of the tree is mainly full with foliage and despite the informal trunk, is most always located directly above the base of the tree.

iii) *Jin* effect (carved remains of dead or unwanted branches to look like dead and rotting limbs of a tree) can also be given with the informal upright style.

c) Slanting

This style is familiar to a tree that slants naturally as a result of strong winds or deep shade during early development, its trunk leans at a definite angle. The stronger roots grow out on the side, away from the angle of the leaned trunk to support the weight.

Species suitable for the slanting style: Most of species suitable for informal style are suitable for this style, as the style does bear similarity to informal upright. Bougainvillea, *Citrus japonica* and Conifers fit particularly well.

This style bears some similarity to informal upright, with the trunk at an angle to either right or left (never to the front), but with the apex not directly over the base of the bonsai.

Considerations for this style

i) At an early age, the bonsai can be trained to an angle by means of wiring the trunk until it is in position. Alternatively, the tree can be forced to grow in a slanted style by putting the actual pot on a slant causing the tree to grow abnormally.

ii) The lowest branches are grouped in threes, and this grouping begins one-third of the way up the trunk. The bottom-most three branches almost encircle the trunk, with two branches thrusting forward, one slightly higher than the other. The third branch, emanating from a point between the first two, is set at such an angle as to make the foliage appear lower than the other two.

d) Cascade

Cascade is the style developed from the trees growing in utmost adverse conditions and getting pulled by the natural forces of gravity. The cascade style of bonsai consists of a growing tip that reaches below the base of a container. The bonsai trunk has a natural taper and branches appear to be seeking the light.

Species suitable for the cascade style: Many species are suitable, if they are not strongly upright like Portulacaria, Bougainvillea, Aralia, Sapota, Ficus *etc.*

To create this style a deep, narrow pot, which will enhance the style and accommodate the cascade style is required.

i) The main trunk should be wired to spill over and down the edge of the pot, with the main focus on the major bend (forming an upside-down U shape).

ii) The branches should be kept uniform and horizontal to the almost directly vertical trunk.

iii) The roots should balance the trunk and branches from opposite side.

iv) The plant should be positioned right into the center of the pot.

e) Semi-Cascade

This is similar to cascade but tip although projects over the rim of the container, but does not drop below its base. The style occurs in nature when trees grow on cliffs or overhang on walls. The exposed roots should balance the trunk.

Species suitable for the semi-cascade style: Many species are suitable, except strongly upright like Portulacaria, Bougainvillea, Junipers, Hibiscus etc.

Technique

The technique and principles for semi-cascade are basically the same as followed for cascade however the tree (growing tip) does not drop below the base of the bonsai pot. Many semi-cascades do not even drop below the edge of the top of the pot. This style is perfect for Junipers.

II. Rock Styles

Rock styles are important as classic bonsai styles. Rocks and trees look marvelous together as bonsai. The rock style bonsai is created in two ways:

Root in rock i.e. a bonsai rooted in a hollow rock:

In the hollow of a rock, a suitable plant is shaped. The tap root, if present, is partly removed and other roots are lightly clipped. The hollow of a rock is filled with soil. The plants are placed in the hollow, covered with more soil and finally with wet sphagnum moss. The entire soil and rock are covered and tied so that the soil does not wash away during watering. The plant is watered thoroughly with a fine hose. The sub moss lasts for nearly a year till the plant is well established.

Root over rock i.e a bonsai on a rock with roots going into the container:

A plant having long roots is shaped, then the long roots area dipped into a thick solution of clay. The plant is placed over a rock, which is placed in a container with some soil. The roots with sticky clay on them are trained over the crevices of the rock and into the container. Then the root system is covered with more clay mixture. Finally, it is covered with moist soil. The whole rock is then covered with moist sphagnum moss and tied. The whole rock, tree and soil in the container are watered with a fine spray.

The root over rock style is described in detail below:

The root over rock'style is widely used and a demanding technique in bonsai. This is when the roots of a plant are made to clingto and grasp a rock - spilling over it and eventually disappearing into the soil. The style has emerged from the trees in nature that grow in small rock crevices and its roots move out and explore in search of nutrients and soil. *Ficus* species are highly suitable for this style. The plant to be trained for this style should first be planted in deep and tall pot to develop long root system.

Materials needed for the rock style

Suitable rock (Calcareous rock or red rock); Plant suited to root over rock style (*Ficus* species); Plastic Grafting Tape; Scissors; Concave Branch Cutters; Bonsai Secateurs; Knife; Root Hook/Fork; Clean and Sharp Sand.

Suitable Species: Most of the *Ficus* species, Junipres, Cidrus, Acer, Punica, Pines, *etc* work well.

Steps to be followed

- Once the roots are long enough (after one year feeding in pot), cut off all unsightly or unnecessary foliage, remove the soil with small fork and wash the soil away from the root ball of the plant - taking care not to damage the roots.
- Place the plant over the rock arranging the roots in the best way. Find the interesting crevices in the rock and implement the roots of the plant into these - to make the finished bonsai look as natural as possible. The thin roots can be overlapped.
- The roots should be fixed on the rock in place by wrapping the grafting tape around the rock reasonably tightly, making sure to cover it completely except for the bottom, where the roots will protrude into the soil. The plant should be placed in a pot and buried in clean sand. The whole root ball along with the rock should be planted with only the bottom part of the trunk of the tree being visible.
- Thereafter, it should be thoroughly watered along with dilute feeding and left in the pot for one or two years for development of the root system.
- After two years, the plant should be removed carefully from the pot. The soil should be removed by scraping and then be washed gently to expose the rock and root-ball.

- The grafting tape should be carefully removed with a pair of sharp scissors to expose roots and rock. Here we see the exposed roots and rock with the grafting tape removed. The roots will appear well developed after a period of two years and truly bonded themselves to the rock in quite an attractive way.
- The plant should be planted back into the pot with the rock above the surface to allow it to rejuvenate for a few months.
- Finally as the tree rejuvenates, it should be planted in a suitable ceramic/ earthenware bonsai pot to display the root over rock style. This style looks elegant particularly with shallow and oval pots.

III. Other Styles

Besides the major five and rock style, there are some other popular styles as described below:

a) Broom Style

This is an upright style with all the branches sprouting from the trunk growing in an upward direction to resemble an inverted broom.

Species suitable: *Casuarina* and *Junipers* work well for this style.

b) Bunjin or Literati Style

Though artistic, this style is not governed by the first rules of bonsai styling. The trunk is interesting with many curves and there are very few branches. Foliage is sparse and more than half the trunk is bare. There is no distinct arrangement of branches. With its powerful profile, the bunjin style is characterized by a rugged miniaturism.

c) Twin or Multiple Trunks

This style consists of two trunks growing from the same root ball. Sometimes, the two trees are placed close to each other to form the twin trunk. The trunks are so placed that one is slightly in front of the other. The branches of the two trunks are arranged as for a single trunk tree. Twin trunks can be trained as slanting, informal, windswept and so on. Multiple trunks are a variant of twin trunks. These extra trunks provide the opportunity to grow a copse of bonsai. There can be differences in the height and thickness of the trunks. Care is needed to ensure odd number of trunks. The group of trunks should have a front and back with spreading sides and back branches coming from the right, and from a suitable angle.

Species suitable: *Ficus benjamina, F. infectoria, F. microphila* and *F. benghalensis* can be well trained for this style.

d) Raft Style

This is an interesting variation that forms a forest. The branches from one side of the trunk are removed and the rest are all wired straight up. While planting, the trunk is laid horizontally with the root ball on one side of the container and the wired branches facing upward. Before it is laid in the soil the side of the trunk which has no branches is given notches, for a flap or bark is removed from under each wired branch, so that it will root. Rooting hormone can also be applied for quick rooting. The trunk is firmly fixed in the container. It is then covered with the soil, watered and looked after. After a period of time, the wired branches develop rooting system.

Species suitable: *Ficus benjamina* and *F. shella*.

e) Group Planting

This style imitates forest look in nature and thus, several trees in group are grown in a container, resembling a forest. The Japanese favour planting an odd number of trees. The main tree should be at the front and a little off centre. The smaller trees go at the back.

Species suitable: *Ixora, Ficus benjamina, Murraya exotica,* Bamboo, *Casuarina*, etc

f) Finger Tip and Mame

These are miniature bonsais measuring 5 to 15 cm (2-6 in) from the base of the trunk to the top of the tree. The training is the same as for other bonsais except that the rules are less rigidly followed. The choice of plant also has to be correct. Plants with very small leaves, flowers and fruit are more suitable. Mames need a lot of attention as they are grown in very little soil.

Species suitable: Portulacaria and junipers work well for this style.

g) Jin Style

It refers to creating of a natural deadwood effect in bonsai. *Jin* or *shari* represents deadwood effect in bonsai and is an effective feature to a design and very realistic feature in many old Pine and Juniper. *Jins* on trees in nature are normally created when branches are broken by natural calamities like strong winds, by the weight of snow or by lightening strikes. When styling and training a bonsai, branches that are not required as part of the foliage mass can instead be used to create *jin*. The use of *jin* on a bonsai implies to the viewer an increased age and shows the struggle of the tree to survive. A common way of creating a *jin* is by carving using power and hand tools; the use of either successfully, requires a great deal of skill, particularly if the wood is featureless, as in the case shown below.

Steps followed for *jin style*

- Remove the bark from the portion of wood that is to be jinned. Any remaining soft tissue is also removed. At this point only the part of the stump that is above the soil level will be worked on; further work will be carried out as the stump is revealed after repotting in the future.
- The first slithers are pulled away from the stump; the top of the wood is either notched to segment it or crushed with pliers. Each segment is then pulled down and away from the stump with jin pliers or ordinary pliers. When pulling away strips of wood, it is important the live section at the sides or the base is not damaged. Strips rarely stop abruptly above the live wood and often need removing near the base with a knife or similar. The pulling down of the segments of wood continues on the first half of the stump.
- The second half of the stump is then pulled away to reveal more of a natural shape.
- After reducing the second half of the stump; the *jin* my reveal good shape and texture.
- As the wood is fibrous, small burrs and pulls are evident. These are best dealt with by gently burning the surface of the wood. It is very important at this point to protect the rest of the tree, in particular the foliage, from not only the flame but also the heat that it generates. It is also important that the *jin* itself is only slightly scorched to remove the fibers.
- Finally, the *jin* is lime-sulphured to whiten it and to stop it from rotting in future. In future, black ink will be added to the lime sulphur when painting the recesses to add a greater illusion of depth. It should be than left to dry and weather naturally.

Bonsai-Containers

A pot for bonsai should be aesthetically appealing and also functional. The pots and containers used for bonsai vary in material, shape and size. Small ceramic or terracotta pots and containers of square, rectangular, oval or round shape are the best for bonsai. Sometimes small cement containers are also utilized for this purpose but these are not convenient to handle because of their heavy weight. Besides, there are glazed pots for bonsai. Unglazed pots require frequent watering than glazed ones, but glazed and hard surface pots tend to have poor drainage. They must definitely be unglazed on the inside. The growing plant never feels comfortable in a pot which is glazed from inside as due to the smooth inside surface, there is hindrance in root development.

The choice of the shape and colour of the container depends upon the style and the type of plant used for bonsai. Usually, terracotta and light colours are preferable. The rectangular and oval shaped containers are ideal for most of the bonsai styles. The shape of the container (round, oval or square) can be selected according to the individual's choice, however it should be shallow. For cascade style, depth of container can be more. Few plants can remain alive if their roots are immersed in soggy soil and so bonsai containers must have big enough drainage holes. There should be one-two drainage holes, and of neutral colours and simple without any decoration.

The pot must not be too decorative, gaudy or brilliantly coloured, as this will steal the show from the bonsai itself. A powerful and rugged looking tree should be planted in a sturdy pot. A tall tree with a slim trunkline and dense foliage goes well in a slim looking pot. With formal upright and slanting styles, it is befitting to use rectangular and thick pots, whereas, with thin and long trunks, the square, hexagonal or even round and rather thin walled pots are good. Twin trunk style or two trees which have a tall and short or thick and thin tree accompanying one another, the best bet would be for a thin rectangular or oval pot. In cascade and semi-cascade, it is important to recreate the feeling of a cliff, from which the tree is trying to survive. For rock plantings, a rectangular or oval, very thin and shallow tray can be used most effectively. As far as the size of the pot is concerned, two standards are followed. The first considers the height of the bonsai. Here the length of the pot should be plus or minus two thirds the height of the tree. Then in relation to the width, the width of the pot should be plus or minus the widest part of the trunk. There are many exceptions such as in cascade or rock planting styles of bonsai. The container should harmonize and compliment the beauty of the tree. For this, the colour of the container plays an important role. The bonsai is always placed on one side in the pot but not in the centre in oval and rectangle pots while in round and square pots it is placed in the centre. In the oval or rectangle pot, plant is placed on right or left side depending on the shape of the bonsai. Evergreens like Ficus, Silver Oak, Juniper or Cypress *etc.* go well in sober, earthy colours like unglazed dark brown, reddish brown or grey pots. White and yellow appear as strong contrast with the plant, blue and light blue do not appear in harmony and the bluish-green pot clash with the foliage colour. For fruit trees, like

Pomegranates or Eugenia or Barbados or *Ficus benghalensis* a cherry reddish brown or blue green pot can be used.

Tools for Bonsai

Appropriate tools are required for development of a good desired bonsai style. Various tools are used in creating miniature trees and tray landscapes. The tools are basically needed for root and shoot pruning, soil removal, separating of entangled roots, for grooving and other techniques.

Purpose

- Root pruning scissors to prune thick entangled roots.
- Rake to disentangle the roots.
- Scissors to cut large sized branches.
- Pruning scissors for small branches.
- Concave scissors for small branches.
- Knob cutter to make deep grooves on the trunk or branch.
- Wire cutter to cut heavy gauge wire without damage to the branch or trunkline.
- Small wire cutter to cut thin wires as well as cut twigs and leaves.
- Giant root or branch cutter for cutting very thick and mature trees.
- Pincer-like tool for crushing the bark of the tree, either trunk or branch for creating effect of *Jin*.
- Fine scissors for cutting twigs and small branches.
- Metallic clamp used for straightening of trunk or branch.
- Broom to create soil contours in a tray landscape or even to clean the rocks after planting.
- A small dainty Trowel is used to level the soil.
- Pincette to extract weeds or removal of unwanted and hurting needles of some trees or even for pinching the process.
- Saws, small and large for cutting branches or a taller unwanted trunk.
- Gravers, commonly used for lino-cutting or wood carving, are useful for carving or smoothening of *jinned* or *shari* effect done on the bonsai.

- Spray bottle to mist or spray water or foliar nutrients on the roots clinging to the rock or even the foliage.
- Grafting knife and blades to make delicate grafts.
- Soil Scoops are useful at various stages of potting or repotting. The angle in the scoops makes the task of putting soil in the crevices or gaps much easier.
- Turn table with lock is useful in turning tree in various positions during training.
- Chopsticks or rods are used very effectively for compacting or prodding the soil.
- Various thickness of copper or enameled aluminium wires to tie or plain aluminium wires to tie or train braches.
- Tool edge sharpening stone or instrument.

Besides, a working table preferably a turntable is the most convenient and required for working on a bonsai.

Protection and Sharpening of Tools

Regular care of tools is must for the long lasting effect and having them in working condition. The tools should be kept clean and rust free. Rusted tools lead to improper and undesirable cuts which can harm the plant health and appearance. The tools should be cleaned after each working session with damp cloth. The resins should be wiped with solvents like terpentine and alcohol. After cleaning of the tools, the tools should be wiped with mineral oil dipped cloth. Regular application of mineral oil in the tools is also desired. This will lubricate and protect the tools from rusting.

A good sharpening stone either an oilstone or a water stone, which is fine on one side and coarse on the other is needed for sharpening bonsai scissors. Spread open and place the surface of one blade on the moistened stone. Move smoothly back and forth, keeping even but gentle pressure and same angle at all times. Never sharpen the coarser side of the stone, switch to the finer side of the stone and a few move back and forth. Finally move the blades on a piece of thick leather for a smooth finish. Then lubricate with oil. Scissor sharpener is also available in the market, which can also be used for the purpose. For the concave or knob or branch cutters, only the upper blade should be sharpened and never the lower blade. When the lower blade comes in contact with the upper blade, a cutting edge is naturally formed.

Bonsai Media

The soil media should have good drainage with some organic mixture. A media comprising of loamy soil, organic manure, leaf mould along with river sand is good for providing good plant hold and drainage. Bonemeal or superphosphate in small quantity can be added to the planting medium. Often the soil in the pot is covered with moss to conserve soil moisture and one or two small stones are placed to give a natural look. A basic bonsai soil mix to use that can be applied to almost all species is: one part loam, two parts sphagnum peat moss, two parts granite grit with some organic matter. All of these ingredients are easily available.

Potting

It is necessary to prepare the plant which is desired to be developed as bonsai for potting in the bonsai pot.

Preparing plant for potting

Remove all the dead branches and all lower branches from the main trunk till one-third of the trunk is bare. The three primary branches to be retained are chosen. They are shortened if wiring is not required, or shortened later after wiring. The cut is always made at a slant just above a growing bud. If braches are growing opposite each other, one of them is removed. If there is a branch growing just above the other, one of them is removed so that the plant is mostly left with alternate branches. Sometimes a branch may be left at the base of the trunk for a while until the trunk thickens. Then, later, it is removed. The desired height of the tree has to be decided so that the plant can be clipped off at a certain height. Sometimes one small top branch can be wired to give the plant a new crown. All excessive foliages are removed from the plant so that the entire branch pattern and the trunk are visible. All the finer branches are pruned back a little.

For starting a bonsai from the natural stunted plant or from a dwarf plant obtained from a nursery, it is necessary to prune the roots. Generally one-third of the roots is cut off and the tap root is pruned if there is an abundant growth of fibrous lateral roots. The unnecessary branches are removed before planting. The basic principle in bonsai culture is to restrict and slow down the growth of the plant by selective pruning of roots and branches. However, at the same time, it is necessary to provide just adequate but balanced nutrition and regulated watering for proper and healthy but slow growth of the bonsai.

Before potting the plant one must decide the style of bonsai to be followed. The method of planting in the pot or container and the training of the plant will depend upon the style of bonsai.

The important steps for potting

- The pot selected for potting must be dry and clean. The drainage holes are covered with pieces of plastic mesh or crocks (broken pieces of pot).
- Spread a layer of coarse soil over the bottom of the pot. This is followed with a layer of main soil. The soil should be moist, neither too dry nor too wet.
- The plant is removed from its pot by holding its base and inverting the pot. The earth is gently shaken off or removed with a pointed stick (or fork) until half or one-third of the original soil is left. If there are any earthworms they should be removed. Now is the time for root pruning. This will depend upon the amount of root growth. Foliage has also to be proportionately reduced so that the scanty roots can support the plant. Look for the tap root (the main root). If there are plenty of fine roots remove the tap root completely. Otherwise, remove just one-third of tap root. Some of the fine roots may need a little trimming so that they can fit into the pot. Remove all dead and diseased roots.
- The plant with its pruned roots is then placed in position over the soil in the pot. If the root ball is very thick, it is held in position by tying it with a piece of wire which is passed through the drainage holes. The large surface roots are spread out and more soil put into the pot till the plant is held firmly in position. With the help of a chopstick or a pointed potting stick the soil is pushed back well into the roots so that no air pockets are left.
- The top of the container can be spread with a very fine layer of top soil which is gently pressed down by hand or with a small trowel. When drastic root pruning is done soil may be built up around the trunk to form a mound of earth.
- Pieces of green moss can be pressed on top of the soil if desired, except on the surface roots to enhance an appearance of age to the bonsai.

Watering

The first watering is the most important. This is done by holding the pot in a basin filled with water to just below the top of the bonsai pot. The water gets absorbed from the drainage holes *via* capillary action. In a short time, the top becomes wet and the plant gets thoroughly watered. Further, bonsai is kept in a place sheltered from the sun and wind for a few days before it is gradually exposed to the nature elements. The second watering should be done when the top soil starts showing signs of drying. This is done with a normal watering can with an attached fine rose.

Repotting

Repotting of bonsai regularly is a must to replace important nutrients, stale soil and allow for new healthy root growth of bonsai. The young bonsai in general requires repotting every year. The old bonsai requires repotting after 2 or 3 years depending upon the plant species and its growth. The repotting is done in the same way as potting.

Repotting time: Early spring or monsoon. Young or small bonsai requires repotting every two or three years and older and larger specimens often require less number of repotting.

Indication of need of repotting: When water takes a long time to drain through the soil or if the roots are crowded around the sides.

The procedure for repotting

- Carefully lift the tree out of its current pot by tilting it to one side and moving it by the base of the trunk. The pot is tapped gently to loosen the rootball and insert a stick through the drainage holes and push the rootball out. Further, remove any moss or accent plants using a chopstick, knitting needle. Try to brush and untangle the roots carefully using a fork. This should be done very tactfully and carefully and avoiding any damage to the important main roots.
- Shake and brush off the soil until about one third to half of the original soil has been removed from the edge and base of the rootball. Spray the roots with water to avoid drying out of the roots and also for ease of pruning of the roots.
- The roots are pruned using very sharp cutters like bonsai clippers or secateurs. Initially, cut the thick, old brown roots that have come close to the edge of the pot and are restricting the growth of the young feeder roots. Around one-third to one-half of the roots are removed. The small feeder roots are not to be pruned. Next, trim the thinner roots which are in excess to the pot size and maintain a suitable shape according to the pot leaving a 1-2 cm (1/2 to 3/4 in) space near the edges.
- Select the pot for repotting or clean the original pot thoroughly. Cover the drainage holes with simple wire mesh. Insert some wire through the drainage holes for anchoring. Add a thin layer of gravel to aid drainage and then a layer of soil-mix.
- Now, decide a basic position for the bonsai moving the tree around and make a small mound over there (usually off-center and slightly to the back of the pot).

- Gently place the bonsai on the mound by spreading its roots out evenly throughout on top of the soil. Now anchor the tree by twisting the wires together (usually with the aid of pliers) over the main root ball of the tree until it is held firmly.
- Now, add more soil up to the base of the trunk but just below the base of the pot. Ensure that the bonsai gets settled firmly with no gaps around the roots.
- After the well settled plant and soil, the features like rock and accent plants and other accessories can be added. The moss should be applied gently and thinly covering the top exposed soil. The moss should be very fine in quality.
- Water the bonsai thoroughly with a fine rose-can/sprinkler. Place the tree in shade with some availability of indirect sun- rays in the morning hours.
- After one month, as the roots have recovered, feeding can be taken up. Note that to balance out the extensive pruning you have just done on the roots, you should prune the branches of the bonsai as well so that it can recover quicker and not be disadvantaged further. Root growth usually does equal to branch growth.

Pruning of Bonsai

Purpose of pruning

Pruning is one of the most important continuous features of bonsai development and maintenance. The purpose of pruning is to give the plant a particular shape and to control its growth by trimming the branches and leaves. The equipments required for pruning are a pair of good scissors for trimming the smaller branches and leaves and a concave cutter for the bigger branches. This is the technique to obtain the right shape of a bonsai and encourage new growth.

Pruning time

To prune correctly, the knowledge of a plant growth habit is essential. Generally, new growth is pruned during the growing season to maintain the shape of the bonsai, while pruning of hard wood (old season growth) is done in mid-autumn.

First pruning

This is drastic and is done at the time of original potting to achieve a certain shape. The plant should normally be left with alternate branches. All branches, which are directly in opposite direction to each other are removed. Excessively long branches should be shortened.

Later stage

Pruning at later stage is done to encourage bushiness keeping the tree in shape. While cutting back a branch, make sure to cut it near a growing bud, checking the direction in which the bud is pointing (and always leaving pair of leaves on the branch). This is important because the new growth will be in that direction.

Care

The cuts on the branch should be cleaned so that the tree can heal quickly. Coniferous trees, particularly pines, have to be pruned very carefully because they will not bud from old wood.

Repeated pruning

Repeated pruning is done in deciduous trees to produce smaller leaves, and in conifers to produce smaller needles.

Pruning for evergreen conifers (Finger pruning)

This is practiced especially in evergreen coniferous bonsai such as junipers and cedars. This involves pinching back new growth which docs not come within the general shape of the bonsai or is at the top of the bonsai - helping to encourage bushy foliage and a more tree-like looking bonsai. To do this, take the growth between your thumb and forefinger whilst holding the branch with your other hand and remove with a twisting movement. Scissors should not be used for trimming conifers as they tend to damage the remaining needles as well as the buds.

Pruning for deciduous trees

For deciduous trees such as maples, the chinese elm and cotoneaster, scissor tip pruning is the best. When trimming outward, trim shoots back to just after the next series of leaves, but dont' cut the foliage as such.

Leaf pruning

Leaf pruning is also known as defoliation. This is practiced for several deciduous and tropical plants such as Ficus or Maples to reduce leaf size, remove unsightly leaves and speed-up growth by causing two seasons' growth in one. The leaves are cut off from each branch leaving a portion of stem intact so that the neighbouring bud is not harmed by mistake. After a fortnight, new leaves appear. In this way the plant is forced to have another crop of leaves in the same season. After pruning, for the next few weeks, the plant should be kept in good position away from extreme climate with adequate water supply.

Trimming or Pinching

Trimming or pinching is a continuous process which had to be done through the life of a bonsai to maintain the quality of the trees' foliage. Pinching back helps preserve the shape of the plant, by inhibiting the natural exuberance of growth. Gradually the smaller branches become finer forming an interesting network. The leaves also become smaller. The number of times the tree has to be trimmed varies depending on the growth of the tree. Trimming should be stopped before flowering. When new growth appears, all branches are pruned back to two or three leaves. During the growing season, plants need a lot of trimming.

Wiring of Bonsai

Wiring of the branches of bonsai is a temporary measure used to hold branches in a desired position to provide desired shape of a tree. Wiring should be checked regularly and removed after some time. It should never be left permanently on a tree.

Wiring depends on the plant species selected for bonsai. Some plants can be very well trained without the use of wiring like Adenium and Jatropha. When wiring, try to imitate the natural curves of trees in nature. Wiring should be done on only those branches that are unlikely to break when pressure is applied to twist the wire around the branch. It is the best to avoid watering a day before wiring.

Mainly two types of wire are available - copper wire and aluminium wire. The copper wire is naturally a better coloured but it is much harder to apply to branches, especially for beginners. Aluminium wire is better and is also available as coloured to look like copper. It is much more malleable than copper wire and has good effect. The size of wire to be used depends on size of branch to be trained and the degree to which the branch is to be modified or bended and the flexibility of the stem. The wire of varied thickness starting from lowest

2.0 mm to 5.00 mm can be used. For pencil-thick sized branches, a gauge of 3.5mm aluminum wire is found good.

Wiring of the trunk

The trunk specially for informal and slanting or cascade needs wiring. For this, metal wire of the right gauge (depending on the branch or trunk) is wrapped around the trunk, upwards from the lower end. Sufficient length of wire to go around the trunk of the tree chosen is selected by cutting. One of the wire is firmly pushed into the soil close to the back of the base of tree. Securing the lower end with one hand, the wire is turned spirally around the trunk till the end of the trunk is reached. Each spiral is held in position with one hand before another spiral is made. In this way the correct degree of tension is maintained. The wire should be tight enough to hold the bends in place. The distance between the spirals should be equal, neither too close nor too far apart. After wiring slowly and gently, apply pressure for bending it to the desired shape by working carefully with both hands. Thus, the trunk can now be trained to shape.

Wiring of branches

The branches are wired in the same way as trunk. One end of the wire is anchored over the fork where the branch grows from the trunk. One turn is taken around the trunk to hold the wire in position then the spirals are wound around the branch in a clockwise direction for right hand branches. The tips of the wires are bent back to finish off. After wiring, the branch is bent into the desired position. When two branches are close together, they can be wired with just one length of wire by commencing from the centre, holding the wire at the centre, first one branch is wired then the other.

After wiring, the tree is kept in shade for two weeks after wiring. Check every few weeks for wire cutting into the bark of the bonsai - particularly during spring and summer.

Period of wiring

All wires must be left in place for a period of 3-6 months after which they should be removed. Ensure by regular checking that the wire does not cut into the bark and leave ugly scars.

Other methods of training

- A piece of wood can also be used to hold two branches apart.
- A screw clamp can be used to give a curve to a trunk.

- Small stones of varied weights tied with thread can be hung from the branches. The weight of the stone can be selected on the basis of desired bend of the branch. This is a good and most easy technique with no risk of scar-formation.

Watering

Bonsai needs plenty of water and nutrients to be able to thrive. Since there are limitations for nutrients and water, it should be supplied sufficiently and regularly. The best time to watering is whenever the top soil starts drying. Shower the foliage thoroughly to wash away any dust and keep your bonsai looking fresh. In the summer water bonsai twice a day. Dormant plants and conifers need less water.

Nutrition Management in Bonsai

To maintain bonsai in good shape and to obtain healthy growth, regular feeding at the right time of year with correct fertilizer is required. Fertilizer must be continually replenished due to a certain amount being washed out each time, the plant is watered.

Fertilizers

Many different types of mixed fertilizers (NPK) are available in the market, which can be used for bonsai. Slow-release fertilizers are also available which will take care of feeding requirements for several weeks at a time. Soluble powder and liquid feeds can be used for rapid effect over short-term periods. These are applied to the soil with a watering can and are used up relatively quickly.

Most commercial fertilizers contain all three main nutrients *i.e.* nitrogen, phosphorus and potassium plus some trace elements.

- **Nitrogen** is an essential ingredient for leaf and stem growth. Too much nitrogen however will make the tree produce long and stringy growth.
- **Phosphorus** encourages healthy root growth and helps the growth of new buds, whilst also protecting against diseases and unfavorable winter conditions.
- **Potassium** (potash) encourages formation of flowers and fruit and is vital in the fight against diseases.

Micro-nutrients mix is also needed in minute quantities.

Time of Nutrition Application

As a general rule, bonsai should be given nutrition from early spring to late summer. A balanced fertilizer should be used with fairly low nitrogen content year round, and one with as little (or no) nitrogen as possible in autumn to aid the tree (so the tree can retain its energy) through the winter. NPK fertilizer with ratio of 0:10:10 is used after monsoon as autumn feed (containing no nitrogen) to harden off the current seasons' growth in readiness for the winter.

Placement of Bonsai

The placement of bonsai is very specific to the species of the bonsai. Bonsai in general should be placed in partial sunlight where they receive good air circulation, relative humidity and where they are easily accessible for maintenance (i.e. watering, feeding and pruning). Bonsai requires lesser light and loves humidity as compared to its original requirement in open conditions in soil.

Display of bonsai inside the house

The bonsai can be brought inside but not permanently and must be ensured to provide extra-special attention. One can keep the bonsai inside for 2-3 days a month at most. The bonsai can be placed for longer duration if some sunlight is available during morning hours. Care must be taken to ensure that they are given the right amount of water (definitely not allowed to dry out) and sunlight as well as humidity, which can be achieved by misting with a general water sprayer.

Display site

The site is the best one possible to get maximum benefit and viewing pleasure from a display of the bonsai

- The site must be suitable for healthy growth of young trees *i.e.* sufficient light, humidity and temperature and should be protected from winds.
- It should be safe in terms of potential for damage by children (frequent cricket matches from next door *etc*) pets or pests.
- The spot chosen must be convenient for watering and routine maintenance.

Bonsai display benches

A good display bench is essential for the outdoor display of bonsai. Once the bonsai is well developed, an appropriate display adds an extraordinary viewing beauty.

The advantages of appropriate display:

- Healthy growth of bonsai is visible.
- Allows regular observation ofbonsai for evaluation and prevents problems from escalating.
- There is ease of maintenance.
- The pleasant view of bonsai is aesthetically and spiritually satisfying.

A well proportioned display bench provides a character to the display area, contributes to bonsai health and even assist in the development of quality specimens. Timber, bamboo or wrought iron (treated) material can be used for construction of the benches. The height of the bench can be decided depending upon the choice. The benches should be in two-three steps. There should be sitting arrangement nearby like swings or garden benches or chairs to rest and enjoy the sight of beautifully displayed bonsai collection.

Seasonal Care for Bonsai

Summer (15th April-15th June)

The following are tips to help your bonsai survive a long and hard summer.

- Care should be taken for frequency of watering and water sprinkling *i.e* the maintenance of soil moisture and relative humidity for the bonsai tree.
- The bonsai plants should be healthy and pest-free. Spray the required insecticides if pest infestations occur.
- Each plant should be given a quarter turn every week. This gives even exposure to the sun and fresh air, and allows to check on the health or disease from all sides.
- Regular pinching should be continued as per the growth. Don't lose the shape you've spent time working on. Thin any tight growth to allow air and light flow.
- Provide shade cloth/net overhead, especially afternoon or site your bonsai under landscape trees or shrubs.
- The plants should be specially protected from the hot west and south-west sun rays.

- Water quality is very important after monsoon. Only good quality water should be used for bonsai. If there is problem of saline water, use only distilled or reverse-osmosis (RO) treated water.
- Regular half strength fertilizer application is important (every 10-15 days).
- In the late summer and throughout the fall, use a high phosphate fertilizer, such as a 10-60-10 blend, for the flowering trees. This allows the plant to build up reserves and rudimentary buds for next spring.

Monsoon (15th June-15th Oct.)

This season is the time of year for re-growth, renewal, repotting and restyling.

Important features:

- **Repotting:** The plants kept in same container for two or more years require repotting and the monsoon showering is the perfect time. The style should be decided and accordingly pot selection and potting and repotting should be carried out. Sharp secateurs and scissors should be used when cutting roots and make clean cuts.
- **Pruning:** This season is also a period of fast new growth. Do not let the new side buds and branches grow too long. Thus, pruning should also be done simultaneously. Cut back and limit the active buds to only a pair or two per branch. The goal is to have compact, ramifying growth. The tools should be cleaned and oiled after each session.

Winter (15th Oct.-15th Feb.)

In extreme climates protecting plants from frost and cold becomes important. This is the time when plants undergo dormant phase. The climate of South Gujarat is moderate and winters are not severe. However, the bonsai need special care for watering.

- Protect trees from harsh winder winds.
- Decrease shading of the bonsai. The morning and evening sunrays should be available.
- If you notice or suspect frost damage to your bonsai, it is the best to wait until the appearance of new growth before trimming any branches.
- Do not fertilize bonsai in the winter. Wait until the appearance of swelling leaf buds before feeding any tree.

Tamarindus indicus - Rock Style

Ficus retusa - Root Over Rock Style

Phyllanthus nivosus-Twin Trunk

Ficus benjamina - Group Planting

Ficus jaquinifolia. Group Planting Style

Differents ferms of Bonsai

- Take time now to clean and oil your tools, clean your empty pots, review your books and magazines.

Spring (15th Feb.-15th April)

The season brings the new flush. The following points are to be taken care off:

- As the buds appear swollen, the bonsai plants should be fertilized with sufficient water immediately.
- Throughout the season, sufficient watering and fertilizing at regular intervals are very effective.
- Pruning and pinching should also be done as per the requirement.

Reference Books

American Bonsai Society –Magazines, U.S.A.
Bonsai — T. Kawamoto & J. Kurihara, Japan.
Bonsai Clubs International –Magazines, U.S.A.
Bonsai Deborah- R. Koreshoff, Australia.
Bonsai in Hawaii –Elaine Okimoto, U.S.A.
Bonsai Techniques I & II –John Y. Naka, U.S.A.
Bonsai Today –Magazines, U.S.A.
Complete Gardening in India –K.S. Gopalawami Aiyangar India.
International Bonsai - Magazine. U.S.A.
Man Lung Artistic Pot Plants –Dr. Wu Yee-Sun, Hongkong.
Nichin Bonsai –Magazines of Bonsai Study Group of Indo-Japanse Association, India.
Nippon Bonsai Kyokai –Bonsai Syunjyu –Magazines, Japan.
Papworth David –Gardeners' Guide to Conifers, U.K.
Practical Bonsai –Paul Lesniewcz & Hideo Katoh, Germany & Japan.
Sekai Bonsai –Magazines, Japan.
The Bonsai Book –Dan Barton, U.K.
The Complete Book of Bonsai –Harry Tomlinson, U.K.
The Japanese Art of Miniature Trees and Landscapes -Yuji Yoshimura & Giovanna M. Halford, U.S.A.
Tropical Garden Plants in Colour-Bose & Chowdhury, India.
Wonderworld of Bonsai and Saikei –Jyoti & Nikunj Parekh
Wonderworld of Tropical Bonsai –Jyoti & Nikunj Parekh
World Tropical Bonsai Forum –Magazines, U.S.A.

Ornamental Horticulture, pp 269-302

New India Publishing Agency, New Delhi - 110034, India

12

Horti-business through Landscaping

S.S. Sindhu

Horticulture and allied sector are integral elements for food and nutritional security in the country. Horticulture is the main segment, while its various sub-segments are fruits, vegetables, aromatic and herbal plants, flowers, spices, tubers, plantation crops and ornamental plants. All these are regarded as essential ingredients of economic security. The wide range of agro-climatic conditions of India is conducive for growing a large variety of horticultural crops, including, root and tuber crops, mushroom and ornamental crops.

The Government of India recognized horticulture crops as a means of diversification in agriculture in an eco-friendly manner through efficient use of land and optimum utilization of natural resources. Horticulture seeks to create ample opportunities for employment, particularly for unemployed youths and women folk. India has maintained leadership in the production of many commodities like fruits, vegetables, flowers and ornamental plants and particularly in landscape business.

India is coming up in landscaping and ornamental horticulture. There is wide scope since large number of MNC and business houses have come up in India. The purchasing power of people has increased to a greater extent.

Thus, over the years, much progress are made for advancement of horticulture and allied sector. Rising investments resulted in increased production and availability of horticultural produces in the rural and urban

Division of Floriculture and Landscaping, IARI, New Delhi

areas. Many schemes and policies were introduced from time to time for upliftment and commercialization of the sector. Large number of investors are taking advantage of the existing potentials in the sector, as well as trying to explore the untapped potentials in the field of horticulture.

Fruits, Vegetables and Flowers

India is the fruit and vegetable basket of the world. India being home of wide variety of fruits and vegetables holds a unique position in production figures among other countries. Over 90% of India's exports in fresh products go to West Asia and East European markets. However, it needs to augment its food and processing industry at a mega scale, according to an agriculture consultant.

Department of Agriculture and Cooperation under the Ministry of Agriculture is the nodal organization responsible for development of the agriculture sector. Ministry of Food Processing Industry is the main central agency of the Government responsible for developing a strong and vibrant food processing sector. It is set up to create increased job opportunities in rural areas, enable the farmers to reap benefit from modern technology, create surplus for exports and stimulate demand for processed food, cut flowers, seeds, tubers and ornamental plants.

In order to remove some of the major constraints affecting fruit, vegetable and ornamental plants and to provide necessary impetus to growth, Ministry of Food Processing Industries have several plan schemes. These include schemes for training of entrepreneurs, creation of infrastructural facilities for fruits and vegetables including mushroom processing. Industrial estates, setting up/ expansion/up-gradation of fruit and vegetales processing units development of backward linkages through contract farming, market promotion through advertisement, promotion of FPO symbol and marketing of small manufacturers' products under common brand name, setting up of and Research & Development in identifying new techniques and practices including those for packaging are mandated projects of Ministry of Food Processing.

Floriculture

Commercial floriculture is of recent origin though the traditional flower cultivation is going on for centuries. Emphasis is shifting from traditional flowers to cut flowers for export purposes. The liberalization of economy since 1991-92 has given an impetus to the Indian entrepreneurs for establishing export oriental floriculture unit under controlled climatic conditions. Maharashtra, Karnataka, Andhra Pradesh and Haryana have emerged as major floriculture centres in recent times. At present, the total area under floriculture is about area 2,32,700 ha. with a production of loose 1729.2 (000MT) of loose flowers and cut 76731.9 lakh (no.) of cut flowers (NHB, 2013).

Department of Agriculture and Cooperation under the Ministry of Agriculture is the nodal organization responsible for development of the floriculture sector. It is responsible for formulation and implementation of national policies and programmes aimed at achieving rapid agricultural growth through optimum utilization of land, water, soil and plant resources of the country. The Horticulture Division in the Department of Agriculture and Co-operation is vested with the responsibility of overseeing the overall accelerated development of the sector. It was set up with a mandate to implement the programmes for improving the production, productivity and utilisation of horticultural crops, support and formulate policies aimed at accelerated growth of horticulture, facilitate availability of disease free planting materials and seeds of horticultural crops, provide thc leadership and co-ordinate activities for promotion of horticulture, *etc*.

Several schemes were initiated by the Government for promotion and development of floriculture sector. Integrated Development of Commercial Floriculture is one such schemes implemented to improve production and productivity of traditional as well as cut flowers through availability of quality planting materials and transfer of technology. Improving human resource capabilities for growing flowers scientifically, promoting production of off season and quality flowers through protected cultivation and improving post harvest handling of flowers are other of activies.

State Governments have set up separate departments for promotion and development of the sector in respcctive states.

Policies and Schemes

Several significant initiatives are taken in recent years by the Government to reverse the downward trend in agricultural production. Some of these important initiatives include:

1. Bharat Nirman
2. National Rural Employment Guarantee Programme
3. National Horticulture Mission
4. Expansion of Institutional Credit to Farmers
5. Establishment of the National Bee Board
6. Establishment of the National Rainfed Area Authority

7. Establishment of the National Fisheries Development Board (NFDB)
8. Watershed Development and Micro Irrigation Programmes
9. Reforms in Agricultural Marketing and Development of Market Infrastructure
10. Revitalisation of Cooperative Sector
11. Agri-business Development through Venture Capital Participation by the Small Farmer Agri-business Consortium (SFAC)
12. Reform and Support for Agriculture Extension Services
13. National Rural Health Mission
14. National Food Security Mission
15. Rashtriya Krishi Vikas Yojana to incentivise states to invest more in agriculture.

Integrated Food Law

1. Legislative Framework for Warehousing Development and Regulation
2. Protection of Plant Varieties and FarmersRights (PPVFR) Act, 2001
3. National Bamboo Mission
4. Knowledge Connectivity through Common Service Centres (CSC) and IT initiatives
5. Integrated Development of Commercial Floriculture.

Objectives

1. Improve production and productivity of traditional and cut flowers through availability of quality planting materials and transfer of technology.
2. Improve human resource capabilities for growing flowers scientifically.
3. Promote production of off season and quality flowers through protected cultivation.
4. Improve on farm post harvest handling of flowers.

Pattern of Assistance

1. Specific attention for development of traditional flowers.
2. Strengthening of Model Floriculture Centres (MFC) for additional facilities and extension of MFC in other potential states.

3. Improvement in productivity and production through higher use of inputs and area expansion
4. Assistance for off-season production of flowers
5. Emphasis on on-farm post-harvest management of flowers.
6. Encouragement for greenhouse cultivation of flowers.
7. Farmer participatory demonstration of cultivator technology

Floriculture Industry for Domestic and Export Market

(a) **Loose Flowers**: Used traditionally for making garlands, worship and decoration purposes in marriages. Marigold, annual chrysanthemum, crossandra, tuberose, jasmine, gaillardia are used for this purpose.

(b) **Cut Flowers and Foliage Plants**: These are used for value added products like bouquets, floral basket and flower arrangements. Foliage plants are also used for decoration purposes. The major cut flowers are roses, carnations, chrysanthemums, gerbera, lilium, tulip, orchid, anthurium, gladiolus and tuberose.

(c) **Plant Nurseries:** include seedlings, rooted cuttings, seeds, bulbs and corms for propagation and supply of live ornamental plants including tissue culture plants.

(d) **Rental Services:** Supply of house plants and arrangements of flowers on annual rent for a specific period, this practice is very common in Govt. offices and corporate houses.

(e) **Value Added Products**: included for making perfume, crude oil, rose water, extraction of natural colours, edible vanilla from orchid, gulkand, gul-rogan *etc*. Mainly these are extracted from different flower crops like roses, chrysanthemum, marigold, jasmine and orchids.

Because of industrialization and increase in population all over the world, floriculture industry is fast emerging as a major venture on the world scenario and is a potential money spinner for third world countries. However, advanced floriculture technology is capital intensive in view of high cost of green houses, equipment, machines, chemicals, cooling, storage, packaging and other infrastructural facilities. Many kinds of ornamental plants are grown for domestic and international trade in developed and developing countries. Today floriculture is recognized as a lucrative profession with much higher potential for return per unit area than most of the field and other horticultural crops.

Business Opportunities

Varied agro climatic conditions in the country and moderate climate control at relatively cheaper cost can deliver quality products at internationally competitive prices. We have skilled manpower to absorb the technology and implement the same at relatively low cost. Proximity to the consumer markets like Europe, Middle East countries and Far East are also added advantages. Vast domestic market with diverse culture and tastes absorb the non-exportable grades of flowers. Most flower usages in India do not require very long stems. Sufficiently large tissue culture facility is available in the country and can meet the requirement of planting materials. There is tremendous scope for growing loose flowers, cut flowers and even for the seed industry. Due to varied types of climate, off season flowers can be grown to meet ever increasing demands in addition to protected cultivation.

(a) Indians are scaling up

Through the past a few years, India is building up commercial floriculture to a sizeable extent to take on growing demand within as well as emerging requirements in foreign markets. Roses grown in Bangalore are much sought after, everywhere. For several years now, they are being exported to the continent during the V-Day, and the volumes are sizeable.

According to people in floriculture trade nearly 4.5 million rose stems were airlifted 10 days before the Valentines' Day. More quantities were on their way to export markets. Big exporters put up 'sold out' boards outside the trading counters.

Indian and Kenyan roses are much in demand. International markets seek them out on two special occasions of Christmas and Valentines' Day. The climatic conditions in Europe during December and February are unsuited for production because of frost and snow. In contrast, there is ample sunshine in tropical countries like India to grow all most all types of flowers.

Given the steep demand-push in Western markets, flower producers in Bangalore perfected a local technique-called peaking to scale up the production over shorter periods to match the demand. Take five acres, where the normal production is about 20,000 to 25,000 stems a day. Says a big exporter, "During the special season, the peaking technique enables an output of over 100,000 stems a day in the sprawling greenhouses."

The normal export price for Indian stem is from Rs.2 to Rs.5 but during the Valentines' Day the rate is upwards to Rs.20/- per stem. Sometimes, retailer asks even more fancy prices depending on the volume of demand. A number of whole sale and retail shops are coming up in all the major cities and even in small towns. The domestic market has grown up many times as compared to the export market.

(b) Markets in UK

UK has a huge floriculture market. Cut flower supplies come not just from Kenya. Also active are India, Israel, Ecuador and Colombia. On the Valentines' Day, British lovers spend an estimated 22 million pounds ($ 31.5 million) on flowers. This trend is increasing even in developing countries like India, Pakishtan, Myanmar, China, Indonesia and Nepal

Thrust on Development of Floriculture

Government of India, Ministry of Agriculture constituted an Expert Group on Development of Floriculture in India in 1988 under the Chairmanship of Mrs. P.P. Trivedi, Secretary (Retd.) to the Govt. of India. The group identified the floriculture production sectors for export oriented production. The group submitted its report in July 1989. It was suggested to follow the approach of intensive floriculture in selected zones on the basis of suitability of climate for growing different varieties of cut flowers.

In keeping with the vast potential offered by the global market and advantages for India, floriculture was identified as a thrust sector for development of exports. An extreme focus group was constituted by Ministry of Commerce under the Chairmanship of Shri Eapen George, Chief Executive, A.V. Thomas Group Co., Cochin. The group reviewed the potential and constraints for the development of floriculture and suggested an action plan in its report submitted in mid 1992.

A study was also conducted by a Dutch consultant commissioned by the Ministry of Agriculture, Govt. of Netherlands in 1992 on the opportunities for Floriculture in India. APEDA was associated in the collection of data and survey of different zones in the country. The study projected a good potential for export of cut flowers from India like roses, carnations, chrysanthemum to West Europe, Japan, Singapore, Hong Kong and West Asia.

In the last two decades or so, floriculture in India received a boost after a number of entrepreneurs realized the export potential and the liberalized government policies made it feasible for them. With the implementation of National Horticulture Mission (2005-2012) by the Govt. of India with the mandate to make the area and production just double of horticultural crops by the end of 2012 through various incentives to the growers in the States under this Central sector scheme.

At present more than 500 units of different kinds are producing flowers for export and the domestic market across the country in addition to the open field cultivation including naturally ventilated green houses. Many farmers have shifted from cereals crops to flower cultivation ranging from small to large scale and getting a very handsome remuneration and drawing their livelihood.

Nutritional Value

Flowers are associated with mankind since time immemorial, as they are used for religious offerings and other social ceremonies. Loose flowers are mostly grown for worship, garland making and decoration. Cut flowers like rose, chrysanthemum, carnation, gerbera, anthurium, orchids and lily are mostly grown for export purposes. Some of the flowers are also edible with high nutritional value e.g. Dahlia tubers, lily, tapioca etc., mainly their roots are rich in starch and even dahlia roots are good to cure cancer.

Floriculture is increasingly regarded as a viable diversification from the traditional field crops due to increased per unit returns and the increasing habit of "Saying it with flowers"during all occasions. Though the art of growing flowers is not new to India, large scale commercial cultivation and protected cultivation in poly houses are relatively new to India. Enormous genetic diversity, varied agro- climatic conditions and versatile human resources offer India a unique scope for diversification into new avenues which were not explored to a greater extent. With the opening up of world market in the WTO regime, there is a free movement of floriculture products worldwide.

Aesthetical Importance

(a) Improving environment: Establishment of fruit orchards helps in maintaining the ecological balance by producing large quantities of bio mass. Fruit trees help in fighting environmental pollutions and also add beauty to the environment (avenue plantation), and thus they are considered as the best component of social forestry. Waste land, salt affected areas and arid semi-arid regions can be utilized for cultivation of suitable fruits for

Landscape View

improvement of surroundings and for high returns. The arboriculture and plantation of seasonal flowers improve the ambience of the locality. Now a days many corporate houses have come up in India and giving due emphasis to landscape horticulture to improve environment and public health. Even on the road side and as avenue trees, many fruit plants like lemon, aonla, jamun, tamarind and mango are planted on road side.

Landscape View of Complex

(b) Beautiful Ambience: With the new concept of housing sector and corporate sector, the flowers and plants are in great demand to improve the ambience and to make surroundings and terraces beautiful to attract business opportunities and appreciation in real estate value.

Economical Prospects

Improving economy: Economy of various States especially, Karnataka, West Bengal, Himachal Pradesh, Uttrakhand, Jummu & Kashmir and NEH regions largely depend upon flower cultivation. They not only get their livelihood and even earn foreign exchange by export. Cultivation of flowers generates employment for rural masses and improves socio-economic conditions. The flower crops besides providing livelihood to a large number of small farmers contribute significantly to economic security. Because of diverse agroclimatic conditions, various types, of flowers are grown in India. Rose, chrysanthemum, carnation, tuberose, tulip, freesia, gerbera, lily, orchid, iris and some of the annual flowers are commonly used as cut flowers. Ficus, dracaena, azalea, kalanchoe, begonia, yucca, saintpaulia, poinsettia and dieffenbachia are common foliage plants.

Flowers in National Economy

Flowers and other floricultural products have great export potential in India in view of the competitive advantage of low cost of labour, tropical and subtropical climate suitable for production of flowers especially during winter (October-March), when there is maximum demand in Europe and their local production is not sufficient to meet the requirement. Keeping this in mind, the scientists of IARI with the assistance of State Trading Corporation of India (STDC) sent the consignment of flowers to Europe market first time in 1969, when other countries like Israel, Kenya, Thailand and Columbia also started developing export trade of flowers. The venture of flower export from India could not succeed because the production was far below the demand and there were no government incentives-exemptions of duty and tax or subsidies and loans. In the present scenario, there is a lot of schemes of Government to promote floriculture in which subsidies are being given in cash and kinds for development of infrastructure and planting materials.

Processing

Many policy and development initiatives including financial incentives lead to higher farm output and remunerative returns and are likely to encourage horticulture farming on a larger scale. Policy and development initiatives including financial and technical support available in horticulture sector provide greater impetus to horticultural development in the country. Higher horticulture growth within the crop husbandry is expected not only to push agricultural growth further. It is also expected to make substantial contribution in the national economy by unleashing substantial value addition through storage, processing, transportation and marketing of horticulture and floriculture commodities and their value added products.

Dry flower and plants of foliage beauty are becoming more popular due to longer shelf life. All kinds of foliage, flowers and part thereof can be used to decorate the interior and are generally used in bouquets *etc.* So far the business of such kind are limited and mainly restricted to southern states and Kolkatta.

Drying Techniques

Several types of dryers and drying methods, each suited for a particular situation are commercially used to remove moisture from a wide variety of food products including fruits, vegetables and flowers. Factors on which selection of a particular dryer/ drying method depends include:

- Form of raw material and its properties
- Desired physical form and characteristics of dried product
- Necessary operating conditions
- Operation costs.

There are three basic types of drying process:

- Solar drying
- Atmospheric drying including batch (kiln, tower and cabinet dryers) and

continuous (tunnel, belt, belt-trough, fluidised bed, explosion puff, foam-mat, spray, drum and microwave)

- Sub-atmospheric dehydration (vacuum shelf/belt/drum and freeze dryers).

The scope is expanded to include use of low temperature and low energy processes like osmotic dehydration.

As dryers are concerned, one useful division of dryer types separates them into air convection dryers, drum or roller dryers and vacuum dryers.

i) **Solar drying:** Sun and solar drying of flowers is a cheap method of preservation because, it uses the natural resource/ source of heat i.e. sunlight. This method can be used on a commercial scale as well at village level provided that the climate is hot, relatively dry and free of rainfall during and immediately after normal harvesting period. The fresh crop should be of good quality and as ripe (mature) as it would need to be if it was going to be used fresh. Poor quality produces cannot be used for natural drying.

ii) **Shade drying:** Shade drying is carried out for products which can lose their colour and/or turn brown if put in direct sunlight. Products having natural vivid colours like herbs, green and red sweet peppers, chillies, green beans and okra give a more attractive end product when dried in the shade.

The principles for shade drying are the same as for sun drying. The material to be dried requires full air circulation. Shade drying is carried out under a roof or thatch that has open sides; it cannot be done either inside conventional buildings with sidewalls or in compounds sheltered from wind. Under dry conditions when there is a good circulation of air, shade drying takes a little more time than is normally required for drying in full sunlight.

Processing Techniques for value addition in floral crops:

a) **Isolation of essential oil:** Generally essential oils of flowers are extracted by hydro-distillation, steam distillation, extraction with low boiling solvents and with liquefied gases.

b) **Hydro distillation:** In this process, the floral material to be distilled is in direct contact with boiling water. The oil and vapours are passed into the condenser to separate the oil from water phase. Water distillation is a slow process but yields superior quality products.

c) **Steam distillation:** In this process, the flower material is placed on a grid in the still. The lower part of the grid is filled with water. The floral material comes in contact only with steam and not with boiling water. Hence with quicker distillation, higher yield and oil of better quality are obtained. The process is suitable for oils with high boiling point constituents.

d) **Extraction with volatile oil solvents:** In this process, the floral material is extracted with low boiling solvents at room temperature. After evaporating the solvent, the concrete is left behind in case of fresh flowers and oleoresin or resinoids in case of dry floral materials. The solvent extracts mostly possess a viscous to waxy consistency and they need to be refined before use. They are found in various forms.

i) **Concrete:** An extract of fresh plant parts by the use of a hydrocarbon solvent is defined as concrete. It is rich in hydrocarbon soluble components and devoid of water soluble components. It is generally a waxy semi solid dark coloured material free from original solvent. The commonly used solvents fro production of concrete are petroleum ether and hexane.

ii) **Absolute:** A concrete not widely used in perfumery, it has to be converted into an alcohol volatile concentrates known as absolute. In this process, the concrete is dissolved in hot (30-60^{0}C) absolute alcohol (ethanol) and then refrigerated to below -5 to -10^{0}C to separate the waxes. From the filtrate, alcohol is removed by vacuum distillation to get alcohol free absolute.

iii) **Effleurage (Extraction with cold fats):** This process involves extraction of perfumery oil from flowers with the help of nonvolatile or fixed solvents, like fat and fatty oils. These solvents bind the essential oils and thereby enable their isolation. The fat loaded with odorous substance is later extracted with ethanol for obtaining perfume oil. With this process, perfume oils of the highest odour and purity are obtained.

iv) **Extraction with super critical and liquefied gases:** In this process, extraction of floral material is carried out with compressed gases (liquid or superior critical fluids) in a high pressure vessel under controlled pressure and temperature. Carbon dioxide (CO_2) and hydrofluoroacarbons (HFC) are being used commonly due to their preferred critical temperature and pressure. Apart from the essential oils, the floral isolates obtained are concretes and absolutes from concrete.

Value Addition

A few ornamentals are known for their typical flavour and taste. If properly utilized, a variety of processed products could be developed which can add to the product range and also help in employment generation, value addition and increasing capacity utilization in agro-processing industry. There are many flower crops being grown for extraction of essential oils as discussed below.

Flower crops for value addition

a) Jasmine (*Jasminum grandiflorum*)

The flower is processed in France, Italy, Egypt, Morocco, Algeria, Turkey, China and India. India has emerged as a dependable source of jasmine isolates and contributes 22% of world jasmine concrete production. The yield of concrete is approximately 0.15% while the yield of absolute from concrete is 55 to 62% depending on month of crop.

b) Rose

i) **Rose water:** Rose water is obtained by water distillation of flowers during flowering season after which it is stored for 2-3 months for maturation of odour. It is then diluted with distilled water in different proportions and bottled. The price varies according to the degree of dilution. The rose water is available in markets in two forms; (i) *Ek-mana-* 20 litres of rose water obtained by distilling 40 kg of rose flowers; (ii) *Do-mana-* 20 litres of rose water obtained by distilling 80 kg of rose flowers.

ii) **Sandelwood based rose attar:** This is prepared by water distillation of rose flowers and passing the distillatc through sandal wood oil.

iii) **Gulkand:** It is a conserve of rose petals, prepared by mixing together equal quantities of rose petals and sugar, and the petals in the marcerated form. This mixture of marcerated rose petals and crushed sugar is kept in clean glass jar in open sun for about a month for maturation.

iv) **Pankhuri:** Dried rose petals are called þankhuri'and are used for the preparation of cold drinks.

v) **Rose oil or Rose otto:** It is obtained by steam distillation of rose flowers. In India, it is the rose water, which is the main product of distillation. Rose oil is not produced on commercial cscale and whatever small quantity of it available in the market is obtained as a by-product of rose water industry. Rose oil, attar and water are much in demand, within as well as outside the country, and both rose water and rose attar are the two main commodities of export.

c) Tuberose (*Polianthes tuberosa*)

The flowers of tuberose emit the most delightful fragrance and are the source of tuberose oil used in high-grade perfumery. The perfume is generally the best extracted by effleurage. The single flowered forms are more strongly scented than double flowered ones and are exploited for extraction of perfume. About 150kg of flowers are required to produce one kg of a brown semisolid absolute of effleurage. After extraction, the flowers contain some natural perfume and are treated with petroleum ether to obtain absolute of chassis as a valuable by product (yield : 1.2 - 1.5%). Extraction of tuberose flowers with petroleum ether yields 0.08-0.11% concrete which gives 18-23% of absolute on treatment with alcohol and contains 3-6% steam distillable oil.

d) Ornamental ginger

i) **Dried ginger:** After removing the dirt and roots from freshly harvested rhizome, it is soaked in hot water to facilitate removal of skin. The skin is scrapped off with pieces of sharpened bamboo. The scrapped produce is washed and dried in the sun for 3-4 days and hand rubbed to remove all remaining bits of skin. Soaking the peeled rhizomes in a 2 per cent solution of lime water for 6 hours before drying gives the dry ginger a good appearance. During drying, the rhizomes loose nearly 70 per cent in weight. Ginger dried for the market contains 7 to 12 per cent moisture. It can be dried artificially by a heating evaporator but the rhizomes loose much of their aroma and become darker with a burnt flavour. The sun drying is still the commercial way of drying. Sun drying also bleaches the produce. The essential oil which gives ginger its aromatic character is present in epidermal cells and excessive or careless scrapping results in damaging these cells leading to loss of essential oils. Therefore, peeling should be done with great care and skill.

ii) **Ginger Oil:** It varies from 0.5 to 3.0%. Oil possesses only aroma and not the flavour of spice.

iii) **Ginger Oleoresin:** It is a blend of oil and resinoides. It is extracted from ginger powder using organic solvents like acetone, ethylene dichloride etc. Its content ranges from 3.5 to 9.5%. Major pungent principle is gingerol (a phenyl ketone).

Trade and Opportunities

The worldwide demand for various floriculture products ranges *viz.* cut flowers, pot plants, flowers, seeds and corms and tubers in addition to various value added products. Many products like perfumes, oleoresins, dried petals and even seeds are being exported to many European and gulf countries. In 1990, it was estimated that some 14 or 15 main countries together accounted for cut flower consumption of over 45 billion (b) guilders and pot plant consumption of over 26 b guilders but what is noteworthy is that the consumption of cut flowers worldwide was raised by 11% since 1985 going from 41 b to 45 b guilders in 1990. This trend continues particularly in the 3 main consuming regions: Western Europe, North America and Japan. The Japanese cut flower market with a strong demand for good quality and new unfamiliar products may grow by 5% per year. Growth of 6% per year is forecast for the American market. Even in European market, the demand is expected up to 15-20% per annum.

With increasing demand, production centres have expanded from traditional centres (USA, Japan, the Netherlands and Columbia) to new centres (Latin America, Africa and Asia). In Asia, India, China, Vietnam and Sri Lanka are moving in intensive floriculture. New consumption centres are also emerging in South-East Asia, Middle-East and Eastern Europe. The scenario provides opportunity to capitalize on the strength and convert weakness into opportunity. However in the consumer-driven market, quality of service and delivery system will play a significant role. Marketing is no more a meeting of buyers and sellers, but it is complex, which is driven by quantity of products and reliability of delivery. New distribution structure with aid of IT is likely to be in place, which would be a virtual market, demanding, branding, cataloguing and quality assurance emphasized. Japan, US, Italy and Netherlands are the worlds' main producers and users of both flowers and pot plants.

On the supply side, Europe plays a great role having more than 76% share of the total world market of cut flowers. The Netherlands are the largest exporting country with a share of 59%. The other major exporters are Columbia, Italy, Israel and Kenya.

The major exporters of floriculture products are Holland, Colombia, Israel, Italy, Spain, Thailand, France, the U.S.A., South America, New Zealand and Equador. The major importing countries are Germany, the U.S.A., France, Italy, Australia, Belgium, Hong Kong and Gulf countries. The important ornamental crops in the international cut flowers trade are carnation, rose, chrysanthemum, orchids, gladiolus, alstroemeria, anthurium, gerbera, gladiolus, gypsophila, liatris, nerine, statice, archilea *etc.*

Environmental Benefits

Temperature and Energy Use

Plants can be successfully used to mitigate heat. Their leaves intercept, reflect, absorb and transmit solar radiation. Trees are called nature's air conditioners because they lower air temperature, shade buildings in summer and block winter winds. Trees can reduce building energy use and cooling costs. A single large tree can produce the cooling effect of 10 room size air conditioners operating 24 hours/day. Dew and frost are less common under trees, because less radiant energy is released from the soil in those areas at night.

Air Quality

Sulphur dioxide, oxides of nitrogen, suspended particulate matter, lead, ammonia and carbon monoxide are the major air pollutants which cause respiratory and other health problems. Plants help in controlling air pollution through acting as biological and physical nets. They cleanse the air by intercepting and slowing dust and other particulate materials causing them to fall out, and by absorbing pollutant gases. Plants increase humidity and thus help in settling of particulates. Increased wind turbulence by plants dispenses pollutants. Fragrance of many flowers masks unpleasant odours, thereby imparting freshness to air.

Precipitation and Humidity

Plants intercept precipitation and slow its decent to soil surface, thus increasing infiltration and reducing run-off and soil erosion. Pubescence on leaves helps in water entrapment. Plants with horizontal branching and rough bark are the most effective. Humidity is increased in hot and dry season by means of transpiration.

Noise Abatement

Vegetation along the highways can screen vehicular noise from reaching adjoining habitations. Plants also create "background" noise of rustling of leaves and wind through branches that can help muffle other noises. Plants having thick and fleshy leaves with petioles which allow higher degree of flexibility and vibration are the best suited for use as noise screens.

Carbon Sequestration

Increase in CO_2 accounts for about 65 percent of the current direct positive radiative forcing due to green house gases produced as a result of human activities. The atmospheric CO_2 concentration has increased 30 percent since pre-industrial times, as a result of increasing emissions from fossil fuel combustion, land conversion and cement production and continues to increase by 0.4 percent per year (Dunne and Harte, 2001).

Water Quality/Erosion

Plants prevent harmful pollutants in the soil from getting into waterways, reduce top soil erosion, slow down water run-off and remove nutrients and sediments, while increasing groundwater recharge. Reducing the flow of storm water reduces the amount of pollution that is washed into a drainage area and also prevents severe clogging. Vegetation along rivers lakes and other water bodies decrease their siltation and reduce cutting of banks during floods.

Wind Protection and Air Movement

Trees perpendicular to wind direction reduce the wind speed up to a distance of 2-5 times the height of the tallest tree on windward side and 30-40 times on the leeward side. Trees can be planted to funnel or baffle wind away from areas - both vertical and horizontal concentration of foliage can modify air movement patterns. Plants can also be used effectively to control snowdrift.

Glare and Reflection Control

Dense foliage absorbs about 70 percent of the suns' rays, reflects 17 percent and transmits about 13 percent. Trees and shrubs help control light scattering, light intensity and modify predominant wavelengths on a site. Trees block and reflect sunlight and artificial lights to minimize eyestrain.

Ecological Benefits

Bio-aesthetic planning aims at enhancing biodiversity along with beautification of environment. Biological diversity has direct consumptive value in food, agriculture, medicine and industry. It also has aesthetic and recreational value. Biodiversity maintains ecological balance and continues evolutionary processes. The indirect ecosystem services provided through biodiversity are photosynthesis, pollination, transpiration, chemical cycling, nutrient cycling, soil maintenance, climate regulation, air, water system management, waste treatment and pest control.

Economic and Social Benefits

Besides savings in the form of reduced energy consumption, plants provide timber, pulpwood, fruits, nuts, mulch, composting material and firewood. Shading by trees extends life of paved surfaces and reduces maintenance cost.

Greenery enhances community economic stability by attracting business and tourists, people linger and shop longer along tree lined streets and apartments/ offices in wooded areas rent more quickly and have higher occupancy rates. Property values are higher in well-planned areas lined with trees. Concept of eco-tourism has gained importance in recent times, and many government and private organizations have come up with camping sites and nature tours to boost tourism in many states. This has resulted in additional revenue and employment generation.

Aesthetic Benefits

Trees and other life forms supported by them have a positive value in scenic quality. Trees change skyline, add variety of form, colour and texture to otherwise dull and monotonous cities. They soften the harshness of concrete structures, complement architectural lines, enframe views and provide background settings to buildings. Dynamic behaviour of vegetation makes the surroundings interesting. Plants can be very effectively used to screen undesirable and disturbing sight lines.

Thunbergia

Integrating Aesthetics and Nature Conservation

India is one of the twelve mega-biodiversity centres of the world with 46,000 plant species and 81,000 animal species, described so far, from 70 percent of the total geographical area surveyed so far. To conserve flora and fauna from extinction, many botanical gardens, national parks and wild life sanctuaries have been created.

Wild Life Conservation

Most of these areas are parts of the natural landscapes of that region declared protected by law. Some of these were hunting reserves of former princely states. Although it is impossible to imitate nature, but based on understanding of the natural processes, one can develop a flexible system that accommodates natural cycles, rhythms and unpredictability of time as a part of the aesthetics of the place. The landscape of Bharatpur Bird Sanctuary is a wonderful example of such process and aesthetics created through generation of a new wetland for recreational purpose. The once scrub land in the brackish area was developed by the Maharaja of Bharatpur, based on the regions bowl shaped topography, potential to collect rain water and attract wild life. Manipulation of water was done for nestling of resident birds and providing feeding grounds for the migratory birds. Emergent wetland species along with aquatic gardens are designed in some parts of the sanctuary.

There will be many such sites in India which, if identified and developed properly, can become great ecological assets - generating revenue through eco-tourism in addition to preserving and enhancing biodiversity. Care should be taken to use only the native species to avoid what is called species invasion like in Lantana. Some of the important national parks and sanctuaries are:

i. Bandipur National Park (Karnataka)

ii. Corbet National Park (Uttarakhand)

iii. Dachigam National Park (J&K)

iv. Dudhwa National Park (U.P.)

v. Gir National Park (Gujarat)

vi. Great Himalayan National Park (H.P.)

vii. Kanha National Park (M.P.)

viii. Kaziranga National Park (Assam)

ix. Keoladeo National Park (Rajasthan)

x. Manas Wild Life Sanctuary (Assam)

xi. Periyar National Park (Kerala)

xii. Ranthambor National Park (Rajasthan)

Botanical Gardens

Many botanical gardens were created for conservation of plant wealth of the country. These gardens serve as centres of collection, maintenance and evaluation of germplasm of important species, apart from education in bioaesthetic planning. Except the ones of national importance and those established by the British, majority have been planned for scientific purpose only and lack beauty. Botanical gardens and arboreta can be made attractive to general public, if aesthetic aspect is integrated with scientific requirement. Properly laid out paths, a few benches and a fountain can create the desired effect. Some of the prominent botanical gardens are:

i. Lalbagh, Bangalore

ii. Royal Botanic Garden, Kolkata

iii. Royal Agri-Horticultural Society Garden, Kolkata

iv. Government Botanic Garden, Ootacamund

v. Lloyd Botanic Garden, Darjeeling

vi. The Botanical Garden of Forest Research Institute, Dehradun

Landscaping Roads

Trees for planting along the roads should be selected with due regard to soil, rainfall, temperature and water level. The tree species should be either truly evergreen, nearly evergreen or be in leaves at least during height of summer. They must not become too wide to interfere with traffic flow and should be capable of developing clear bole up to a height of 2.5-3.5 m from ground level. They should be fast growing and wind firm, and should not be thorny, or drop too many leaves. Trees should be deep rooted. Traditional roadside trees like jamun and seedling mangoes should be preferred which have long life and are well adapted.

Highways

Bio-aesthetic planning for highways should begin at design stage, so that it could be a part of the master plan. Design objectives of highways should be to fit the road into the existing terrain, and harmonise it with general landscape of the region. Existing features along the road alignment like natural wooded areas, streams, ponds, rock outcrops, scenic vistas, historical buildings, monuments, gardens, *etc*. contribute significantly to general landscape and every effort should be made to preserve, incorporate and integrate these into the landscape. In the construction stage, removal of vegetation and cutting of trees should be done judiciously and limited to minimum possible extent, so that the least habitat loss occurs. Plantation of ornamental and economically important species should be carried out to a sufficient distance on either side of the road. Functional planting applies to such problems, as protection of slopes against erosion, screening of unsightly views, reducing headlight glare and providing shade in summer. Trees serve as biological reference points along border of the road and increase drivers' perception of orientation. Arranged in varied forms and colours, they reduce drivers' fatigue and boredom. Besides pollution abatement, planting blends the road into the surrounding countryside and enhances the overall beauty. Vegetation also provides habitat for many species of birds, rodents etc

Tree planting along highways can be done according to the following schemes

- Front row of flowering trees and rear rows of utility trees: Front row should be planted with flowering trees of different species, each for 3-5 km and the subsequent rows should be planted with species of economic importance such as timber, fruit and fuel wood.
- Mass effect of flowering trees: Flowering trees should be planted as large groups for about 0.5 km after every 5 km of plantation of economically important species. Such groups should be arranged alternately on both sides of the road. This will break the monotony of singly planted species.
- Roadside amenities like rest areas, truck parks and off strip parking places should be created on all highways for safety, convenience and enjoyment of the travellers. Truck operators in particular have to undertake long journeys, which are often exhausting and boring. Roadside rest areas and truck parks should be planted with shady trees. Such areas should fit into the natural surroundings and become a part of general road landscape.

City Roads

Roads and streets are lifelines of urban areas. City dwellers spend considerable part of their outdoor time on them. Without plants, these structures seem to pierce through the body of the city and create restlessness. To make them more enjoyable and safe, they should be lined with shady and flowering trees. An outer row of shady trees and inner row of flowering trees, with cycle track/walkway between the two, will be a good arrangement for wider roads. This ensures pedestrian safety and reduction in number of accidents. Species of flowering trees should be sufficiently varied to avoid monotony. The central verge, wherever available, should be planted with small to medium sized shrubs to cut off headlight glare.

Wherever single row is to be planted, one of the following schemes could be adopted (Arora, 2001):

- **One kind of flowering tree on both sides:** In this scheme, when the trees come to bloom, there is riot of colour and the road looks beautiful for a particular period. However rest of the time the road looks dull. Maintenance is easy in this type of arrangement, e.g., *Bauhinia variegata*, *Cassia nodosa*, *Lagerstroemia speciosa* etc.
- **Two kinds of flowering trees blooming at one time on both sides:** Here, the two species are planted alternately on both sides of the road and different colours at same time avoid monotony, e.g., *Grevillea robusta* (Yellow) - *Jacaranda acutifolia* (Blue).

- **Two kinds of flowering trees blooming at different time on both sides:** In this case, one tree flowers in one part of the year and the other in another part of the year. Roads remain colourful for longer period, e.g., *Cassia fistula* (Yellow in May-June) -*Jacaranda acutifolia* (Blue in March-April).
- **Shady trees only on both sides:** This type of arrangement is suitable for roads with heavy pedestrian traffic like approach roads of bus stops, railway stations and government offices, e.g., *Ficus infectoria*, *Mimusops elengi*, *Alstonia scholaris* etc.

Rural Roads

Rural roads are the most neglected throughout the country as for tree plantation is concerned. Here, preference should be given to planting of multipurpose trees which besides providing shade and flowers, also provide economic products such as fuel wood, timber, fruits, fodder, etc. Roads should be planted with small to medium sized trees preferably on the southern side, so that fields are not shaded.

Landscaping Urban Centres

Proper town planning is a pre-requisite for bio-aesthetic treatment of our cities and towns. This can be seen only in new towns with wide roads and well laid out public parks, rather than in congested old towns with narrow crooked streets. Older parts of our cities like Delhi, Amritsar, Lucknow, Varanasi and Kolkata have ugly, ill ventilated houses, joined together along narrow lanes. Deficient in open places, they offer little scope for bio-aesthetic planning. The few existing trees in such areas should be preserved, as far as possible and wherever possible some open spaces should be created and converted into parks. Some open spaces are, however, available around religious places and historical buildings, which can be used to give a healing touch by planting flowering trees and shrubs.

Trees for Urban Landscaping

In case of individual buildings, large trees help architectural composition and provide the required background and in fact look like structural members of the scheme. Shrubs and climbers also serve their purpose by creation of various decorative patterns. Colourful flowering plants in window boxes add to beauty of the building.

Hospital Compounds

Flowering trees, shrubs with fragrant flowers and shady trees should be planted in compounds of hospitals. Medicinal plants like neem, which also provide good shade should be preferred. This improves microclimate of the hospital complex and the beauty of flowers will make the otherwise dull environment, somewhat interesting.

Institutions/Offices

Maximum possible ornamental and economic species should be planted in educational institutions to educate students about importance of plants in everyday life. Students should be involved in maintenance of these plants

Flowering shady trees and shrubs should be planted in courts, government offices and post offices, so that visiting public may rest under the shade and enjoy their beauty while waiting. Railway stations, bus stands and airports, which are gateways to the cities, should be meticulously planted to accord welcome to the commuters and make the long waiting period interesting. Proprietors of commercial buildings should be encouraged and given all assistance in the planning and planting of ornamental plants.

Industrial Areas

Industrialization is one of the primary causes of environmental deterioration. Nevertheless, it has to persist and proliferate for continuous development of mankind. All that can be done are to make the industries more eco-friendly. Bio-aesthetic planning is thus very important for industrial areas.

Government of India has issued certain guidelines for industrial sites according to which industries should be located at least 25 km from ecologically sensitive areas, such as national parks, sanctuaries, lakes, biosphere reserves; archaeological and historic places and from major settlements (>3 lakh). Green belts should extend 500 m from the boundary of such areas. However, current scenario of the industrial areas demands strict enforcement of such guidelines for minimizing loss to the environment.

Bio-aesthetic plan for industrial areas should include creation of buffer zones around the areas with mixed plantation of economically important species, to serve the dual purpose of revenue generation and environmental amelioration. A green belt of suitable trees having a width of 50 to 100 m between residential and industrial zones decreases dust level up to 52 per cent and noise pollution by 40-60 percent. Five hundred metre belt can mitigate about 76 percent SO ^{2} and 67% NO from the atmosphere. Premises of the industrial units should be planted with pollution tolerant flowering plants. This helps in improving working conditions for the industrial workers and increase their output. Deciduous species remove air pollutants in a better way by accumulating them in their leaves, which are all shed during dormant season.

Historical Buildings

We have a rich legacy of historical buildings and heritage sites, which besides their archaeological importance are prime attraction for domestic and foreign tourists. Some of these have well planned gardens around them, like The Taj, while most of them have drab surroundings. With the intended promotion of heritage tourism in new tourism policy, the surroundings of these icons of our glorious past, need to be revamped, for presenting their true grandeur. Such sites should be beautified, keeping in view their significance, and type of vegetation historically associated with them be preferred. Approaches to these sites should also be ornamented with flowering trees and shrubs as a welcome gesture.

River Banks

Rivers like Ganga and Yamuna are regarded sacred and many temples and ghats are seen along their banks. These should be planted with sacred trees like kadam and ashoka, which will greatly enhance their sanctity and also add colour and charm to them. Apart from beautification of banks, plants help in controlling erosion and silting of rivers, and also help in reducing cutting of banks during rainy season. Planting of riverbanks is not meticulously undertaken, and on most of the area, general agriculture and forced vegetables are grown. A national level body should be formed to undertake riverside plantation and serious efforts should be made for enhancing their beauty and maintaining ecological balance. Many wetlands of international importance are identified along rivers e.g. Bhoj wetland, Bhopal (M.P.); Deepor Beel, Guwahati (Assam); Harike lake and Ropar headworks in Punjab and Pong Dam lake in Himachal Pradesh. Such areas have immense tourist potential, which can be exploited by providing proper facilities and aesthetic improvement.

Canal Banks

Banks of canals generally should be planted with moisture loving trees. Most of the tree planting alongside the canals were done by Forest Department and economically important trees were planted to generate regular income. Wherever they pass through near a city or town or a highway crosses them, flowering plants should be preferred. Headworks of canals should be developed into pleasure resorts and water sport facilities should be created wherever possible. Canal side plantations also serve as important habitat for birds and small animals as they are not much disturbed by human activity.

Airport

New airports are coming up in a big way in the country and recently Delhi, Bangalore, Hyderabad and Kolkata airport were upgraded with the latest state of art and technology. Their maintenance part has ample opportunity for horti business.

Landscaping Rural Areas

Our villages had a tradition of planting trees along field bunds, panchyat lands and around wells. Although economic considerations always dominated over beauty but there used to be sufficient green cover anyway. Panchyats usually assembled under Pipal or Bargad trees, which used to be centres of village activities. Intensive agriculture has taken its toll and commercial plantations of Eucalyptus and Poplar have replaced long living fruit or timber trees at many places. Village ponds, which used to be prominent features of rural landscape, are fast vanishing or being converted into sewage and wastewater dump and every effort should be made not only to conserve them but also to make them attractive. They can be used as water harvest tanks especially in low rainfall areas, where they can provide drinking and irrigation water during hot summer months.

Planting flowering trees on the approach roads, village schools, panchayat ghars and religious places, can beautify villages to a great extent. Common lands of the villages usually under the supervision of panchyats, provide ample opportunity for beautification, with a variety of flowering plants. A scheme was launched in Punjab for planting flowering trees, shrubs and fruit trees near the tube wells in villages and plants were provided to the farmers free of cost. Such schemes can greatly improve rural landscape and should be followed in whole of the country.

Role of Various Organisations

Institutions

Of late, most of the agricultural universities and horticultural institutes in the country either have a separate Department of Floriculture and Landscaping or it is a part of Horticulture Department. Basic knowledge of ornamental horticulture is given to all the students at undergraduate level or through special traning programmes. Eight universities are offering post-graduate courses and five are offering doctorate courses in Floriculture and Landscaping. Students are given thorough training in science and art of ornamental horticulture, with special emphasis on growing and maintenance of ornamental plants and their use in landscaping of parks, open spaces, buildings, roads, etc. Services of such trained personnel should be utilized while planning residential, commercial, industrial and other areas suitable for bio-aesthetic planning.

Peoples' Participation

In a India it is practically impossible for any government to undertake bio-aesthetic planning on its own. In a situation, where potable water and electricity are not available so far to all the people, it is futile to think of beautification by public funding alone. Peoples' participation is, therefore, very essential for realising the dream of a beautiful and colourful India.

Peoples' participation in any programme is driven by the benefits they get from it. Urban people will involve in bio-aesthetic planning primarily for their recreational needs and for integrating with nature. Increasing awareness about pollution has also forced the people to think positively about greening of cities. They should be further educated regarding other benefits like climatic amelioration, water quality improvement, decrease in energy consumption, *etc.*

Educating the people is very essential to ensure their participation and the best method is to educate the children. Excursion tours to nearby parks and botanical gardens with a few tips on environment, could be a great fun besides useful information for them. Organizing theme based science fairs and quiz competitions can infuse love and respect for nature in them. Ministry of Environment and Forests has introduced Urban Environmental Information System in Agra, Kanpur, Patna and Bhubaneshwar to provide information to the public in the form of local environmental report to be prepared by the municipalities on the status of socio-economic development and environment of the cities. Such efforts should be extended to whole of the country so that people could be made aware of the current state of their surroundings. NGO's, agri-horticulture societies, Eco-clubs *etc*. should be involved in such campaigns for effective dispersal of information.

Nursery as Industry

A nursery can be a part of a diversification strategy to make a farm more profitable, or it may be only an enterprise. Nursery production usually involves reduced levels of synthetic fertilizers and pesticides, use of IPM systems and focus on building the soil to promote plant health. The plant material available in most of our nurseries today unfortunately is out dated or substandard. Lack of quality planting material is the major hindrance for not realizing the full potential of floriculture in India. Plant material of various kinds (seedlings, budded plants, rooted cuttings, bulbs, tubers, corms, annual seed, etc.) is required for commercial flower production, pot plant production (and their rentals), for adding to home garden and for landscaping (corporate landscaping, bio-aesthetic planting, etc.).

Most of the nurseries still follow the same old practices of production, management and marketing of nursery materials. An attempt was made to focus some of the strategies for Hi-tech nurseries for future growth of nursery industry in India which include:-

- Production of quality planting materials i.e. through ***satellite nurseries***
- Development of infrastructure
- Container plant production and rentals
- Plug plant production
- Corporate landscaping
- Plant tissue culture
- Seed production

This segment may not require much research but requires adequate extension for transferring the already developed technologies. On the research side, the use of PGR's can further be exploited in propagation and seed production of ornamentals.

Value Addition

In floriculture trade value addition takes place at every step to create a novel product to cater demands of heterogeneous consumers. The trends in floriculture keep pace with trends in fashion industry and hence evolution of floriculture is much more rapid than other industries. The flower trade across the continents is valued at US$ 6 billion but in reality the trade is much more in magnitude as the value added products and their sale start after sale of fresh flowers at auction houses.

Extraction of essential oils: The major flower crops important for essential oil extraction are rose, jasmine, tuberose, vanilla etc. The major producers of essential oils are Brazil, China, USA, Egypt, India, Mexico, Guatemala and Indonesia. The demands for the essential oils progress at a rate of 7-9% per annum and offer unique scope for large scale cultivation of essential oil bearing crops and their processing using both conventional and ultra modern vacuum distillation/ solvent extraction/ liquid CO_2 methods. Essential oils and perfumery from natural sources are in great demand. In India, flower crops grown for essential oil production are limited and include mainly rose, jasmine, tuberose, etc. *Rosa damascene* is exclusively cultivated for extraction of essential oils, rose water, attar, gulkhand, etc. in certain pockets of Rajasthan and U.P.

Research should be focused on development of varieties with higher oil content and standardizing distillation methods for higher oil recovery. Further, identification of more crops and standardization of production technology need to be included in the research agenda. Promotion of this sector encourages ancillary industries like steam distillation and use of indigenous technical knowledge (ITK) for making value added products.

Rock Gardens and Water Garderns

Rock Gardens and Water Gardens

Versatile Gardens

Annual Gardens

Flowers in this garden include purple fountain grass, other ornamental grasses, red impatiens and yellow icicle pansies

Flowers in this garden include ornamental grasses and impatiens

Flowers in this garden include ferns, hostas, geraniums, lilies and daisies

Flowers in this garden include lavender, dusty miller, marigolds, and impatiens

Flowers in this garden include petunias and ageratum

Flowers in this garden include mass plantings of impatiens

Perennial Gardens

Flowers in this garden include ornamental grass, geraniums, hostas, hydrangea, allium and euonymus

Flowers in this garden include ornamental grasses and shrubs

Flowers in this garden include daisies, hostas and ornamental grasses

Flowers in this garden include ferns, salomon and seal

Border Gardens

Flowers in this garden include ivy gournd cover, ferns, and fuchsia in the container

Flowers in this garden include ageratum, celosia, and marigolds in the top border

Flowers in this garden include dusty miller, impatiens, marigolds and lavender

Flowers in this garden include and hostas

Container Gardens

Flowers in this garden include purple and yellow icicle pansies, Winter Garden Collection conifers and heather, and ornamental kale

Flowers in this garden include purple and yellow Icicle Pansies, Winter Garden Collection conifers and heather, and ornamental kale

Flowers in this garden include purple Icicle Pansies, and Winter Garden Collection conifers and heather

Flowers in this garden include purple and yellow Icicle Pansies, Winter Garden Collection conifers and heather, and ornamental kale

Flowers in this garden include geraniums and dracaena spike

Flowers in this garden include geraniums and sweet potato vine.

Flowers in this garden include ferns and elephant ears.

Flowers in this fall container garden include icicle pansies

Summary

Bio-aesthetic planning not only beautifies the environment but also makes it more livable and ecologically stable. It improves microclimate, mitigates pollution, creates habitat for many animal species, adds variety of form, colour and texture to the environment and brings about positive changes in human behavior. Bio-aesthetic plans should be drawn, as part of the master plans, to realize its full potential. Peoples' participation through Park Management Committees, Kitchen Garden Societies, Eco-clubs etc. is very essential at all the stages from planning to management, for making the country beautiful as a whole. The government must ensure rigid enforcement of environmental laws and incentives should be given for developing effective environmental management systems

References

Anonymous, (2001). National Climatic Data Center Website. http://www.ncdc.noaa.gov/img/climate/online/doe/Indiaanntemp.gif

Arora J.S., (2001). Introductory Ornamental Horticulture. Kalyani Publishers, Ludhiana, p 204.

Boden T.A., Kanciruk P. and Farrell P. (1990). Trends 90, A compendium of Data on Global Change, Carbon Dioxide Information Analysis Center, Oak Ridge, Tennessee, USA, p. 104

Dunne J.A. and Harte J. (2001). Greenhouse effect. In: *Encyclopedia of Biodiversity* (Ed., Lewin S.A.), Academic Press, New York, 3:277-293.

Kalidasa, Miller B., Miller, B.S.. The Plays of Kalidasa (1999). Published by Motilal Banarsidass Publication.

Mahawer L.N. and Mishra R.L. (1996). Trees for combating dust and noise pollution. In: Eco-friendly trees for urban beautification (Ed., Khosla P.K.), Indian Society of Tree Scientists, Solan, 285-290.

Randhawa M.S. (1964). The Cult of Trees and Tree-worship in Buddhist-Hindu Sculptures. All India Fine Arts & Crafts Society, New Delhi, p. 66.

Randhawa M.S. (1978). Beautifying Cities of India. Indian Council of Agricultural Research, New Delhi, p 15.

Ornamental Horticulture, pp 303-322

New India Publishing Agency, New Delhi - 110034, India

13

Wealth of Bougainvillea

S.S. Sindhu and B.K.Banerji

Introduction

Bougainvillea is a popular ornamental plant in the tropics and sub-tropics of the world. The plant is mainly grown for its attractive bracts in wide array of colours which splash colour to the surrounding where it grows. Wide adaptability to different agro-climatic conditions and easy multiplication has made it a popular ornamental plant of the world. The mass effect of the colourful bracts of Bougainvillea can alter the face of any given place in a magnificent way which could not be achieved by punctuating any other plant. The flowering of Bougainvillea is compared with the Azalea of temperate countries and very truly called as Glory of the Tropics.'

The popularity of Bougainvillea has now crossed from tropical to temperate countries and successfully being grown as pot plants in green houses. Selection of cultivars of *Bougainvillea spectabilis* and *B. peruviana* for temperate region is recommended for better result. Availability of a large number of cultivars under horticulturally important species, provide a unique scope to the Bougainvillea lovers for wide choice according to their need. Single bracted cultivars with large bracts in different colours are obvious choice for every landscape designer while double bracted cultivars produce mass effect of colour as specimen plant. Considering its ornamental and commercial importance in nursery trade, R & D work were taken up by different national research institutes (IARI, IIHR, NBRI, BARC), agricultural universities, Agri-Horticultural Societies (Kolkata and Chennai) and even by progressive nursery men. As a

[1]Division of Floriculture and Landscaping, IARI, New Delhi
[2]Floriculture Section, National Botanical Research Institute, Lucknow, U.P.

result, a large number of new cultivars have been developed in India. Considering the contribution made by India, The international registration authority for new cultivars lies with The Indian Agricultural Research Institute (IARI), New Delhi which is an international recognition conferred upon by International Society for Horticultural Science.

Recently, lot of development taking place for further popularization and exploitation of Bougainvillea around the world by way of development of new cultivars, standardization agro-technology suited to different zones. Countries like Thailand, Singapore, Malaysia and India have been excelling in this area. Establishment of American Bougainvillea Society is another landmark in Bougainvillea cultivation.

Origin and History

Bougainvillea is a native of South America and was first collected by Commerson, a French Botanist, at Rio de Janeiro, Brazil. From its place of origin, Bougainvillea first entered to Europe in 19th century and later introduced to Asia, Africa, America, Australia, New Zealand. At present, it is available in almost all major countries where horticultural development has taken place.

In India, *Bougainvillea spectabilis* was introduced in 1860 from Europe and its improvement was started later on in early 20th century with the introduction of few more cultivars. However, the popularity really started with the introduction of cv. Mrs. Buttfrom Royal Botanic Garden, Kew to The Royal Horticultural Society, Kolkata. Percy Lancaster, the renowned British horticulturist, has been credited for the development of first cultivar of Bougainvillea - Scarlet Queen'in 1920 and subsequently another excellent bi-coloured cultivar Mary Palmer'developed by him paved the way for Bougainvillea cultivation in India.

As the popularity grew, Bougainvilleas got lot of attention from horticulturist and garden lovers. Subsequently, R & D work was started in different Horticultural Societies particularly in Kolkata, Chennai followed by agricultural universities and research institutes like IARI, New Delhi, IIHR, Bangalore, NBRI, Lucknow. The main emphasis was collection of germplasm at one place and improvement by employing different methods. These initiatives gave birth many new cultivars and popularization of Bougainvillea.

Bougainvillea is an unusual natural shrub likes dry climate. Its native place is tropical reason of South America. Bougainvillea belongs to the family Nactanginanceae. It has a special quality to withstand, survive and flourish profusely in varying range of soils and climatic conditions. It has natural and genetic constitutional gift in it for adaptability with the leading quality of growth and development behaviour. In fact in distress and harness situation alongiwth hard pruning, it survives and flowers profusely which vary according to variety in plantations.

It is well known fact that rockey places and rockery or a construction made in a surface of ups and downs and tapering slopes form a suitable ground for planting it. Rockery is made with varying size of stones, rocks, soils, manure mixture etc. materials. Different types of plants are planted which includes foliage, xerophytic, very dwarf shrubs, edging etc. plants. Some trained in dwarf shape i.e. like bonsai height. Bougainvillea plants can be planted in dry sides or upper corners. Very dwarf varieties of bougainvillea, well trained in limited height may be planted with the condition that such rockey places and rockery should not have streams/water falls.

Classification

The genus *Bougainvillea* belongs to the family Nyctaginaceae having four horticulturally important species namely *B. spectabilis, B. glabra, B. peruviana and B. x buttiana* . These species have large and showy bracts. There are numerous cultivars under these species arisen as a result of bud sports, seedling variation, mutation, hybridization and polyploidy.

B. spectabilis –N ative of central and eastern Brazil; plant vigorous, leaves elliptic to ovate; floral tube swollen at the base; bracts develop all along the branch. Representative cultivars are –Lateritia,' Speciosa,' Splendens' etc.

B. glabra – Native of northern Brazil; leaves elliptic with short hairs, bracts elliptic in purple, mauve, rose and white; old bracts persistent. Representative cultivars are - Cypheri,' Dream,' Formosa,' Glabra,' Magnifica,' Sanderiana' etc.

B. peruviana – Native of Equador, Peru and Colombia; climbing; leaves broadly ovate; bracts crinkled, magenta-pink. Representative cultivars are –Arjuna,' Lady Hudson,' Equador Pink,' Mrs. H.C. Buck' etc.

B. x buttiana - Native of Comombia, Trinidad, a natural hybrid between *B. peruviana* and *B. glabra*. Climbing in nature; leaves broad, tip slender, hairy on both surfaces; bracts broadly ovate, crimson to mauve. Representative cultivars are –Mrs. Butt,' Cherry Blossom,' Los Banos Beauty,' Los Banos Variegata' etc.

Single bracted Cultivars: Blondie,' Dr. H.B.Singh,' Lady Mary Baring,' Splendens,' President,' Dream,' Poultoni Special,' Palekar,' Mrs. H.C. Buck,' Dr. R.R. Pal' etc.

Double bracted Cultivars: Archana,' Cherry Blossom,' Los Banos Variegata,'Mahara,'Mahara Variegata,'Marietta'and Roseville's' Delight'

Cultivars with variegated leaves : Archana,' Arjuna,' Hawaiian Beauty,' L.N. Birla,' Louise Wathen Variegata', Mahara Variegata,' Manohar Chandra Variegata,' Nirmal,' Scarlet Queen Variegata,' Thimma,' 'etc.

Cultivars with white bracts: Hawaiian White,' Jennifer Fernie,' Shubhra,' Shewta 'and Sova'etc.

Cultivars with yellow bracts: Enid Lancaster,' Golden Glow,' Lady Mary Baring, Suverna 'and Yellow Queen 'etc..

Cultivars with red bracts: Dr. R.R.Pal,' Flame,' Isabel Greensmith,' Mahara,' Mrs. Butt,' Palekar,' Scarlet Glory,' Scarlet Queen 'etc.'

Cultivars with purple bracts: Baron Hado,' Elizabeth,' Formosa,' Glabra,' Glabra Magnifica,' Refulgens,' Splendens 'etc.

Cultivars with mauve bracts : Dr. H.B. Singh,' Dream,' Enid Walker,' Lavender Glory,' Prolific,' Trinidad 'etc.

Cultivars with pink bracts: Alick Lancaster,' Cascade,' Los Banos Beauty,' Mahatma Gandhi,' Poultoni Special,' Princes Margaret Rose,' Spring Festival 'etc.

Cultivars with orange bracts: Lateritia,' Louise Wathen,' Mrs. Mc Clean,' Roseville's' Delight,' Tetra Mrs. Mc Clean,' Zakiriana 'etc.

Cultivars with double /multi-coloured bracts: Begum Sikandar,' Cherry Blossom,' Chitra,' Mary Palmer,' Mary Palmer Special,' Odissi,' Thimma,' Wajid Ali Shah 'etc.

Cultural Practices and Uses

1. **Site/location:** Since bougainvillea requires sunny and open area for the maximum time of days in different months of the year. In small places, rockeries, background area of it can be planted bougainvillea varieties where these varieties in such situations will add background beauty with colour and foliage by back portion of the surrounding area.
2. **Selection of Varieties:** All about rockery, rockey places are situated in different conditions. Surroundings are of rockery, climate, sun-direction etc. should be kept in mind while considering the requirement of the plants. Bougainvillea varieties, its plants' age, variegation of foliage, colour of flower/bracts etc. should be taken in to consideration. In colour effect annuals, some other suitable plants are planted, so the colour

arrangement in contrast and suitability should always be taken into consideration. Mary Palmer, Tomato, Rosevilles' Delight, Subhra, Thimma varieties may be taken according to site, situation.

3. **Preparation of Pits:** In the rockey and rockey background surrounding area, pits preparation should be done as per the general methods. Mixture of manure, soils etc. material should be filed in the pits. Here general, vigourous, desired coloured varieties should be planted in pits. After planting it should be irrigated and it is followed according to need and site, direction of sun and shade. In both sides of border lines from ground to upper portion of the rockery few a very short height plants of desired coloured bracts may be taken. Pits at desired places should be filled in with the mixture of less quantity of manure and soil. Mixture should contain very little quantity of F.Y.M. i.e. 1/4 parts and 3/4 parts normal sandy loam soil.

4. **Planting:** Plants of bougainvillea should be planted in already prepared pits. After planting, these should be irrigated. After that watering may be done as and when required because bougainvillea needs very less quantity of water in general cultivation. Its planted soil should be dry which induces flowering.

5. **Intercultural:** It needs very little in rocky places and rockery area. However, weeds, grasses, undesired plants, insect-pest affected plants or plant parts should be removed as and when required. This operation should be done in the rainy season and after rainy season.

Some of the promising varieties used in different purposes are given below:

Bush – Jayalakhshmi, Dr. R.R. Pal, Golden Glow, Mahara, Shubhra, Joe, Enid Lancaster, African Sunset. Wajid Ali Shah, Tetra Mrs. Mc Clean, Shweta, Nirmal.

Standard – Archana, Arjuna, Chitra, Mary Palmer Special, Asia, Begum Sikander, Gokul, Jayalakshmi, Orange King, Rose Queen.

Hedge – Dr. B.P. Pal, Parthasarthy, Alic Lancaster, Bhabha, Partha, Elizabeth, R.R. Pal, Dream, Los Banos, Beauty, Snow Lady.

Pot Plant – Dr. H.B. Singh, Enid Lancaster, Flame, Gopal, Begum Sikander, Golden Glow, Cherry Blossom, Shrubhra, Scarlet Queen.

Topiary – Krumbeigal, Lord Willingdon, Mahatma Gandhi, Jakiriana, Trinidad, Star, Cherry Blossom, Formosa, Boisde Rose, New Red, Happiness, La Jolla, Srinivasa, Treasure.

Bonsai – Sharma, Soundarya, Srinivasa, Rosevilles' Delight, Tomato, Thimma, Shubhra, Begum Sikander, Mrs. H.C. Buck, Blondie.

Climber – Surekha, Manohar Chandra Variegata, Hawaiian Beauty, Pallavi, Los Banos Variegata, President, Chitra, Jubilee, Louise Wathen, Lady Mary Baring, Easter Parada.

Umbrella Shape – Superlia, Thimma, Marry Palmer, Meera, Mrs. G.S. Randhawa, Mrs. H.C. Buck, Partha, Rao, Rosea Fuchsea, Sensation, Wajid Ali Shah, Archana, Dr. R.R. Pal, Meera.

Variegated Bougainvillea

Bougainvillea is an ornamental plant belonging to the family Nyctigenaceae. It has a short history of domestication. It is a native of winter dry tropical region of South America . The genus includes the showiest ornamental of tropics and subtropics and now cultivated in tropical area where they have become an important part of the tropical horticulture. Growers in South Florida, South California, Kenya , Mauritius , Trinidad , Australia , Ceylon and India , have selected various forms of bougainvillea. Bougainvillea was introduced in India from Europe in 1860.In our country it is flourishing well in all the metropolitan cities like Bangalore, Bombay, Calcutta, Chennai and Delhi. It is generally used as bush, bonsai, and cascade, climbers, hanging basket, espalier, pot plant and topiary work. Bougainvillea are in great demand by soft landscape architects, for development of home gardens, factory gardens, municipal gardens, Multinational companies indoor and outdoor landscape, slopes of river bank, sides of railway track, railway station, airport surroundings, five star hotels, holiday resorts, newly built townships, multistoried flats, historical monuments and temple surrounding and road side plantation and gazebos. This hardy and thorny plant needs low nutrition and water for its vegetative and floral growth. Its wide adaptability into different agro climatic conditions blended with broad spectrum of recurring blooming habit has made bougainvillea commercially important plant in the nursery trade. It thrives well in saline and alkaline soils. Bougainvillea is drought and pollutant resistant plant.

The most complete modern botanical monograph of the genus is by Heimerl, who recognize ten species only, three of which are of Horticultural importance viz. *B.spectabilis* willd, *B.glabra* choisy and *B.peruviana* Humb & Bonp as they have colourful bracts. Garden bougainvilleas have originated from these three elemental species. Bougainvillea generally blooms thrice a year. The first flush can be seen in first quarter of the year followed by second and third in last quarter. On the basis of the bract type bougainvillea are divided into two main groups i.e. Single bracted and Double or Multibracted.In single bracted bougainvilleas flowers are small but the bract round the flowers provides magnificently colourful effects which ranges from white through shades of pink flushed mauve to terracotta and orange and then to scarlet and crimson and ending mauve to terracotta and orange and then to scarlet crimson and ending mauve and magenta with dark margin magenta purple.

Bud sports and induced mutations have played an important role in origin of many new forms of variegated single and multibracted bougainvillea and many of the present day chlorophyll variegated bougainvillea cultivars are the result of it. Induction of chlorophyll-variegated mutants by gamma irradiation in single bracted bougainvillea has been reported (Abraham and Desai, 1977; Gupta and Shukla, 1974). Gamma rays has been successfully utilized to develop chlorophyll variegated mutants in multibracted bougainvillea (Banerji et.al.,1983, Banerji and Datta,1986, 1987a,b, Banerji et.al.,1987, Banerji and Datta,1988, Datta and Banerji,1994, 1997, Banerji,2007,2008). Chemical mutagen has been used successfully utilized to induced chlorophyll mutation in multibracted bougainvillea 'Los Banos Beauty' (Jayanti et al., 2006, 2007). These variegated bougainvillea looks very attractive even in the off seasons when plant is devoid of colourful bracts. These mutants are ideal for pot culture and bed. 'Pallavi' and 'Mahara Variegata' abnormal leaf mutant' is ideal for topiary culture. 'Los Banos Variegata' and 'Mahara variegata' is suitable for road side planting as it breaks the monotony of green and the mutant gives contrast attractive looks. These mutants are also suitable for rock gardening as their foliage adds beauty to the surrounding background.

These chlorophyll variegated bougainvillea are in great demand in soft landscape as it adds new dimension to the nature by exhibiting their multicoloured leaves which breaks monotony of the green surrounding especially during the off season when the plant is devoid of blooms. These bud sports and induced mutants with variegated leaves can be used for soft landscaping and topiary work. Morphological descriptions of variegated bougainvillea along with their origin are given below:

Archana: It is and evolved as a bud sport from multibracted bougainvillea cultivar 'Roseville's Delight.' Bract colour is apricot orange to venaceous orange. Leaves are variegated. Combination of cream yellow and green tissues is clearly visible on leaf surface. Dr.G.S.Srivastava (1975) has reported this sport from National Botanic Garden, Lucknow .

Arjuna: It is a gamma ray induced mutant of 'Partha.' Leaves are variegated. Central portion of the leaf is green and creamish tissues surround this portion. Bract colour is pinkish purple. Gupta and Shukla (1974) reported this mutant from National Botanic Garden, Lucknow .

Bhabha: It is a bud sport of Louis Wathen.' Leaves are variegated. Cream and green tissues spread all over the surface of leaf. It is reported from Lal Baugh Botanic Garden, Bangalore in the year 1960. Leaves are variegated. Cream and green tissues spread all over the surface of leaf.

Brilliant Variegata: It is a hybrid of bud sport B.buttiana cv. brilliant. Leaves are variegated. Centre of the leaf is deep green and a cream white tissue surrounds it. Young shoot is pinkish. Bract colour is like Brilliant. The cultivar is introduced by Lal Baugh Botanic Gardens, Bangalore in 1975 from Peter Green Smith Nursery, Kenya.

Carmentica: It is also known as Gladalan Red'and Philippines Red. It is a bud sport of bougainvillea x *buttiana*. It was originated at Philippines and patented in U.S.A. under the name Carmencita.' R.E. Holttum (1970) mentioned about it in book entitled Flowering wines of the world'edited by Edwin A. Neningar (pages233-245). References are also available in the bougainvillea growers' handbook written by Jan Ire dell (1990). Bracts are persistent and lavender mauve in colour.

Cinderella: This bougainvillea cultivar is a bud sport of single bracted bougainvillea cultivars *Mrs. Butt'* . Jan Ire dell (1990) mentioned about this cultivars in her book entitled the bougainvillea growers handbook. The colour of the bract is double red and its leaves are variegated. The combination of green and cream variegation gives very attractive look to these cultivars. Bracts are persistent.

Coconut Ice: It is variegated sport of multibracted bougainvillea cultivar '*Cherry Blossom*'. Jane Ire dell from Australia for the first time reported about it in 1986. Bract colour is white with pink.

Daya: It is a bud sport of Partha.' Leaves are variegated. Centre of the leaf is green and this tissue is irregular. Surrounding leaf margin is light yellow in colour. Bract colour is faint pink. V.K.Palakar & Palekar Com. from Bombay reports it in the year 1966.

Gangamma: It is a bud sport of Lady Mary Baring.' Leaves are variegated. Margin of the leaf is white and its center is green. Mr.Y.Muniverkatapa a nurseryman from Bangalore in the year 1974 reports it.

Gangaswamy: It is a bud sport of Lady Mary Baring.' Leaves are variegated. Centre of the leaf is green and white tissues in the margin portion surround it. Bract colour is like Lady Mary Baring. Flowering is sparse. It is reported from Lal Baugh Botanic Gardens, Bangalore in the year 1974.

Glabra variegata: It is a bud sport of B.glabra.' Plant growth is bushy and compact. Leaves are variegated and its size is small. Internodes are short. Thorns are thin and small. Leaf margin is white and cream with two shades of green. It looks very attractive. Juvanile bracts are whitish purple which turns purple in mature stage which looks alike B.glabra 'Sanderiana.' The cultivar has been introduced to India by Lal Baugh Botanic Gardens, Bangalore from Kenya , (Africa) in the year 1961.

Godrej Cherry Blossom: This new cultivar of multibracted bougainvillea is evolved as a bud sport of *B. spectabilis* var. 'Cherry Blossom.' Plant growth habit is intermediate. Leaves are variegated and its base is broadly ovate. Bracts are triangular and persistent and their colour is cream yellow.

Harissi: It is reported from Sri Lanka by Rodney Jonklass. Foliage is small. Leaves are variegated. Patches of green and white tissues make the variegated leaves attractive and beautiful. Bract colour is purple.

Jayalaxmi Variegata: It is a bud sport of 'Jayalaxmi.' Leaves are variegated. Young leaves are coppery red and mature leaves are creamish yellow. Bract colour is brick red. Abraham and Desai from BARC, Trombay Bombay reports it in the year 1977.

Jawahar Lal Nehru: It is a bud sport of bougainvillea cv. 'Lal Baugh.' Leaves are variegated. Young leaves are coppery and mature leaves are green with irregular patches in the centre and prominent minosa-Yellow variegation along the margin. Bract colour is scarlet rose. It is reported from Indian Institute of Horticulture Science, Bangalore in the year 1975.

Kuvempu: It is a bud sport of 'Rao.' Young shoots and leaves are reddish. Mature leaves are variegated. Colour of variegated leaf is creamy white and green. Bract colour is red like Rao. Sometimes it is reverted back into little extant. It is reported from Lal baugh Botanic Garden , Bangalore in the year 1960.

Lady Hudson of Ceylon Variegata: It is a bud sport of 'Lady Hudson of Ceylon.' Leaves are variegated. Pale yellow tissues spread all over the surface of leaf. Abraham and Desai reported it from Bhabha Atomic Research Center, Trombay, Bombay in the year 1979.

Laxmi Narayan: It is a bud sport Of 'B. spectabilis.' Leaves are variegated. Variegation in leaves comprises of combination of green and cream tissue. It is prominent in older leaves. Bract colour is bright rosy purple. It is reported from Lal Baugh Botanic Garden Bangalore in the year 1961.

Los Banos Variegata: It is a gamma ray induced mutant of multibracted bougainvillea cultivars Los Banos Beauty.' Banerji and Datta reported the mutant from floriculture Section, NBRI, Lucknow in the year 1987. Leaves are variegated and has excellent pattern of colour combination. It looks very attractive due to presence of green irregular centre and white margin which is more than 1cm. Bract colour remains the same i.e. mallow purple. In extreme cases few albino shoots are emerged in old plant with pink stem. Very old plant produces few branches with Rhodamine purple bracts, which are Mahara type. These branches have green leaves. The cultivar has been registered with International Bougainvillea Registration Society, New Delhi.

Los Banos Variegata silver margin: It is a gamma ray induced mutant of multibracted bougainvillea cultivars Los Banos Beauty.' Banerji reported the mutant in 2002 from floriculture section, NBRI, Lucknow . Leaves are variegated and have combination of green and silver margin. It looks very attractive due to presence of green regular center and silver white margin with of which is 1-2mm. variegated leaves are slightly incurved and in extreme cases leaves are spoon shaped. Mutant is not clearly visible from a long distance as its variegation is not very contrast like Los Banos Variegata. Bract colour is mallow purple just like Los Banos Beauty and Los Banos Variegata.

Los Banos Variegata Jayanti:' This is a chemical mutagen (Ethyl Methane Sulphonate {EMS} 0.02% aqueous solution) induced chlorophyll variegated mutant of bougainvillea cv. Los Banos Beauty.' The mutant differs from original variety only in foliage colour without in any change in bract colour. Mutant showed two types of variegated leaves. One type was mosaic which, has four visible colours i.e. Olive green (Yellow Green Group 146B, Fan-3), Pale (Yellow Green Group 146D, Fan-3), Deep Green (Yellow Green Group 146 A , Fan-3) and Light Yellow (Yellow Green Group 154D, Fan-3), mixed with different shades of green. The second type of leaf have green centre with creamish white margin. Foliage of this type has three major colour combinations i.e. Light Yellow (Yellow-Orange Group 19C, Fan-1)'Green (Green Group 147B, Fan-3) and Light Green (Yellow-Green Group 148C , Fan-3) (Jayanti and Datta, 2006, Jayanti *et.al.* 2007 and Banerji, 2008). Bract colour is mallow purple just like Los Banos Beauty and Los Banos Variegata and Los Banos Variegata silver margin.

L.N.Birla: It is a bud sport of Maharaja of Mysore.' Leaves are variegated. Variegation in leaves comprises of combination of green and yellow tissue. Bract colour is just like Maharaja of Mysore. It is reported from Agri Horticulture Society of India Calcutta.

Louise Wathen Variegata: It is a bud sport ofLouise Wathen.'. Leaves are variegated. Variegation in leaves comprises of combination of green and cream yellow tissue. Bracts are like Louise Wathen. It is reported from Royal Horticulture Society of India Calcutta in 1935-40.

Mahatma Gandhi Variegated: It is a bud sport ofMahatma Gandhi.'. Leaves are variegated. Variegation in leaves comprises of combination of green and white tissue. Bract colour is like Mahatma Gandhi. It is reported from Pratap Nursery & seed store Dehradun in the year 1969.

Manohar Chandra Variegata: It is a bud sport ofManohar Chandra.'Young leaves are coppery and old leaves have green center and yellow margin. S.C.Sharma and Bisaria report it in the year 1985 from National Botanical Research Institute, Lucknow .

Mataji Agnihotri: It is a bud sport ofMary Palmer.' Leaves are variegated. Variegation in leaves comprises of combination of green and yellow tissue. Green tissues are present in the center of leaf and it surrounded by cream tissues. M/S K.S.Gopalaswamy Iengar & sons report it from Bangalore in the year 1982.

Mrs. McClean Nirmal: It is a bud sport of Mrs. McClean.' Leaves are variegated. Young leaves are green to coppery white. Variegation in leaves comprises of combination of green and white tissues in patch form. Mr.G.S.Srivastava reports it in 1982.

Mahara Variegata: It is a gamma ray induced mutant of multibracted bougainvillea cultivars '*Mahara*.' Datta and Banerji reported the mutants in the year 1994 from floriculture section, NBRI, Lucknow . Leaves are variegated and their pattern is not fixed. Variegation varies from leaf to leaf with combination of creamish yellow and various shades of green overlapped patches. Young leaves are coppery while the old leaves have dull yellow and faded green colour. Bract colour is Rhodamine purple which remains unaltered. It has been registered with International Bougainvillea Registration Society, New Delhi.

Mahara Variegata abnormal leaves: It is a gamma ray induced mutant of multibracted bougainvillea cultivarMahara.'Banerji reported this mutant from floriculture Section, NBRI Lucknow in the year 2002.Leaves are variegated and are of peculiar type. Margin is unevenly undulated and leaf lamina is asymmetrical in shape. Variegation in various shades of green and light green pattern haphazardly spread all over the surface of leaves giving it a curio look. Old leaf texture is leathery. Bract colour remains unaltered but its shapes and size reduced. Most of the bracts are asymmetrical with acute tip. This is the first mutant of bougainvillea where leaf variegation is associated with leaf and bract shape and size changed.

Marietta: It is a bud sport of Bougainvillea buttiana var *Mahara'* . It was raised by Mrs. Marietta H.Raymundo Malabon, Rizal of Philippines and reported by Dr.J.V.Pancho in 1967, college of Agriculture, Laguna, Philippines. The bract colour is Rhoda mine purple, which remains unaltered and just, like Mahara but foliage is variegated. The variegation is a combination of white and green. Margins of young growing leaves are pinkish. Central green and marginal white colour combination is very attractive and charming. Bracts are persistent.

Mrs. Butt Variegata: It is a bud sport of Mrs. Butt' Leaves are variegated. On leaf surface irregular patches of yellow and green tissues are clearly visible. Rodney Jonklaas Jaela from Sri Lanka reported this sport.

Mrs.Eusnia Rajasinghe: The cultivars evolved as a bud sport of Mahara.' Don Justin A. Rajasinghe reported it from Srilanka. It is variegated cultivars with central portion of the leaf is green and spreading into the margins with ash colour. Leaf margin colour is dominated by yellow colour just like Marietta. Young and old leaf of this cultivar is same while in case of Marietta the younger leaves are bronze in colour. Bract colour is same like Marietta.

Muni Venkatappa: It is a bud sport of Mary Palmer.' Variegation in leaves comprises of green tissues in the center and white tissues are present in margin. It is almost like a Thimma but variegation pattern is entirely different. Y.Munivenkatappa, a nurseryman from Bangalore reported it.

Pallavi: It is a gamma ray induced variegated mutant of '*Roseville's' Delight'* evolved by Banerji et.al. (1987) at Floriculture section of National Botanical Research Institute, Lucknow . The Variegated mutant has been released in the name of Pallavi. The mutant has been registered with International Bougainvillea Registration Society; New Delhi. Variegated leaves are having light green, pale green combination in various shades. Variegation is very prominent in young shoots and it masks with age and in older shoots it is not very clear. No Change in bract colour was noticed however in older plants shoot having Mahara types of bract bunches developed which are very common as it is commonly found in the old Roseville's' Delight growing plants. Bracts are persistent. Variety is ideal for pot culture and topiary work.

Parthasarthy: It is a bud sport of Partha.' Leaves are variegated. Leaves have light green colour in the center with cream white surrounding margin. Some of the leaves exhibit entirely cream colour. Bract colour is like Partha. It is reported from Dr.S.C. Sharma from National Botanical Research Institute, Lucknow in the year 1974.

Phoenix: It is a bud sport of Mrs. Butt.' Plant growth is vigorous. Stem have straight thorns and their size is small. Leaves are variegated. Centre of the leaf is green and it is surrounded by golden and cream tissues. Young leaves have golden tip. Bract colour is dark red. It is reported by Jan Iredell from Australia in 1986.

Poultoni Variegata: It is induced mutant of Poultoni.' Leaves are variegated. Leaves have light yellow colour and green patches spread all over the leaf surface. Some of the leaves have no green tissues and clearly visible among the variegated leaves due to their prominent light yellow colour. Abraham and Desai in the year 1986 reported the mutant from Bhabha Atomic Research Center, Trombay, Bombay.

Preeti: It is a bud sport of Partha.' Leaves are variegated. Leaves have dark green colour in the center and it margin have light yellow tissues. Bract colour is dark pink. V.K.Palakar & Co., Bombay in the year 1970, reported this sport.

Rao: It is a bud sport of Mrs. Butt.' Leaves are variegated. Leaves have green center and it margin is white. Young shoots and leaves are pink. Bract colour is red. It is reported from Lal Baugh Botanic Garden Bangalore in the year 1954. (CFTRI, Mysore).

Refulgens variegata: It is a bud sport of bougainvillea spectabilis cv. Refulgen.' Leaves are variegated. Variegation is in the form of streak and its colour is white. Gopala Swamy Iyengar sons, from Bangalore reported this sport.

Scarlet Queen Variegata: It is a bud sport of Scarlet Queen.' Leaves are variegated. Leaves have a cream and green tissues combination, which spread all over the surface of leaf and provides it beauty. Bract colour is deep carmine scarlet and it is like a Scarlet Queen. S.Percy Lancaster reports it from Calcutta in 1935-40.

Soundarya: It is a bud sport of Mrs.H.C.Buck.' Leaves are variegated. M/S Soundarya Nursery, Madras reported this sport.

Sharma: It is a bud sport of Mrs. Fraser.' Leaves are variegated. Variegation pattern is diffused type. It is a combination of green and cream tissue along with red patches of various sizes, which spread all over the leaf surface. Bract colour is terracotta. The cultivar is named after Shri M.D.Sharma, floriculturist from Lal Baugh Botanic Gardens, Bangalore in the year 1961.

Silver Top: It is an induced mutant of Versicolour.' Leaves are variegated. Leaves are crinkled and boat shaped. Old leaves have narrow white border. Adaxial surface of the leaf is silver Mosaic. Bracts are narrow. Abraham and

Desai reported this mutant from Bhabha Atomic Research Center, Trombay, Bombay in the year 1986.

Singhe: It is a bud sport of Mahara off White. Leaves are variegated. Green tissues are present in the center of leaf and ash colour tissues surround it. Foliage resembles with Marietta. Young and old leaves are of same colour. Don Justin Anaeletus Raja Singhe reported this sport.

Spectabilis Variegata: It is a bud sport of B.spectabilis. Leaves are variegated. Cream and white tissues spread all over the surface of leaf and its pattern quite resembles with cv. Rao.' Bract colour is fuschia purple. S.Percy Lancaster reported this sport from National Botanic Garden, Lucknow in the year1986.

Surekha: It is a bud sport of Scarlet Queen Variegata.' Leaves are variegated. Pale yellow patches of different shape and size spread all over the surface of leaf. Bract colour is carmine to fuschia purple. Bract margin is crinkled. S.C.Sharma reported this sport from NBRI, Lucknow in the year1981.

Suvarna: It is a bud sport of Lady Hudson of Ceylon Suvarna.'

Thimma: It is a bud sport of Mary Palmer.' Leaves are variegated. Central portion of the leaf is yellow and it is surrounded by green tissues upto margin. Young stem colour is yellow. Bract colour is like Mary Palmer. It has been reported from Lal Baugh Botanic Garden Bangalore in the year 1960.

Vellayani: It is a bud sport of Mrs.Butt.' Leaves are variegated. The colour of the leaf is prominent green and cream. Bract colour is crimson red. It has been reported from Lal Baugh Botanic Garden Bangalore in the year 1950.

Villayani Variegata: It is a seedling selection from B.buttiana. Leaves are variegated. Irregular and diffused tissue of green and cream colour is distributed on leaf surface. Bract colour is like 'Asia' and its colour is magenta crimson. Shri Vittal Superintendent, Madras Municipal Corporation Garden reported it.

Vishakha: It is a bud sport of Mrs. H.C. Buck.' Leaves are variegated. Colour of the leaf is dull green and its margin is white. Bract colour is spinal red. It is reported from Indian Agricultural Research Institute, New Delhi in the Year 1976.

New Cultivars (2015)

Dr. P.V. Sane

A new variety developed by NBRI, Lucknow 'Dr. P.V. Sane"is bud spot of Dr. R.R. Pal, Possess erect plant habit, straight thorn & leaves are Variegated - Green with yellowish green speckled (Fan 3, Green group 137-C & Fan-3 Yellow-green group 144-B) with profuse flowering, ovate type of bracts and its is good for pot culture & standard.

Leah Nagpal

Leah Nagpal: Originated as seedling of Verna Nagpal, Plant habit erect thorns are small, colour of the leaves are light green with yellow with a glabours texture, size is small and the shape is almost elliptic (widest in middle), flowering is profuse, bracts are small. This variety is suitable for Shrub and Pot culture.

Geoffrey Nagpal: Originated as bud sport of Manila Hybred, Plant habit is erect, leaves are medium green, with glabours texture, size are small, shape is broad, flowering is profuse, bracts are large. The central star is not prominent and remains at unopened flowering stage. This is good for shrub as well as pot culture.

Other Suitable place for Bougainvillea

Planting along the road

Roads are the important sites of the urban areas which contribute significantly in mitigating the pollution by planting trees on both sides. Unfortunately, in most of the old Indian cities and towns, there is hardly any provision of sufficient space for the same. However, it is necessary to study the type of road, overhead electrical cables, space available on both sides, central verge, traffic triangles, round-about and other open space available before taking up any plantation. It has been observed that trees and shrubs which are drought and frost resistant generally tolerate pollutants. Selection of trees is another important task. Before selecting any plant species, it is necessary to consider following characters: agro-climatic conditions, evergreen or deciduous, height and spread, canopy architecture, growth rate and habit (straight or branched trunk), aesthetic effect (foliage, conspicuous and attractive flower colour), pollution tolerance and dust scavenging capacity.

Central Verge

Central verge of the two way roads in the cities and towns are often found neglected and devoid of any planting. It is recommended that this area should be well utilized by planting dwarf trees and shrubs. This will not only serve aesthetic purpose but also functional being physical barrier for the glare of head lights of the vehicles which is essential for effective and safe operation of the roads during dark hours. Planting may be done either in single or double row depending upon the space available. Since these plants are more close to the automobile exhaust, their capacity for pollution tolerance should be considered before selection. Following plant species have been reported as pollution tolerant and recommended for plantation: *Bougainvillea Cvs.* 'Chitra,' H.C. Buck,'Lady Mary Baring,'Mary Palmer Special,'Partha,'Shubhra are most suitable.

Traffic Island

Traffic islands at the main road intersections vary in shape and size from square, triangle to round. Whatever may be the shape, these islands should be properly planted with the dwarf trees, shrubs and ground covers recommended for planting along the road and central verge which will contribute effectively in mitigating the air pollution. The ideal are Bougainvillea Cvs. Dr. H.B. Singh and Thimma, Subra and RR Pal.

Area under the flyovers

The evergreen house plants are very ideal for planting under the shade of the flyovers provided there is arrangement for daily watering of these plants.

Greenbelt

Greenbelt is defined as the mass plantation of pollution tolerant trees and shrubs in an area for the purpose of minimizing air pollution by filtering, intercepting and absorbing pollutants in an effective manner for improvement of the environment. Every town and city must have certain area earmarked for the development of greenbelts. The effectiveness of the greenbelt depends on the several factors viz. climatic conditions, design, selection of plant species and it

Use of Bougainvillea in Plant disease control

Leaves extract of *B. spectabilis* inhibit tomato spotted wilt tospovirus on Capsicum annum in laboratory test (Balasarawasthi et.al., 1998). *B. spectabilis* leaves extract were highly effective in reducing Okra yellow vein mosaic virus infection of Okra (Pun et.al., 1998). Antivirus protein was already characterized by Balasarawasthi et.al., 1998.

Use of Bougainvillea in antimicrobial activities

Solvent extract of leaves showed maximum inhibitory effect of tested bacteria. The photochemical and antimicrobial activities of various solvent extracts of *B.spectabilis* leaves were tested against Gram positive and Gram negative bacterial strains by observing the zone of inhibition. The bacteria used in the study were *Staphylococcus aures, Bacillus subtilis, Streptococcus faecalis, Micrococcus lutes, Escherichia coli, Pseudomonas aeruginosa, Salmonella typhii, Klebsiella pneumoniae, Proteus vulgaris, Serratia marcescens, Shigella flexneri* and *Vibrio cholerae.* It was observed that ethanolic, methanolic, chloroform and ethyl acetate foliage extracts showed maximum inhibitory effect on all tested bacteria except Vibrio cholerae compared to other solvent extracts. The alcoholic extracts were more active against all Gram positive and Gram negative bacteria. The antibacterial action of various extracts of *B.spectabilis* leaves may indicate their high potential as antibacterial herbal remedies.

Use of Bougainvillea in Antidiabetic medicines

Diabetes mellitus is a metabolic syndrome characterized by an increase in the blood glucose level. Treatment of diabetes is complicated due to multifactorial nature of the disease. *Bougainvillea spectabilis* is reported to have medicinal values including antidiabetic properties. Aqueous and methanolic leaf extract of *B. spectabilis* showed good tolerance and significantly reduced intestinal glucosidase activity with regeneration of insulin-producing cells and increase in plasma insulin. *B.spectabilis* leaves contains D-pinitol (3-O-methylchiroinositol) and is claim to exert insulin-like effect (Narayanan et.al., 1987). It is reported to have medicinal values including antidiabitic properties. *B.spectabilis* chloroform, methanolic and aqueous extracts were investigated for biochemical parameters important for controlling diabetes.

Traditional Use

Traditional practitioner in Mandsaur uses the leaves of bougainvillea for a variety of disorders like diarrhea and reduces stomach acidity. It is also used in treatment of cough and sore throat. A decoction of dried stem, 10g in 4 glasses

Bougainvellea: A Versatile Plant for Landscaping

of water is recommended dose for treatment of Hepatitis. For treatment of blood vessel disorder and leucorrhea a decoction of dried flowers 10g in 4 glasses of water is recommended.

References

Anonymous. 1966 R.H.S. Colour Chart, The Royal Horticultural Society London in association with the Flower Council of Holland.

Banerji, B.K., Nath, P. and Gupta, M.N. 1983. Gamma ray induced somatic mutation in Bougainvillea cv. 'Roseville's Delight': Induction of chlorophyll mutation. In: *National Symposium on Recent Development in Nuclear and Allied Techniques and their application in Agriculture, Biology & Animal Science'*, May 5- 7, G .B. Pant Univ. of Agric. & Tech., Pantnagar, pp. 41.

Nath, P., Banerji, B.K. and Gupta, M.N. 1983. Spontaneous and induced mutations in Bougainvillea. News Letter, Bougainvillea Festival, Bougainvillea Society. Pp.1-6.

Banerji, B.K. and Datta, S.K. 1986. Improvement of ornamental plants by induced mutation: Bougainvillea. *Floriculture*, 3: 4-9.

Banerji, B.K. and Datta, S.K. 1987. Gamma ray induced chlorophyll variegated mutants in Bougainvillea cv. 'Los Banos Beauty.' *Journal of Nuclear Agriculture & Biology,* 16(1): 48-50.

Banerji, B.K. and Datta, S.K. 1987. Gamma ray induced genetic improvement of double bracted Bougainvillea. Paper presented at 'All India Seminar on Advances in Botanical Research in India during the last ten years.' Nov. 1-3, 1987, Dept. of Botany, Dungar College , Bikaner .

Banerji, B.K., Nath, P. and Datta, S.K. 1987. Mutation breeding in double bracted Bougainvillea cv. 'Roseville Delight' . *Journal of Nuclear Agriculture & Biology,* 16(1): 45-47.

Banerji, B.K. and Datta, S.K. 1988. A review on variegated Bougainvillea. *Floriculture,* 8: 1-6.

Datta, S.K. and Banerji, B.K. 1994. Improvement of double bracted bougainvillea through induced mutation. National Symposium on Frontiers in Plant Science Research, N.S.A. Khan Center of P.G. Studies and Research A.U. College, Mallepally, Hyderabad-1, Feb. 13-14.

Datta, S.K. and Banerji, B.K. 1994. 'Mahara Variegata' A new mutant of Bougainvillea. *Journal of Nuclear Agriculture and Biology,* 23(2): 114-116.

Datta, S.K., Dwivedi, A.K. and Banerji, B.K. 1995. Investigations on gamma ray induced chlorophyll variegated mutants. *Journal of Nuclear Agriculture and Biology*, 24 (4): 237-247.

Banerji, B.K. and Datta, S.K. 1996. Bougainvillea mein utparivartan prajanan. *Vigyan Vani*, 2: 51-53.

Datta, S.K. and Banerji, B.K. 1997. Improvement of double bracted Bougainvillea through gamma ray induced mutation. *Frontiers in Plant Science*, 64: 395-400 (Editor Dr. Irfan A. Khan)

Jayanti, R. and Datta, S.K. 2006. Improvement of Bougainvillea varieties using Chemical Mutagens. In: '*National Conference on Bougainvillea*,' April12-13, 2006, NBRI Lucknow, Pp17.

Jayanti, R., Datta, S.K. and Banerji, B.K. 2007. Los Banos Variegata-Jayanti,' Applied for Registration in International Bougainvillea Registration Authority, IARI, New Delhi , May 25, 2007.

Banerji B.K.2007. Multibracted Bougainvillea. In *Hi-Tech Floriculture*. (Eds.) Satyanarayan Reddy, B.S., Janakiram, T., Balaji, S.Kulkarni and Misra, R.L. Indian Society of Ornamental Horticulture, IARI, New Delhi. pp: 190-192.

Ornamental Horticulture, pp 323-333
© Editor, S.S. Sindhu, 2016
New India Publishing Agency, New Delhi - 110034, India

14

House Plants

S.S. Sindhu[1] and Sapna Panwar[2]

House plants play an integral role in providing decorative value and aesthetic sense to the place. They are important component of interiorscaping and provides softscape. House plants are also considered as natural air filters as they provide freshness and many house plants are reported to absorb many house pollutants. The demand of house plants are generating many employment opportunities like plant rental services, interiorscaping etc. The house plants are preferred in homes, offices, hospitals, multiplexes etc.

Placement

These attractive house plants rightly deserve their place as interior decorations. "Shade loving" is an adjective often used to describe house plants, but this is a misnomer, Plants may be shade tolerant but they do not actually seek poor light. Diffused light through the glasses or windows' can be effective you can supplement the sun's rays with artificial light in the form of a "warm white" fluorescent tube. Then may be shifting them after frequent intervals in the open for a short duration in morning hours to supplement the sunlight during winter months and leaving them out overnight and bringing them after exposing them to early morning sunlight during hot months exposure for a desired time frame in the monsoon are some of the effective methods to make your house plants feel happier.

[1&2] Division of Floriculture and Landscaping, IARI, New Delhi

Growing media

The growing media must be of best quality. The media should be porous so that it gives good root aeration. At the same time, it should be well drained and should have good retention of both water and nutrients. The commonly used media should be locally available like soils but nowadays artificial media containing inert material are also being used for growing house plants. The growing media should possess the following properties:

- It should be easily available
- It should be cheap
- It should provide good aeration
- It should be inert in nature
- It should be well drained
- It should be free from diseases and pests
- It should retain appropriate moisture content
- It should posses good physical and chemical properties

Most common components of growing media are

(i) Vermiculite

Vermiculite is a light weight, sterile, micaceous product having good amount of K and Mg. The pH is usually 7.0 to 7.5. The main benefit of using is its tendency to compress and loss drainage enhancing quality with time.

(ii) Perlite

Perlite is sterile, light weight, porous and light rock material of volcanic origin. It is white in colour. The pH usually ranges from 7.0 to 7.5. It does not hold water and nutrients. It is essentially high expanded aluminium silicate rock and improves aeration.

(iii) Cocopeat

It is prepared from the coconut husk. It has very high porosity and very good drainage quality. It is free from pathogen and provide good aeration.

(iv) Sand

Sand should be sufficiently gritty with a good particle size. The main disadvantage of sand is that it is very heavy and evaporates moisture too quickly. It is mainly used to provide aeration.

(v) Soil

Soil is commonly used growing media. It should be rich in nutrients. Soils have generally higher water retention and cation exchange capacity. The major disadvantage is that it is too heavy and contain various harmful microorganism that causes diseases. So, soil sterilization is needed prior to planting

Media Preparation

A good medium for house plants should contain most commonly locally available media components. Recommendations that are generally followed are garden soil-2 parts, organic matter-1 part, sand-1 part and leaf mould-1 part. The ratio of media components can be changed depending upon plant type and environmental conditions. Nowadays most commonly used media is 3:1:1 of coco peat, vermiculite and perlite. Some of the media that can be prepared especially suitable for particular crop.

House plant mix-1 (1:1:1): 1 part peat moss, 1 part vermiculite and 1 part perlite. This media is suitable for growing succulents that don't require extreme dryness.

House plant mix-2 (2:1:1): 2 part peat moss, 1 part vermiculite and 1 part perlite. This is good for non-woody foliage plants.

House plant mix-3 (3:2:1): 3 part sphagnum moss, 2 part vermiculite and 1 part perlite. This is best for shrubbery, woody plants and most vines.

Containers

Clay or terracotta pots as are commonly used and are best to grow house plants. Fiberglass pots available these days are other successful kind that can be used for growing house plants indoors. Earthenware pots are most commonly containers used in form of dishes, pots, shallow bowls and troughs. A variety of potting containers available in varied designs and colours. Metal receptacles of brass and copper are also used to enhance the aesthetic look.

Potting

Potting mixture suitable for these plants should consist of two parts of garden soil, one part of well decomposed organic manure which may be vermi compost as well and one fourth of sand. The basic mixture should be thoroughly mixed. Additional requirement for pots of 25 to 30 cm size is about 150 gms of coco peat or leaf mould powder 100 gms of Neem cake powder 100 gms of insecticides may be anti termite powder or malathion or any other branded insecticide to prevent soil borne infections. While filling the pot, cover the drainage hole in the base with some crocks (broken pieces of earthen pot). Take out the plant along with the root ball of earth and put in the soil mixture, making a hole to accommodate the plant and press properly. Fill the edges with growing mix and tap the pot to shake it down well, and firm it with your fingers. Finally water the plant carefully. Keep the newly potted plant in shade and semi shade for about a week and if it shows any signs of wilting, spray its leaves.

Propagation

The method of propagation vary from plant to plant. The house plants can be propagated by sexual or asexual method of propagation. The following are the examples of the plants propagated by various methods

Seeds: Asparagus, Cacti, Azalea, *Impatiens, Kalanchoe, Calceolaria, Cyclamen, Gloxinia, Coleus, Schefflera, Araucaria, Begonia*, Aralia, African violet, Palm, *Philodendron, Cineraria*, etc.

Stem cuttings: They are futher divided into subcategories depending upon texture of plant tissues

i) **Herbaceous cuttings**: *Portulaca, Kulfa, Pothos, Coleus.*

ii) **Soft wood cuttings**: *Euphorbia splendens, Pentas lanceolata, Pedilanthus, Eranthemum, Tradescantia, Oilea, Alternanthra.*

iii) **Semi hardwood cuttings**: *Acalypha, Crotons, Hamelia patens, Jatropha, Dieffenbachia, Crinum asiaticum, Dracaena.*

iv) **Hardwood cuttings**: *Ficus lyrata, Ficus elastica, Dracaena, Bougainvillea, Hibiscus, Duranta.*

Through leaves

i) **Leaf cuttings**	*Leaf bud Kalanchoe, African violet, Rex begonia, Begonia, Peperomia. Sansevieria, Bryophyllum* etc.
ii) **Leaf petiole**	*Begonia, Peperomia*, Scindapsus, Gloxinia and African violet,
iii) **Leaf lamina**	*Begonia, Peperomia,*
iv) **Leaf Bud**	*Kalanchoe, Poinsettia, Hydrangea* etc.

Cane or nodal cuttings: Dieffenbachia, Aglaonema, Monstera, Dracaena, etc.

Air layering: Rubber plant, Hedera helix, crotons

Watering

More house plants are killed by kindness then by neglect. And the most common sort of kindness is to give them too much to drink. The amount of water a house plant needs depends on the conditions you keep it in and on the weather. The role of plant roots is to forage for soil water and the nutrients that are dissolved in it. But the growing roots also need air. As the soil becomes wetter and wetter, the water drives air from the space between soil particles and the soil eventually becomes waterlogged. When this happens, roots will first stop growing, then rot.

Ways of watering: There are generally two ways of watering

1. Top watering
2. Bottom watering

Top watering

In this method, the pot is watered thoroughly with a hose from top. Here the excess water flows out of the container through the drainage hole. The main advantage of this method is that salts that are trapped in the soil are leached out with the excess water.

Bottom watering

In this method soil absorbs water from below surface of the pot. The main advantage is that medium absorbs appropriate quantity of water thus reduces wastage of water. Generally followed by professionals. The main drawback is that salt build up is inevitable in this method. The examples of bottom watering are through capillary mats, Wick watering and Pots with attached saucers.

Feeding

A regular feeding at least once in a month with a handful of Neem cake powder or vermicompost or coco peat in rotation provides regular diet to the plants. Feeding a dosage of about 250 gms. Mixture of all the three organic manure with a tablespoon ful of Bone-meal added to it that during spring and monsoon keeps the house plants in their full health.

Diseases

The most common fungus which attack the pot plants are *Alternaria, Cercospora, Collectotrichum, Fusarium, Phytophthora, Rhizoctonia Sclerotinia* etc. The fungus attack both foliar and underground parts. Various fungicides are available in market to control the fungus like Dithane M-45. The most common bacteria that attack house plants are *Erwinia, Xanthomonas and Pseudomonas*. Generally Streptomycin 100-200 ppm is found suitable bactericide.

Insect pests

Aphid: House plants susceptible to aphid infection are *Dieffenbachia, Aphelandra, Brassaia*, etc.

Control: Water spray can be effective. Spray of Systemic fungicide like Monocrotophos will be useful in case of severe infestation

Mealy Bugs

Mealy bugs infest a wide range of house plants like *Ficus, Dieffenbachia Asparagus, Aphelandra, Crassula, Crotons, Dracaena*, etc.

Control: Washing the plants with insecticidal soap can be effective. Spraying any systemic fungicide like monocrotophos in case of severe infestation can be effective.

Scales

The foliage plants affected are *Adiantum, Aphelandna, Ardisia, Asplenium, Nephrolepis* and *Zamia.*

Control : Washing with insecticidal soap can be effective or Horticultural oil can also be used for control

Thrips

The common house plants affected by thrips are *Dieffenbachia, Aphelandra, Syngonium., Ardisia, Brassaia, Philodendrons* etc.

Control: Frequent misting can be effective. Systemic insecticide like Rogor can be sprayed in case of severe infestation

Araucaria spp.

Codiaeum variegatum

Coleus spp.

Dieffenbachia spp.

Dracaena spp.

Dracaena reflexa

Ficus spp.

Iresene spp.

House plant at the corner of house

Scindapsus aureus

Syngonium spp.

Mites

Major foliage plant hosts are Brassaia, Calathea, Chamaedorea, Chrysalidocarpus, Cissus, Cadiaeum, Cordyline, Dieffenbachia, Dracaena, Fastia, Hedera, Maranta and Polyscias.

Control: Frequent misting prevents severe infestation. Spraying Kelthane or Vertimac effectively controls the mites.

Slugs and snails

Control: Collect and destroy trapped slugs and snails every day or two, look for them under pots and pot rims. A commercial slug and snail bait containing Metaldehyde or Methiocarb can also be used for effective control.

Nematodes

Meloidogyne spp. is the most commonly nematode affecting the house plants like Ardisia, Asparagus, Brassaia, Caladium, Calathea, Dieffenbachia, Maranta and Sansevieria.

Control: Nematodes can also be controlled effectively by carbofuran and aldicarb.

Precautions

Success with your house plants depends on how you choose them and on how you treat them from the first moment on. Make sure when selecting plants from the nursery that they are 'hardened off' to cope with the ups and downs of the prevailing climatic requirement and specific household environment. Look for plants that are sturdy and make sure there are no signs of damage or disease. Most expensive plants are not necessarily the best specimen. Ensure that in the process of shifting or transportation, they are not subjected to shock or suffer from drought and subjected to direct sunlight and take care not to over water them.

Selection

1) **Aglaonema** - are decorative plants with mottled leaves. It can grow in poor light conditions as well. It needs regular watering.
2) **Aspidistra** - a hardy house plant which can grow with ease in shade.
3) **Areca palm** - a very decorative house plant popular and graceful which can with stand indoor conditions reasonably well.

4) **Spider plant** - (chlorophytum): This is one of the easiest and most popular of all house plants. It is practically indestructible, and reproduces itself freely, a ready source of gifts for friends. It will tolerate a wide range of temperatures.

5) **Draceana varieties** - a number of exciting foliage plants come under this name, different species of the same genus. With wide range of selection this can give you much colourful and green foliage.

6) **Philodendrons**: The most philodendrons make excellent house plant. It can be grown in a pot of any size and can be trained on a moss stick and hydro-ponic Another similar looking but with dark green leaves is sweet heart Vine also being known as Oxycadium. Grows best in shade.

7) **Ferns**: An excellent and easy plant to grow indoors. This is very attractive with different kinds of green colour and foliage.

8) **Umbrella tree** (*Scheffelera actinophylla*): Can grow tall with broad leaves but very hardy indoor normally called Brassia. But quite a few species of Scheffelera makes lovely and hardy house plants including the one with variegated leaves.

9) **Mother in laws' tongue** : (Sensevieria) despite its name this is a very pleasing house plant. It has long narrow leaves. Very tolerant of most environments A very good house plants for cleaning of indoor air pollution.

10) ***Hedera helix*:** Common Ivy as is known can grow indoors. A climbing by habit, it can be trained and supported over the frame or even grown in pots. To fill the pots. An ideal plant for hanging baskets or cylindrical containers for being arranged as hanging house plants.

11) ***Begonia rex:*** The plant is very attractive in its look**.** Leaves are marbled, mottles and blotched in various patterns in green, purple, red, brown or silver colours.

12) ***Scindapsus aureus***: Known commercially as "pothos". It is vine and climbs tall by rootlets; leaves heart shaped, waxy dark green, yellow variegated.

13) ***Euphorbia pulcherrima***: It is a branching shrub grows up to 3 m height, The decorative value of the plants is the terminal shoots forming dark red, velvet, lanceolate bract leaves surrounding the tiny yellowish flowers.

14) ***Spathiphyllum***: It posses glossy green , thin leathery, oblong lanceolate leaves, inflorescence on reed like stems. The ovate spathe, white turning

green with age, 8-10 cm long; maize like white spadix also looks attractive.

15) ***Monstera:*** It is woody stemmed, climber with hanging aerial roots having hanging thick leathery leaves which are glossy green, pinnately cut and perforated with holes. The leaves of a young plant are heart shaped and without holes. They often overlap and clinig closely to a tree trunk.

16) ***Codiaeum variegatum:*** Crotons are beautiful tropical ornamental shrubs. The ornamental value is of leaves. Leaves are thick leathery, glabrous, ovate to linear, entire or lobed or spirally twisted, 10-30 cm long, variegated in beautiful colours. The green and yellow colour of leaves often change to shades of red.

References

Bose, T.K., Maiti , R.G., Dhua, R.S. and Das P. 1999. Floriculture and Landscaping published by Naya Prakash, Calcutta.

Janakiram T., Raju, D.V.S., Prasad, K.V., Singh, K.P., Namita, Swaroop, K. and Jain R. 2012. Horticulture for Environment and Ecotourism published by Division of Floriculture and Landscaping, IARI, New Delhi.

Randhawa G.S and Mukhopadhyay A. 1986. Floriculture in India published by Allied Publisher private Ltd. New Delhi pp 656.

Singh Kanwar P., Raju, D.V.S., Swaroop K, and Singh Krishan P. 2008. Practical Manual on Indoor Plants published by Division of Floriculture and Landscaping, IARI, New Delhi.

Swarup V. 1997. Ornamental Horticulture published by Macmillan India Ltd., New Delhi, pp.462.

Ornamental Horticulture, pp 335-348
© Editor, S.S. Sindhu, 2016
New India Publishing Agency, New Delhi - 110034, India

15

Seasonals

S.S. Sindhu

Seasonals or annual flowers refers to the group of herbaceous plants which complete their life-cycle i.e. both vegetative and reproductive phase within one year or one season. They are most popular among garden lovers and amateurs as they are easy to culture and moreover they also provide a beautiful and instant display of colours in the garden. Seasonals are mostly preferred for growing in pots, beds, borders, window boxes, hanging baskets or as cut flower for interior decoration. They enhance the aesthetic value of a place within a short span of time and create the instant landscape. Seasonals are generally grown in beds either individually with or without perennial plants in borders. They are also grown for cut flowers for interior decoration or for loose flower purpose. Some of the seasonals are valued for their naturally dried flowers like straw flower, paper flower *etc.* Some of annuals the annuals also provide sweet fragrance to the garden.

Classification

I. According to growing season, the seasonals are categorized into following three groups

a) Winter season annuals

b) Summer season annuals

c) Rainy season annuals.

Division of Floriculture and Landscaping, IARI, New Delhi, India - 110 012

Winter season annuals: The winter season annuals are sown in the month of September-October. Examples are Antirrhinum, Californian Poppy, Larkspur, Marigold, Calendula, Stock, Sweet William, Dainthus, Linaria, Cineraria, Sweet Pea, Sweet Alyssum, Sweet Sultan, Nasturtium, Phlox *etc.*

Summer season annuals: The summer season annuals are sown in the month of February and transplanted in the field during late March. The common summer season annual flowers are Coreopsis, Cosmos, Gaillardia, Kochia, Marigold, Portulaca, Sunflower, Tithonia, Zinnia *etc.*

Rainy season annuals: These rainy season annuals are sown during May and they become ready after 30-45 days of transplanting. This group includes annuals like Amaranthus, Balsam, Celosia, Gomphrena and Torenia.

II According to usage of seasonal, they are categorized into various groups which are as follows

Bedding plant: Annuals flowers like carnation, dahlia, pansy, verbena, sweet william, sweet sultan, petunia, zinnia, marigold, balsam, gomphrena, gaillardia etc. are suitable for bedding purpose.

Loose flower: Annuals flowers like marigold, chrysanthemum, China aster, zinnia, gaillardia, gompherena, sunflower, calendula, paper flower etc. can be utilized for loose flower purpose.

Cut flower: Many annual flowers can also be utilized for cut flower purpose like antirrhinum, corn flowers, lupin, carnation, sweet william, sweet sultan, China aster, larkspur, lisianthus, paper flower etc.

Pot purpose: Annual flowers also serves as the pot plant and many annuals are grown in pot for example antirrhinum, marigold, China aster, petunia, carnation, ice plant, dimorphotheca etc.

Hanging baskets: Annual flowers like nasturtium, ice plant, verbena, phlox, sweet alyssum, daisy, portulaca etc. are utilized for hanging basket purpose.

Shady situations: Some of the annual flowers flourish well under shady situations like salvia, cineraria, clarkia, verbena, alyssum. iberis, ageratum etc.

Screening purpose: Annuals like sweet pea, hollyhock are used for screening purpose.

Rock garden purpose: Annuals best suited for rock garden are ice plant, nemesia, gamolepis, phlox, verbena, nasturtium, ageratum etc.

Dry flower purpose: Some of the annual flowers are also utilized as dry flower for example statice, acroclinum, helichrysum, nigella, lady's lace etc.

Green filler purpose: Annuals such as gypsophila, kochia, mollucella are utilized as filler.

Fragrance purpose: Some of the annual flowers also possess sweet fragrance like sweet pea, sweet sultan, sweet william, sweet alyssum, mignonette, carnation, stock etc.

Pollination behaviour and breeding system

The breeding system of plants fall mainly into 2 classes- those which are self-pollinated e.g. balsam, sweet pea, antirrhinum, lupin, salvia, etc. Outbreeders are those crops which produce seeds only after cross pollination e.g. Ageratum, Cineraria, Verbena, Hollyhock, petunia etc.

Climate

The annual flowers or seasonals are grown successfully in all parts of the country. They are grown all round the year in one or other parts of India. In north Indian plains, winter season annuals are grown during winter seasons, while summer and rainy ones are grown during summer and rainy seasons respectively. In northern hills, all seasonal or annual flowers are grown during summer season. In the places having milder climate particularly in Bangalore and Pune, mostly all seasonals or annuals flowers can be successfully grown.

Soil

Seasonals or annuals thrive best in well-drained, sandy loam soil. The soil should be rich in organic matter. The pH of the soil should be maintained between 6.0 and 7.5 and it should neither be too acidic nor too alkaline. The acidic soils can be reclaimed by adding lime. It is recommended to add organic matter in clayey soil to make it porous and add sufficient organic matter in sandy soils for improving its texture.

Propagation: Almost all annuals can be propagated easily by seeds. However, marigolds can also be propagated through cuttings especially in places with high humid conditions.

Sowing Time: Sowing time of seasonals depends upon climate of the place. It varies in different parts of India. In the northern plains, winter season annuals are sown during September-October. Summer season annuals are sown during January-February, while rainy season annuals during May-June. In south India, September-October is ideal time for sowing annuals. In northern hills, March-April is most appropriate time for sowing annual flower seeds, whereas March-May in southern hills.

Planting and transplanting

The seasonals are more commonly grown by raising the nursery. The seeds are sown thinly so as to avoid dense planting which lead to damping off disease. The seeds of annual flowers are also sown in earthen pots, seed pans or wooden seed trays. The seedlings are ready for transplanting after 25 days of sowing. The right stage of transplanting is 4-leaf stage of the seedlings. Before transplanting, seedlings are hardened off by withholding water for 1 or 2 days or by exposing them gradually to sunlight. Transplanting is generally done in the evening so that seedlings got well established stand. Transplanting in evening is good as the night cool temperature is beneficial for the establishment. Light watering is recommended for every day either in early morning or late in the afternoon for about a week. It is required for proper establishment of the seedlings.

Planting method: The seeds of annual flowers are sown in nursery beds, pots, seed pans or wooden trays. The seeds of a few annuals having bold seeds like sweet pea, morning glory, lupin nasturtium and hollyhock can be sown directly at permanent places. The seed compost should consist of one part each of garden soil, coarse sand, farmyard manure and leaf-mould. For preparing the nursery beds, the soil should be dug up thoroughly and sufficient farmyard manure, should be mixed in soil. Raised nursery beds of convenient size (normally 60cm wide and 15cm high) should be prepared. If soil is heavy, recommendation is to add some quantity of sand. In order to raise healthy planting material, it is better to sterilize the soil of nursery bed or earthen pots with 2% formalin or seeds should be treated with Cerasan (0.2%) and Captaf (0.2%) to prevent the seedlings from damping off disease. The seeds should be sown thinly and evenly as thick sowing causes damping off of seedlings. Mixing of fine sand in very small seeds is advisable for even sowing. In nursery beds, the seeds are sown in rows spaced 6 cm apart. Then, they are covered with finely sieved leaf-mould. Watering is to be done with a watering can having a fine rose both in beds and pots. In beds, when germination is over, water is given for proper moisture. There-after, the beds should be kept weed-free.

Planting distance

Dwarf annuals: They should be planted at a distance of 30cm x 30cm e.g. daisy, gaillardia, gazania, candytuft, stock, mesembryanthemum, pansy, petunia, phlox, portulaca, mignonette, French marigold, torenia, nasturtium, venidium, verbena, viola etc.

Medium-tall annuals: They should be planted at a distance of 45cm x 45cm e.g. amaranthus, antirrhinum, calendula, celosia, sweet sultan, cornflower, cineraria, clarkia, coreopsis, cosmos, larkspur, carnation, digitalis, dimorphotheca, gomphrena, gypsophila, balsam, kochia, limonium, lupin, bells-of-Ireland, nicotiana, nigella, garden poppy, rudbeckia, salvia etc.

Tall annuals: They should be planted at a distance of 60cm x 60cm e.g. hollyhock, chrysanthemum, sunflower, helichrysum, heliotrope, pimpinella and zinnia etc.

Watering

Watering is an important operation that is responsible for the overall growth and development of the plants. The frequency and amount are the important factors to be kept under consideration. Regular watering is done as most of the annuals do not develop long deep root system. Waterlogging should be avoided as they deprive the roots of the aeration. The water requirement of the annuals varies with respect to crop, season, location etc. In lighter soils, more frequent irrigation is needed than that in heavy soils. The season of planting also determines the frequency of irrigation. During summer season, irrigation should be done at weekly intervals in beds, while at 10-12 days intervals in winters. Irrigation during rainy season depends upon prevailing weather conditions. Potted plants need daily watering during summer, whereas on alternate days in winter.

Manuring and fertilization

Both organic and inorganic fertilizers are essential to exploit the best quality and production of the seasonal. The farmyard manure or compost @ 3kg/m^2 is mixed in the soil at the time of planting. Chemical fertilizers-20g urea, 60-120g superphosphate and 30-60g muriate of potash/m^2 should also be added. It is recommended to add half dose of urea and full dose of both superphosphate and muriate of potash at the time of bed preparation. The remaining quantity of urea must be applied one month after transplantation. Spraying plants with 2% urea twice or thrice is beneficial for good growth and flowering. Nowadays, some readymade pot-mixtures are available in the market. The liquid manure is recommended for pot grown plants and it is prepared by fermenting l-2kg each of fresh cowdung and oil cake in 10 litres of water in a drum for one week. It is diluted to tea colour and sieved with the help of a muslin cloth. It is applied @ 500-l000ml/ pot at 7-10 days intervals.

Growth retardants

Growth retarding or promoting substances play a major role in getting dwarf plants or higher flower yield. Growth retarding chemicals- CCC (1,000-

2,000ppm), B-Nine (2,000- 5,000ppm) and SADH (1,000-3,000ppm)- are used to control growth and flowering in hollyhock, arctotis, sweet sultan, coreopsis, cosmos, phlox and pansy. These chemicals retard plant height, produce more number of leaves andbranches and improve flowering, Similarly, GA_3 (100-400ppm) gives beneficial results in respect of growth and flowering in African marigold, China aster and antirrhinum. Application of these growth substances is more effective at vegetative stage.

Aftercare

Pinching: Pinching is a common practice to break apical dominance in plants so as to promote axillary branching. It is done by removing the terminal portion of the seedlings. Pinching is done as soon as seedlings are established in beds. It is essential for making the plants bushy. It increases flower production and slightly delays the flowering. The optimum time of pinching is 20-25 days after transplanting. It is commonly practiced in crops like carnation, marigold etc.

Disbudding: The main objective is to divert cell sap to the few flowers so as to obtain big size bloom instead of more number of small sized flowers. It is desirable in cases where big size blooms are required or in those flowers which produces flower buds at an early stage, therefore buds are removed as soon as they are at pea size stage.

Weeding and hoeing : Weeding should be done regularly so that annuals grow and flourish well without any competition for water, sunlight, nutrients. 3-4 manual weeding are preferable for the entire cropping period. It is recommended to remove the weeds before they start setting seeds. Hoeing loosens the soil by the breaking the hard soil crust which allows better aeration to the roots and also removes the weeds. Avoid deep hoeing as it will cut the fine roots.

Roguing : It is essential to remove "off-types"from the field so as to avoid contamination from mixtures. It is done to maintain the purity of the germplasm. For this purpose, monitoring and inspection is required at regular intervals.

Staking: Staking is done for annuals with weak, slender or strag-gling stems e.g. sweet pea, carnation, morning glory and nasturtium, antirrhinum etc. The support is given when they are 15-20cm tall. The stakes prepared from split bamboos.

Alyssum

Annual Chrysanthemum

Antirrhinum

Calceolaria

Calendula

Dahlia

Dianthus barbatus

Dimorphotheca

Gazania

Group display

Linaria

Mesembryanthemum

Phlox

Tagetes sp.

Verbena

Common Insect pests

Insects

Insect	Nature of damage	Control
Aphids	Sucks sap from young foliage and buds. It resulted in growth retardation and flower quality is affected	Endosulfan or Rogor @ 1.0- 1.5 ml/lt. of water
Thrips	Sucks sap from leaf undersurface and flower buds. Growth and quality is affected	Rogor or monocrotophos @ 1ml/lt. of water
Catterpillars	Damage seen on flowers buds and foliage.	Endosulfan or Malathion @ 1.0- 1.5 ml lt. of water
Leaf hoppers	Sucks sap from foliage and stem	Rogor @ 2ml/lt. of water
Mites	Sucks sap from the foliage and plants show webby appearance	Kelthane (Dicofol) @ 1ml/lt. of water

Diseases

Disease	Symptoms	Control
Damping off and root rot	Infected plants shows damaged root system	Soil sterilizationDrenching of nursery beds with Captaf (0.1%) and Bavistin (0.1%). Seed treatment with Captaf (0.1%)
Botrytis	Brown colour spots developed on flower petals under humid conditions	Spary Mancozeb @ 2g/lt. of water
Leaf spots and blight	Appearance of small brownish spots which later on results in blights	Spray Dithane M-45 @0.2%
Powdery mildew	White powdery spores appear on leaves and in extremes covers the entire plant	Spray Karathane (0.1%)
Inflorescence Blight	Burnt appearance of inflorescence in extreme cases	Spray Dithane M-45 @0.2%

Harvesting

Most of the seasonal or annual flowers are grown for garden display purpose. However, the commercial annual flowers which are grown for cut flower or loose flower purpose should be harvested when they are fully open and are should be sold in the fresh form in the local markets in short time. For

Description of some Important Annual Flowers

S. No.	Botanical Name	Common Name	Family	Flower colour range	Remarks
Summer and Rainy Season Annuals					
1.	*Amaranthus caudatus*	Love-Lies-Bleeding	Amarantaceae	Catkins of purple crimson, pale green or white flower	Mostly grown in groups or in rows. Also grown in pots for indoor decoration
2.	*Celosia* sp.	Cock's Comb	Amarantaceae	Deep crimson, deep ink cardinal red to deep orange	Suitable for pot culture
3.	*Gaillardia pulchella*	Blanket flower	Asteraceae	Yellow, orange, scarlet, copper or combinations	Withstand dry conditions and mostly grown for bedding purpose. Also good cut flower
4.	*Gomphrena globosa*	Globe amaranth	Amarantaceae	white, purple, magenta, pink and rose colour	Bedding purpose
5.	*Helianthus annus*	Sunflower	Asteraceae	Yellow, golden orange	Bedding purpose
6.	*Impatiens balsamina*	Balsam	Balsaminaceae	Red, pink, violet, rose, and white	Early flowering and mostly grown in beds
7.	*Kochia scoparia* var. tricophylla	Summer cypress or burning bush	Chenopodiaceae	Foliage: green	Mostly grown for foliage beauty in pots
8.	*Portulaca grandiflora*	Sun plant or Rose moss	Portulaceae	White, yellow, red, orange, pink, lavender, crimson, purple	It is most ideal for beds, pots and hanging baskets
9.	*Tithonia rotundifolia*	Mexican sunflower	Asteraceae	Deep orange to scarlet colour	Bedding purpose
10.	*Zinnia elegans*	Zinnia	Asteraceae	White, green, red, pink yellow, orange, salmon, lavender, mauve	Bedding purpose
Winter Season Annuals					
11.	*Acroclinum roseum*	Paper flower	Asteraceae	White, pink	Bedding purpose
12.	*Ageratum houstonianum*	Floss flower	Asteraceae	Blue, white, pink	Bedding purpose

S. No.	Botanical Name	Common Name	Family	Flower colour range	Remarks
13.	*Althaea rosea*	Hollyhock	Malvaceae	White, crismson, pink yellow and red	Screening purpose, herbaceous border
14.	*Alyssum maritimum*	Sweet Alyssum	Cruciferae	White, lilac	Bedding purpose, pot purpose
15.	*Anchusa capensis*	Anchusa	Boraginaceae	Blue	Bedding purpose
16.	*Antirrhinum majus*	Snapdragon	Scrophulariaceae	White, cream , yellow, rose, salmon, pink, mauve, red, magenta, orange, bronze	Mostly grown as a pot plant. Bedding and good cut flower
17.	*Arctotis stoechadifolia*	African daisy	Asteraceae	White with lilac centre	Suitable for pot, bed and herbaceous border
18.	*Bellis perennis*	English daisy	Asteraceae	White, pink	Pot purpose
19.	*Brachycome iberidifolia*	Swan river daisy	Asteraceae	White, light pink, blue	Bedding purpose
20.	*Calendula officinalis*	Pot marigold	Asteraceae	Yellow, orange	Bedding purpose, pot plant, cut flower
21.	*Callistephus chinensis*	China aster	Asteraceae	White, yellow, blue. Lavender, red, salmon rose, magenta, scarlet	Loose flower as well cut flower crop
22.	*Centaurea cyanus*	Corn flower	Asteraceae	White, pink, blue	Bedding purpose, pot plant
23.	*Centaurea moschata*	Sweet sultan	Asteraceae	Mauve, white, yellow	Bedding purpose, pot plant, cut flower
24.	*Chrysanthemum coronarium*	Annual chrysant hemum	Asteraceae	White, creamish white, creamish yellow, yellow	Bedding purpose, loose flower crop
25.	*Coreopsis tinctoria*	Tickseed	Asteraceae	Yellow	Bedding purpose
26.	*Cosmos bipinnatus*	Cosmos	Asteraceae	White, pink, yellow, orange, magenta	Bedding purpose
27.	*Dahlia variabilis*	Dahlia	Asteraceae	White, cream, yellow, orange, scarlet, pink, mauve, purple, crimson	Excellent pot plant
28.	*Delphinium hybridum*	Larkspur	Ranunculaceae	Deep blue, rose pink, rosy scarlet, white	Good cut flower. Suitable for mixed annual borders

S. No.	Botanical Name	Common Name	Family	Flower colour range	Remarks
29.	*Dianthus barbatus*	Sweet William	Caryophyllaceae	Crimson, red, maroon, salmon. White, purple	Suitable for bedding, borders, pot and cut flower purpose
30.	*Dimorphotheca aurantiaca*	African daisy	Asteraceae	White, yellow, orange, salmon	Bedding and pot plant
31.	*Eschscholtzia californica*	California poppy	Papaveraceae	Yellow, white, orange, crimson, pink, red	Suitable for bedding purpose
32.	*Gamolepis tagetes*	Gamolepis	Asteraceae	Yellow	Suitable for bedding, edging, pot purpose and also for rock garden
33.	*Gazania splendens*	Gazania	Asteraceae	White, pink, yellow, orange, red, brown	Suitable for bedding, edging, pot purpose
34.	*Gypsophila elegans*	Baby breath	Caryophyllaceae	White, pink	Bedding plant, Filler in bouquet arrangement
35.	*Helichrysum bracteatum*	Straw flower	Asteraceae	Yellow, orange	Suitable for bedding and rock garden
36.	*Iberis* spp.	Candytuft	Cruciferae	White, purple	Bedding, edging, pot and herbaceous border purpose
37.	*Lathyrus odoratus*	Sweet pea	Leguminosae	White, light blue, lilac, mauve, pink, rose, red, maroon	Screening purpose
38.	*Limonium sinuatum*	Sea lavendar	Plumbaginaceae	Blue, lavender, rose	Suitable for beds, borders, cut flower
39.	*Lupinus hartwegii*	Lupin	Leguminosae	White,blue, pink	Cut flower, bedding, border and pot plant
40.	*Mathiola incana*	Stock	Crucifeare	Red, white, yellow	Good cut flower
41.	*Mesembryanthemum criniflorum*	Ice plant	Aizoaceae	Pink, white, yellow	Suitable edging, hanging baskets, pots plant and rock garden
42.	*Nigella damascena*	Love in a mist	Ranunculaceae	White, light blue, rose	Pot plant and cut flower purpose
43.	*Petunia hybrida*	Petunia	Solanaceae	White, ram, yellow, pink, blue, purple, mauve, salmon	Suitable for bedding, pots, hanging baskets, window boxes

S. No.	Botanical Name	Common Name	Family	Flower colour range	Remarks
44.	*Phlox drummondii*	Star flower	Polemoniaceae	White, cream ,pink, pale yellow, pink, salmon, rose, blue, red	Suitable for pot, bedding, border, window boxes
45.	*Reseda odorata*	Mignonette	Resedaceae	White, light yellow, red, crimson.	Suitable for pots, beds, window gardens
46.	*Rudbeckia lacinata*	Cone flower	Asteraceae	Yellow	Bedding purpose
47.	*Saliva splendens*	Sage flower	Labiatae	Scarlet, red white, purple	Suitable for shady areas
48.	*Schizanthus wisetonesis*	Butterfly flower	Solanaceae	White, yellow,pink, lilac, red	Pot plant and cut flower purpose
49.	*Senecio cruentus*	Cineraria	Asteraceae	Blue, purple, salmon, crimson, red, pink, white	Suitable for shady areas, good pot plant
50.	*Tagetes erecta*	African marigold	Asteraceae	Yellow, orange	Bedding, loose flower crop

cut flowers purpose, Antirrhinums are cut when one-third of the florets are open and China aster flowers are cut along with their stems when they develop their original colour. However, loose flowers are harvested when they are fully open e.g. marigold, gypsophila etc. Harvesting should be done either late in the afternoon or very early in the morning.

Flower yield: In African marigolds, yield of 20-22 tonnes of fresh flowers is obtained from one hectare crop, whereas in French marigolds 10-12 tonnes/ha of fresh flowers is obtained. China aster gives the yield of 10-12 tonnes/ha of fresh flowers.

Seed yield: Seed yield per unit area depend upon the nature of flowers and climatic variations. In pansy, seed yield is estimated around 15-20 kg/acre, Salvia -20 kg/acre, Verbena –40 kg/acre, African marigold- 50 to 60 kg/acre, French marigold- 60 to 75 kg/acre and Cineraria- 25 kg/acre.

References

Bhattacharjee S.K. 2004. Landscape Gardening and design with plans published by Aavishkar Publishers, Rajasthan, pp 512

Bose, T.K., Maiti , R.G., Dhua, R.S. and Das P. 1999. Floriculture and Landscaping published by Naya Prakash, Calcutta.

Bhanu L. Desai 1962. Seasonal Flowers published by Indian Council of Agricultural Research, New Delhi pp 180.

J.S. Arora. Introductory Ornamental Horticulture. 1992. Published by Kalyani Publishers, New Delhi, pp 188.

K.L. Chadha. 2006. Handbook of Horticulture, ICAR, New Delhi, pp 1031.

Randhawa G.S and Mukhopadhyay A. 1986. Floriculture in India published by Allied Publisher private Ltd. New Delhi pp 656

Swarup V. 1997. Ornamental Horticulture published by Macmillan India Ltd., New Delhi, pp.462.

Ornamental Horticulture, pp 349-388
New India Publishing Agency, New Delhi - 110034, India

16

Production of Loose Flowers

Kanwar Pal Singh, Gunjeet Kumar and Sapna Panwar

Introduction

Floriculture is a fast emerging, rapidly expanding industry in the present scenario. The importance of flowers has been realized throughout the world and now it has been developed into an intensive form of agriculture. Flowers play an integral role in life of each and every human being. They are the part and parcel of day to day activities. There is common saying with flowers that a man is born with flowers, live with flowers and dies with flowers. This shows the importance of flowers in our lives since time immemorial. The flowers create an atmosphere of beauty, happiness and harmony. With the fast changing lifestyles, floriculture can become a lucrative business in India in view of low labour cost, varied agro climatic conditions and large collection of plant wealth.

Among flowers, loose flowers hold an important place in terms of high demand for usage in various religious and social ceremonies. They are mainly used for worshipping, making floral garlands, floral decorations in wedding, religious functions, interior decoration, self adornment etc. The earlier trend of using flowers for aesthetic value is changing slowly into business with the realization of its vast economic potential as value addition of flowers by drying, extraction of essential oils, colouring pigments, antioxidants etc. is also gaining importance.

In India, mostly loose flowers dominate the domestic market. Marigold, jasmine, tuberose, crossandra and china aster are the major loose flowers grown in India. Production technology of these flowers is described as:

Division of Floriculture and Landscaping, Indian Agricultural Research Institute, New Delhi

Marigold

Marigold (*Tagetes* sp. Linn.), is an important cultivated loose flower crop mainly used for floral decorations, religious offerings and also for making garlands and flower baskets. It is native to South and Central America, especially Mexico. The genus *Tagetes* comprises of about 33 species of which *Tagetes erecta, Tagetes patula and Tagetes minuta* are commonly cultivated species. Today, it is one of the most important commercial flowers grown worldwide and in India it accounts for more than half of the nations' loose flower production. Major marigold growing states are Karnataka, Gujarat, Maharashtra, Haryana, Andhra Pradesh, Uttar Pradesh, Chattisgarh, Odisha, Jammu & Kashmir, Puducherry, Andaman Nicobar, Arunachal Pradesh, West Bengal, Tamil Nadu, etc. It occupies an area of 42,880 hectares with production of 3,60,210 metric tons loose flower (NHB database, 2012-13). Marigold has gained popularity because of its adaptability to various soil and climatic conditions, longer blooming period, beautiful flowers which have good shelf life.

Uses

Marigold flowers are mainly grown in India for loose flower production and being used extensively for worshipping, making floral garlands, floral decorations in wedding, religious functions, interior decoration etc. Marigold flowers are rich source of carotenoids and being grown on commercial scale for extracting these carotenoids which are used extensively in food colouring as natural source. Flower petals contain xanthophylls which are major carotenoids fractions and lutein forms 80-90 percent of the total xanthophyll content. Carotenoids are the major source of pigment for poultry feed used for intensification of colour of egg yolk and boiler skin. Marigold species especially *T. minuta* oil is the most valuable and precious for using in high grade perfumes and cosmetics. Marigold leaves and flowers possess a good insect repelling properties, therefore, marigold oil is being used on commercial scale for formulation of insect repellent.

Species and Cultivars

There are about 33 species of the genus *Tagetes*. The characters of important species are given below

***Tagetes erecta* (African marigold)**: The plant is an annual. It grows to a height of 70-100 cm. The plant type is erect. The leaves are pinnately divided and leaflets are serrated. Flower type varies from single to double. The flowers are gobular. Flower colour varies from sulfur yellow to orpiment orange. African marigold can be cultivated throughout the year.

Tagetes patula **(French marigold)**: The plant is an annual which grows to a height of 30-45 cm. The stem is reddish and the foliage is dark green. Leaves are pinnately divided and leaflets are serrated. Flowers are yellow to brick red. In most of the cases, brick red coloured blotches are seen on yellow background.

Tagetes tenuifolia **(Signet marigold)**: This species is an annual with branching habit. Leaves are pinnately divided. Flowers are yellow and numerous.

Tagetes lucida **(Sweet scented marigold)**: The plants of this species are perennial. Leaves are sessile, small and lanceolate. Flowers are usually 2-3 rayed, produced in dense terminal corymbs.

Tagetes minuta: This species is an annual producing small white flowers in dense terminal clusters. Commercially oil is extracted from this species.

Varieties

In marigold, many varieties and strains are available which vary in plant height, growth habit, flower shape and size. Mostly local varieties are being cultivated by the farmers which are generally low yielders. The Division of Floriculture and Landscaping, Indian Agricultural Research Institute, New Delhi has developed high yielding varieties of marigold which are as follows.

African Marigold

Pusa Narangi Gainda: The plants of this variety are of medium stature, vigorous and uniform, foliage-dark green which grows to a height of 70-80 cm. It flowers in 125-136 days after sowing. The flowers are orange coloured, carnation type, double and compact in nature. The variety is most suitable for loose flowers and carotenoid extraction.

Pusa Basanti Gainda: It is little late maturing variety which takes 135-145 days. The plants are 60 -70 cm tall, vigorous and uniform with dark green foliage. Flowers sulfur yellow coloured, carnation type, double and compact. It is ideal for loose flower production and also for growing in pots and beds in the garden.

French Marigold

Pusa Arpita: The variety produces tall, bushy and vigorous plants. Plant height ranges from 90-100 cm . The foliage colour is dark green and stem colour is greenish purple. The crop establishment is very good and the plants make good bush with strong stems. The variety produces medium sized, light orange, compact flowers. The variety is most suitable for loose flowers and carotenoid extraction.

The other important varieties of important species are listed below:

Tagetes erecta: Giant Double African Orange, Giant Double African Yellow, Cracker Jack.

Tagetes patula: Red Brocade, Rusty Red, Butter Scotch, Valencia, Sussana.

Tagetes tenuifolia : Golden Gem, Lulu, Pumila.

Soil and Climate

Marigold can be cultivated in a wide variety of soils. A deep fertile, friable soil having good water holding capacity and aeration is suitable. Sandy loam soil with a pH of 7.0-7.5 having good drainage is favourable. Marigold needs plenty of sunshine, so it should be grown in open sunny conditions. In winter, seedlings and plants are damaged by severe frost so nursery should be sown under moderate conditions. The environmental conditions markedly influence the growth and flowering of marigold. Marigold is a quantitatively short day plant so flowering is accelerated by short days whereas long days may delay flowering when combined with higher temperatures. Mild temperature during growing period greatly improved flowering. The flowering of marigold can be taken throughout the year by taking crops in succession in order to ensure their regular supply to the market.

Propagation

Marigold is propagated by both sexual and asexual propagation. However marigold is commercially propagated by seed. Plants raised from seeds are vigorous, healthy and establish well. It is also propagated by cuttings to obtain uniformity in growth and flowering traits. Male sterile lines are also propagated and maintained by cuttings.

Seed Propagation

Seeds of many cultivars are easily available and germinate quickly. The seed rate is 800-1000 grams per hectare. Marigold seeds germinate well at temperatures ranging from 18 to 30°C. Nursery beds are prepared after thoroughly mixing well rotten farmyard manure. Raised nursery beds of convenient size (normally 1 m wide and raised to 15 cm) are prepared. Before sowing, the seeds are treated with Captaf or Bavistin. The nursery bed may also be drenched with 0.1% Bavistin. Seed should be finely covered with farmyard manure or leaf mould. Watering should be given with the help of a fine rose can. Formation of hard pan should be prevented for good emergence of the seedlings. The sowing should not be too dense to prevent the damping off of the seedlings. The seeds can be sown from mid October to mid November. The seedlings get ready for transplanting in 25-30 days after seed sowing.

Vegetative Propagation

Marigold is also propagated through cuttings to produce true to type plants and to maintain male sterile lines. Cuttings of 6-10 cm length can be planted in sand. The base of the cuttings may be treated with commercial formulations of Indolebutyric acid (IBA) to promote the rooting.

Planting

The seedlings are transplanted in the field after one month of seed sowing. A spacing of 45 x 45 cm for African marigold and 30 x 30 cm for French marigold has been found to be optimum for commercial cultivation. For vigorously growing varieties such as Pusa Arpita optimum spacing is 60 x 45. The seedlings at the time of transplanting should be disease free and should have 3-4 true leaves. Light irrigation should be given after transplanting.

Nutrition

For marigold, well rotten farmyard manure @ 30 tonnes should be incorporated per hectare well in the soil at the time of soil preparation. In addition to farmyard manure, it is advisable to apply 120 kg/ha of nitrogen and 80 kg/ha each of phosphorus and potash for getting good vegetative growth and flowering. Whole quantity of phosphorus and potash should be applied at the time of land preparation. Nitrogen is to be applied in two split doses. The first dose should be applied 20 days after planting and the second dose 40 days after transplanting. It will be better if two foliar sprays of 0.2% urea are done at an interval of 15 days.

Pinching

In African marigold, because of apical dominance the plants grow tall and lanky with very few branches. Such plants are susceptible to lodging and the flower production will be less. Pinching is done to break apical dominance and to encourage axillary branching and thereby increasing flower production. French marigold does not require pinching. Pinching slightly delays flowering but improves flower production. Pinching 40 days after transplanting has been found to be optimal.

Growth Regulators

The cultivars of African marigold usually grow taller and growth retarding chemicals have been recommended for reducing the height and improving the appearance as potted plant. SADH 2500 ppm reduced plant height and improved the foliage colour. When 2-5% solutions of CCC were sprayed, there was a marked improvement in growth and flowering. The treated plants were dwarf in nature. Repeated applications of GA_3 slightly improved height and flowering.

Irrigation

For all stages of vegetative growth and flowering, sufficient amount of moisture is essential. Moisture stress at any of the vegetative and flowering stages markedly influences growth and flowering. The frequency and quantum of irrigation is dependent on species, variety, soil and season. During the summer months, the plants are liable to dessication due to hot winds. Therefore the plants should be irrigated according to the atmospheric conditions. Winter crop requires irrigation at an interval of 8-10 days.

Weed Control

In marigold, weed control is laborious and it involves lot of cost. If weeds are not controlled in time, growth, flowering and productivity are drastically affected. Atleast 5-6 manual weedings are required for the entire cropping period. Alachlor @2.5-3.0 kg/ha is effective in controlling the weeds during the entire cropping period.

Harvesting and Packaging

The flowers are harvested when they are fully open and have attained full size. Field should be irrigated before plucking as flowers will retain freshness for longer period after harvest After harvesting, the flowers are packed in small and big bamboo baskets. They are then transported to nearest or distant markets by buses or lorries. It is better to harvest the flowers in the morning and to store them in a cool place before packing.

Diseases

Damping off (*Pythium* sp., *Rhizoctonia* sp.)

This disease affects the seedlings. Before emergence brown coloured spots appear on radicle and plumule causing pre-emergence death of the seedlings. After emergence, brownish necrotic spots appear on the collar portion of the seedlings. These necrotic spots in severe cases form a girdled portion leading to the collapse of seedlings. The disease can be controlled by prophylactic drenching of nursery beds with Captaf and Bavistin. Seed treatment with 0.1% Captaf will also reduce the incidence of the disease.

Pusa Arpita

Pusa Basanti Gainda

Pusa Narangi Gainda

Leaf spot and blight (*Alternaria* sp., *Cercospora* sp., *Septoria* sp., *Colletotrichum* sp.)

Various pathogens cause leaf spot and blight in marigold. The symptoms appear as small brownish spots on leaves. In case of *Alternaria* infection, the spots show concentric circles. In severe cases, the infection may spread to inflorescence. In case of *Cercospora,* the spots show bird eye appearance with dark border and light coloured centre. *Septoria* infection is marked by black spots. The blight of marigold is caused by *Colletotrichum* sp. The infected plants show burnt appearance. These leaf spot diseases can be controlled by Dithane M-45 0.3% spray.

Inflorescence blight: (*Alternaria zinnae*)

Initially minute brown spots appear which enlarge into blotches. The spots show zonations or concentric circle (Target board) pattern. Inflorescence gives a burnt appearance in severe cases. This disease can be controlled by Dithane M-45 (0.3%) spray.

Flower bud rot: (*Alternaria dianthi*)

This disease appears on inflorescence or young buds which show discoloured appearance in initial stages. The buds show dry rotting in case of severe infection. Bud rot can be controlled by Dithane M-45(0.3%) spray.

Powdery mildew: (*Oidium* sp., *Leveillula taurica*)

Whitish tiny spots appear on leaves initialy. The entire plant shows powdery appearance in severe cases. The disease can be controlled by Karathane (0.1%) sprays.

Viral diseases: (Cucumber Mosaic Virus, Aster Yellow Virus)

Infected plants show chlorotic streaking and yellow mosaic appearance. Infected plants get dwarfed. Roguing of diseased plants and insecticidal sprays to control vector population prevent the spread of the disease.

Insect Pest

Red spider mite: (*Tetranychus* sp.)

Mites infect old plants. The diseased plants give dusty appearance. The plants show webbed appearance. The disease can be controlled by spraying Kelthane (Dicofol) 0.1%

Hairy caterpillar: (*Diacrisia obliqua*)

In the initial stages they eat away the foliage later stages they eat away the ray florets of the inflorescence. This can be controlled by spraying Nuvan(0.2%).

Leaf hopper: (*Empoasca fabae*)

The hoppers suck sap from the leaves and stem. The infected plants show cupped or rolled leaves and give wilted appearance spraying of Metasystox (0.2 %) controls the hoppers.

Jasmine

Jasmine is one of the oldest fragrant flowers cultivated by man. The flower is used for various purposes viz., making garlands, bouquet, decorating hair of women, religious offering etc. It is also used for production of Jasmine concrete which is used in cosmetic and perfumery industries. More than 80 jasmine species are found in India, of which only three species are used for commercial cultivation. They are *Jasminum sambac* (Gundumalli / Madurai Malli), *J. auriculatum* (Mullai) and *J. grandiflorum* (Jathimalli / Pitchi). The first two species are mainly cultivated for selling as fresh flowers whereas the last one is cultivated for concrete extraction. The main beauty and uniqueness of Jasmine is its odour, which cannot be imitated by any known synthetic aromatic chemical, still has a unique status in the perfume world. ancient Literature of Sangam' period (500 B.C. to A.D. 200) clearly indicates that all these three species have been mentioned in those ancient literatures and therefore, South India could well have been an important center of origin for many of the Jasminium species.

In India, jasmines are cultivated throughout the country but the commercial cultivation is confined to Coimbatore, Madurai, and Dindigul(Tamil Nadu); Bangalore, Bellary, Mysore and Kolar (Karnataka); Knnauj, Jaunpur and Gazipur (Utter Pradesh); Udaipur, Jaipur, Ajmer and Kota (Rajasthan); Ranaghat, Kolaghat, Pancskura (West Bengal); parts of Andhra Pradesh and Maharastra.

Tamil Nadu is the leading producer of jasmine in the country with an annual production of 77247 t from the cultivated area of 9360 ha. The flowers produced in the state are being exported to the neighbouring countries viz., Sri Lanka, Singapore, Malaysia and Middle East countries. The major jasmine producing districts of Tamil Nadu are Dindigul, Salem, Madurai, Tirunelveli, Virudhunagar, Trichy, etc. Since the crop requires lots of manpower for harvesting and other operations, only small farmers are cultivating the crop. It is an ideal crop for small farmers whose land holdings are less than 1 acre.

Uses and Importance

Widely cultivated for its flowers, jasmine is enjoyed in the garden, as a house plant, and as cut flowers. Its flowers are used for worship, making garlands, gajra, veni, etc. The delicate jasmine flower opens only at night and may be plucked in the morning when the tiny petals are tightly closed, then stored in a cool place until night. The petals begin to open between six and eight in the evening, as the temperature lowers.

Jasmine flowers are used for extracting essential oil . Its flowers are either extracted by the labour-intensive method of enfleurage or through chemical extraction. It is expensive due to the large number of flowers needed to produce a small amount of oil. The flowers have to be gathered at night because the odour of jasmine is more powerful after dark. The flowers are laid out on cotton cloths soaked in olive oil for several days and then extracted leaving the true jasmine essence. Some of the countries producing jasmine essential oil are India, Egypt, China and Morocco.

Species

Jasmine belongs to the family Oleaceae and more than 100 species of jasmine are reported . Some of the important commercially grown species of India are described below.

Jasminum grandiflorum (Mysore Mallige)

Popularly known as royal or spanish jasmine. Mysore Mallige is the most well-known variety, which derives its name since it is grown mostly around Mysore city in Karnataka state. Mysore Mallige, mostly grown in and around Mysore city is a viable crop for small farmers. Farmers reap two crops of this seasonal flower. Apart from the local market, the flower is in demand in parts of Kerala and Tamil Nadu. The principal aromtic components are Indol, Jasmone, Benzyl Acetate, Benzyl Benzoate, Methyl Anthranilate, Linalool & Geraniol.Its modern day applications are in perfumery, cosmetics, Incense, Aroma Therapy and Ayurveda.It is also used externally to soothe dry and sensitive skin. The essential oil content in the flowers ranges from 0.24 to 0.42 per cent.

Jasminum auriculatum (Hadagali Mallige)

Hadagali Mallige is known for its rich fragrance and shelf life. in Bellary district of Karnataka. Plant is small shrub with little scandent nature. Leaves are simple, thick, margin folded upside, slightly pubescent. Flowers are borne in axillary cymes. Flower with long corolla tube around 1 cm (0.39 in) long. Petals 7, spreading, white in colour.

The sandy red soil prevailing in this region is ideal for cultivation of Hadgali Mallige. The dry weather and good water supply also adds to the congenial environment for the cultivation of this crop. It is propagated mainly through cuttings. Planted directly in the months of July-August on the onset of monsoon. The flowering season spreads up to six months. Flowers are highly scented and possess good essential oil recovery (0.24 to 0.42%). The flowers are used for extraction of essential oil.

Jasminum sambac **(Udupi Mallige)**

It is commonly known as Arabian jasmine, Tuscan jasmine, bela and mogra. The variety grown as Udupi Mallige. Plant is a small and bushy with leaves of light green with yellowish shade, nerves slightly raised beneath, acute at both ends, entire, ovate-lanceolate. Flowers in cymose inflorescence and borne in the axils and also terminal.

Propagation is mainly by cuttings. Planting is done in the month August-September.

Jasminum multiflorum **L.**

Commonly known as kunda, kundo and kakada. This is one of the resistant jasmine, not scented and very ornamental. Long duration flowering beginning from December to August.

J. bessianum

Deciduas or semi ever green climber. Leaves simple opposite, ovate and lanceolate.flowers are fragnant, rosy pink in colour.

J. arborescens

In local language it is known a muta bela and neba mallinga. Flowers are white and have strong fragrance. It is grown in West Bengal and Uttar Pradesh.

J. calophyllum Vahl

It is known as Pandal malli in local language. It is profuse bloomer and is generally grown in home garden with flowers which are scented. Flower colour is white. Flower appears round the year. Plants are healthy free from pest and disease attack.

J. flexile Vahl

It is also known as (Syn. J. caudatum Wall.). It is a profuse flowering species grown widely in home garden for its fragrant flowers. Plant blooms throughout the year. Cultivar is free from insect, pest and disease.

J. humile L

It is also known as Semmalligai and pelli chameli in local language. It produces yellow flower and is also known as yellow Jasmine. Flowers are fragrant and are used in the perfume industry.

J. pubescens Wild

It is known as Kundo and Kunda in local language and it is called as J. multiflorum Burm. It is an ornamental plant with slightly fragrant, flower colour is white and it blooms throughout the year. Flowering is in its peak during winters. It is grown in north India and its performance is better during cool winter climate.

Important Varieties Developed in India and Their Characteristics

Parimullai

This variety belongs to the species *Jasminum auriculatum* (Jui). It has a medium round bud with flowering duration of about 9 months/year. It is resistant to gall mite. The average yield is 8t/ha,

CO 1

This variety belongs to the species *Jasminum auriculatum* (Jui). Flowers of this variety have a long corolla tube and thus easy to harvest. The average yield is 8.8 t/ha

CO 2

This variety belongs to the species *Jasminum auriculatum*(Jui). Flower buds of this variety are bold with long corolla tube. It is tolerant to phyllody disease. The average yield is 11.1 t/ha.

CO-1 (Pitchi)

This variety belongs to the species *Jasminum grandiflorum*(Chameli). It is released by T.N.A.U., Coimbatore. It is suitable for loose flower production and oil extraction. The average flower yield is about 10 t/ha/year while the estimated concrete yield is 29 kg/ha.

CO 2

This variety belongs to the species *Jasminum grandiflorum* (Chameli). The variety produces bold pink buds with long corolla tube. The average yield is 11.68t/ha.

Arka Surabhi (Pink pin)

This variety belongs to the species *Jasminum grandiflorum* (Chameli). It is released by IIHR, Bangalore. The average flower yield is 10 t/ha.

Gundumalli

This variety belongs to the species *Jasminum sambac* (Mogra). Flowers are round with good fragrance. The average yield of flowers is 7-8 t/ha while the estimated concrete yield is 15kg/ha.

Double Mogra

This variety belongs to the species *Jasminum sambac* (Mogra). The flowers have 8-10 whorls of petals with excellent fragrance resembling that of white rose.

Flowering

Flowering in Jasmine is almost throughout the year in *J. grandiflorum, J. flexile, J. pubescens, J. calophyllum* and *J. regidum*. Other species flower from February to October. Flower opening in Jasmine is not uniform in all the species and it varies from species to species. On the basis of flower opening time Velusamy et.al. (1976) grouped the species in to early, medium and very late.

Cultural Practices

Soil and Climate

Jasmine can be grown on a wide range of soils. Well-drained, rich loamy soil with a pH ranging from 6.5-7.5 is ideal for their cultivation. Jasmine prefers mild and tropical climate. Jasmine is commercially grown in India under open field conditions. The ideal requirements for successful cultivation of jasmine are mild winter, warm summer, moderate rainfall and sunny days. Jasmines grow well upto 1200 m. A well-distributed annual rainfall of 800 to 1000 mm is optimum for growth and development.The dry and warm weather with low humidity is good for the crop. Flowering starts during March-April and continues up to June-July, April -May being the peak season.

Season of planting

The ideal time for planting in North India is during July-August and from the end of January-February, while in South India planting is done any time between July-December.

Spacing and planting

Tamil Nadu Agricultural University, Coimbatore has recommended a planting distance of 1.5 x 1.5 m for *J. auriculatum* and *J. sambac* and a spacing of 2.0 x 1.5 m for *J. grandiflorum* . The soil is well pulverized and weeds are removed before planting. One or two initial ploughings are required to remove the weeds present in the land, which is followed by digging of pits. Pits of 45 cm3 are prepared one month before planting and are exposed to sunlight. A few days before planting each pit should be applied 10 kg of Farm Yard Manure (FYM) before filling the pits; pits are filled with FYM, fresh soil and coarse sand in the ratio of 2:1:1. Addition of BHC (10g/pit) to the soil in the pit helps to prevent the attack of termites. Pits are watered to settle the soil compost mixture. Well-rooted, healthy and strong seedlings obtained from cutting/layering are planted in each pit. A hole is dug in the centre of the pit sufficient enough to accommodate the soil ball of the seedling. The soil ball is placed in the centre of the pit and the soil is firmly pressed around the seedlings. The plants are then immediately watered.

Propagation

Seed

Seed setting and Germination Seed setting and its germination is a rare phenomenon, however it depends on genotype and surrounding environment. Extensive studies on seed set and germination has been carried out at IIHR, Bangalore. Under natural pollination seed setting was almost negligible in *J. grandiflorum* Linn. strain Pink Pin and Pink Thrum and *J. sambac* Ait var Gundumalli 30% in *J. calophyllum* Wall and 50% in case of *J. auriculatum* Vahl var. Parimullai. *J. grandiflorum* is sterile in nature. Meiosis in Pollen sterility and non viability may be reason of sterility. The diploid species *Jasminum auriculatum and Jasminum grandiflorum* can arise plant from seed and havin 75-84% germination ability. Higher temperature and humidity hasten the germination. Maximum germination occur in saw dust and sphagnum moss.

Layering

Layering is done during June-July in North India and from June to December in South India. For preparation of layers, well matured, one year old shoots are selected and are buried in the soil 10-15 cm deep after making a shallow, slanting cut in the portion that is to be buried. The root formation occurs in 90-120 days.

Cutting

It is the easiest method of propagation of jasmine *J.grandiflorum* and *J.sambac* are best propagated by apical cuttings while *J.auriculatum* is propagated by semi hardwood cuttings. Normally 22-25 cm long cuttings with 3-4 nodes are planted in rooting media.

Cuttings taken during April-September has highest percentage of rooting with maximum rooting in June planted cuttings. The basal portion of softwood cuttings is treated with growth regulating substances (IBA 400ppm and IAA @ 1000 ppm) before planting. The cuttings are buried more than 5 cm deep in the rooting medium and are spaced 7cm apart. The cuttings are ready for transplanting into the main field after 4 to 5 months of planting in the rooting media.

Grafting and budding

Approach grafting in Gundumalli (*Jasminum auriculatum*) on Mullai (*Jasminum sambac)* was successful. Besides patch budding and side grafting practices proved that *Jasminum grandiflorum* is compatible with *Jasminum sambac* cv. Madhaban.

Cultural operations

Manuring & fertilization

Many commercial growers use organic manure @10 kg/plant/year. A fertilizer dose of 100g N, 150g P_2O_5 and 100g of K_2O per plant over a basal dose of 100 kg FYM /plant/year is ideal for *J. grandiflorum* at Bangalore. Further Mg (40 kg/h), Zn (10kg/h) and B (5 kg/h) gives good result for yield. In *J. auriculatum* 120g N, 240g P_2O_5 and 240g of K_2O per plant is required. While for *J. sambac* 90g N, 120g P_2O_5 and 240g of K_2O per plant is beneficial.

Weeding

Weed growth within the jasmine crop competes with the main crop for nutrients. Therefore it is essential to remove the weeds at the appropriate time. After pruning, the soil 15-30 cm from the main stem around the bushes is stirred to a depth of 15 cm. This should be repeated every two to three months. Though manual weed control is effective, but it is very expensive. Mulching also reduces the weed population considerably. Chemical weed control is comparatively economical, convenient and efficient in eradicating weeds with one or two applications.

Intercropping

In the initial years when there is sufficient space between the plants, vegetable crops and ornamental plants can be grown as intercrop.

Pruning

Pruning is an important activity as it influences growth, flower-bud initiation, differentiation and, ultimately, the flower production in Jasmine. Usually, irrigation is stopped before pruning and plants are pruned to half their original length. All the leaves are stripped off after pruning. All the cut ends are smeared with Bordeaux paste to prevent entry of pathogens. Pruning is done at 45cm from the ground level. Pruning done during mid of December to mid January results in maximum number of branches and higher yield of flowers

Irrigation

Adequate moisture in the soil is necessary for proper growth and flowering in jasmines. Plants are irrigated by flooding once a week in the summer months. After flowering, no irrigation is normally required till after the next pruning and manuring.

Harvesting

The plants start flowering six months after planting. However the commercial yields are taken from the third year onwards. For fresh flowers, fully developed, unopened flower buds are hand-picked early in the morning. The flowers should be handled carefully during harvest and transport.

Post harvest management

Flower should be graded according to their shape, size and freshness. Corrugated fibre boxes are the best for packing before sending them distance market. Bamboo basket can also be used for local market.

Stage of harvest and pre and post harvest handling influence the flower shelflife. Application of boric acid, alluminium sulphate and silver nitrate keep the flower fresh for longer duration compared to soaking the flower in normal distilled water.

Yield

The flower yield in jasmines varies considerably according to the species and cultivars and management practices. The flower-yield of various cultivars in kg/ha are given below:

J. auriculatum - 4733 to 9152 kg/ha

J. sambac - 739 to 8129 kg/ha

J. grandiflorum - 4329 to 10144 kg/ha

The life of a jasmine plantation is between 10-15 years. It is advisable to replant the whole plantation after every 15 years.

Diseases

Leaf-blight (*Cercospora jasminicola* and *Alternaria jasmini*)

This disease occurs in a severe form on *J. grandiflorum*. Reddish-brown, circular spots are produced on the upper surface of the leaves, spreading rapidly in the rainy season. The infected leaves curl and start drying from the margins. Even the young shoots dry up. In severe cases of infection, vegetative buds and young branches dry up. The flower production is very much reduced in infected plants. Spraying of 0.4% Benlate solution, 0.2% Dithane M-45 or 0.1% Bordeaux mixture has been found effective in control of the disease.

Rust (*Urornyces hobsoni*)

Rust occurs on all the aerial parts of the plants including flowers. Yellowish orange coloured pustules appear on the lower side of the leaves and also on young twigs and flowers buds. The infected parts become distorted.The disease can be controlled by pruning the branches or spraying Copper Oxychloride (0.3%) or Mancozeb (0.2%). Sulphur at the rate of 20-25 kg/ha is also useful.

Wilt (*Fusarium solani*)

The disease occurs in patches and the roots turn black. In the case of sclerotial wilt, in addition to the above symptoms, white mycelia are found generally girdling the roots and the sclerotia are found adhering to the roots of the wilted plants. Drenching the soil around the plant with 1% Bordeaux mixture controls this disease.

Insect- pests

Bud Worm (*Hendecasis duplifacialis*)

It is a greenish larva with a black head, which bores into immature jasmine buds and feeds on floral structures and, in severe cases, webbing of buds. A basal application of Carbofuran (40g/plant) is recommended for control.

Gallery Worm *(Elasmopolpus jasrninoghagus)*

It is a serious pest, which causes webbing of terminal leaves, shoots and towers. The plants should be sprayed with Malathion (0.2%) to control these insects.

Mites

The mites attack the undersurface of leaves, which become yellow and drop off. Severe puckering and discoloration of leaves are caused by the gall mite in *J. auriculatum*. The variety Parimullai released by the TNAU, Coimbatore, is resistant to gall mite. Wettable Sulphur (0.3%) can be sprayed on the infested plants to control this pest.

Root-Knot Nematode (*Meloidogyne incognita*)

It causes severe stunting of plants, branches become dry with yellow leaves which drop prematurely. Application of Neem cake at one t/ha or Carbofuran at 2.5kg/ha effectively suppresses the nematode population.

Tuberose

Tuberose is an important bulbous ornamental plant known for beauty and elegance of its flowers. It is cultivated commercially for its varied uses. The loose flowers are used for making artistic garlands, floral ornamentals, bouquets and buttonholes. The flowers remain fresh for a long time and can withstand long distance transportation. The long flower spikes are excellent cut flower. The delightful fragrance of flower is an excellent source for extraction of essential oil. The fresh flowers yield 0.08 to 0.11 percent concrete of which 18 to 23 percent is ábsolute.' The crop can be successfully grown in pots, beds and borders. The tuberose bulb contains an alkaloid lycorine and has number of medicinal properties.

Owing to its strong fragrant flowers with longer keeping quality, ability to withstand long distance transportation, versatile uses of flowers, reasonably good recovery of pure concrete, low production cost and high profit and negligible attack of pests and diseases, the area under tuberose cultivation is on rise.

Tuberose is native of Maxico from where it spread to other parts of the world. The name tuberose is derived from tuberosa, this plant being the tuberous hyacinth as distinguished from the bulbous hyacinth and hence the name tuberose. Over 12 species of tuberose have been reported among which *Polianthes tuberosa* is popular and cultivated. Presently, it is cultivated on large scale in France, Italy, South Africa and in many tropical and sub tropical areas. In India, It is grown commercially in Karnataka, Tamil Nadu, Maharastra and West Bengal spread over an area of more than 20,000 hectares.

Cultivars

There are mainly two production types in commercial cultivation.

Single

In this type flowers have only one row of corolla segments and are more fragrant than other types. The cultivars Calcutta Single, Mexican Single, Phule Rajani, Prajwal, Rajat Rekha, Shringar, Khahikuchi Single, Hyderabad Single, Pune Single and Arka Nirantra belong to this group. Two hybrids Shringar and Prajwal have been released by IIHR which are superior in quality and yield when compared to local ‘single’ type. While ‘Shringar’ possess pinkish buds ‘Prajwal’ has white buds. The hybrids yield 40% higher loose flower yield as compared to local type. A gamma induced mutant called ‘Rajat Rekha’ has been released by NBRI, Lucknow.

Double

In this type, flowers have more than three rows of corolla segments. The flowers are less fragrant and generally spikes are cut and used for decoration in vases. Calcutta Double, Hyderabad Double, Pearl Double, Swarna Rekha, Suvasini and Vaibhav are the cultivars in this group. Recently two hybrids ‘Suvasini’ and ‘Vaibhav’ have been released by IIHR. Number of flowers in spike in these hybrids is more and open boldly as compared to local types in which flower opening is incomplete. The yield of cut flowers is also 25% higher than the local double type. At NBRI , Lucknow a gamma induced cultivar ‘Suvarna Rekha’ has been released.

Soil and Climate

Tuberose can be grown in wide range of soils, however loam and sandy loam soil having a pH range of 6.5 to 7.5 with good aeration and drainage are found to be most suitable for its cultivation. It can be grown successfully under high saline alkaline conditions by following suitable cultural practices. The soil should be rich in organic matter and retain sufficient moisture for proper growth.

The tuberose grows well in topics and subtropics. The commercial cultivation is mainly confined to warm and humid areas with temperature range of 20^0 to 35 °C. At high (>40 °C) and low temperature (<10 °C) it is reported to reduce spike length, weight and quality of tuberose flowers. A temperature range of 20-30 °C is found to be optimum for maximum bulb production. In North Indian plains it blooms from April to November.

Propagation

Tuberose is mainly propagated vegetatively through bulbs. It is essential to select good planting material for obtaining higher yield and good quality flowers. It is better to avoid fresh bulbs as it will result in more vegetative growth. Bulbs stored for 4-6 weeks after harvest are ideal for planting. Seed propagation is a slow process and is not used for commercial propagation.

The size of bulb plays an important role in growth and production. Although larger bulbs causes slight delay in sprouting, results in early flowering, higher yield of spikes and loose flowers. In general bulbs having diameter between 2.0 and 3.0 cm are suitable for planting.

Selection of Site and Land preparation

The site selected should have plenty of sunlight. Excess shade results in lanky plant growth and poor quality of flowers. Good drainage is must as water logging damages the roots and thus affects growth and flowering. Land has to be thoroughly ploughed two to three times and soil has to be brought to fine tilth. Approximately eight tonnes of FYM, 80 Kg each of phosphorus and potash should be thoroughly mixed in the soil during field preparation.

Planting

The planting time varies from place to place depending on the climate conditions. It is planted during February –March in northern plains and April –May in hills. In West Bengal the ideal month for planting is first week of April. In southern India it is done during July –August. In places with mild climatic conditions like Bangalore, April –May is found to be ideal for obtaining good growth and flower production.

Planting on flat bed is found to be better for obtaining higher spike and rachis length and also for longer duration of flowering as compared to planting on medium or high ridges. It is better to follow staggered planting at 15 days interval to obtain continuous flowering. An increase in planting depth delays sprouting. Depending on the size also depth of planting varies. Larger size bulbs are planted more deeply as compared to smaller bulbs. Therefore the depth of planting varies from 3-10 cm from the soil surface. Deeper planting of 6 cm however results in better plant height, number of spikes, spike lenth, flowering duration and bulb production. In sandy soil planting is deeper as compared to clayey soil. Plant growth, yield and quality bulb production is markedly influence by density of planting. It is suggested to follow a spacing of 20 x 20 cm or 25 x 25 cm thus accommodating about 40 to 50 thousand bulbs per acre. A closer spacing of 20 x 10 cm showed early flowering and higher yield while wider spacing of 25 x 25 cm resulted in better vegetative growth and larger number of bulbs and bulblets.

Nutrition

Depending on soil type and climatic conditions, the nutrients requirements of crop varies from place to place. It is essential to get the soil tested before planting for pH, EC and other nutrients. This helps in supplying the nutrients in required quantities for better growth and production. While deficiency of Nitrogen and Phosphorus results in poor growth and flowering, excessive nitrogen result in tall and weak spikes thus making them vulnerable for breakage due to winds and also more susceptible to diseases and pests. Calcium deficiency results in cracking and in acute cases even bud rot is observed. Deformed and stunted inflorescence is found to be due to deficiency of boron.

At the time of planting recommended dose of farm yard manure, 1/3 rd N and full dose of P and K are applied while the remaining N is further split in to two doses and are applied 30 and 60 days after planting. The split application of nitrogen helps in better availability without being lost in leaching. Foliar spray of 1.60 N, 0.60 P_2O_5 , and 0.64 g of K_2O / I improves essential oil recovery.

Irrigation

The frequency of irrigation is dependent on soil type, stage of growth and weather conditions. It is better to irrigate before planting and avoid further irrigation till the bulbs have sprouted. Excess moisture at the time of sprouting results in rotting. Irrigation can be done once in a week or 10 days depending on weather conditions. During summer when temperatures are high, irrigating twice a week is essential. It is reported that drip irrigation at 2000-3000 m^3 per hectare gave the best yield of bulbs of planting size (30-32 bulbs / m^2).

Inter cultural operations

To check the growth of weeds and for loosening of soil, it is important to do earthing up and weeding regularly so that growth and productivity of the crop is not affected. Removal of weeds helps in reducing the competition for nutrients with the main crop. Pre plant application of atrazine at 3.0 kg a.i. per hectare is beneficial to control weed population.

Harvesting and Yield

Flowering starts 3-4 months after planting. Loose flowers are harvested for making garlands and other floral ornamentals while for bouquets and vase arrangement, long spikes are cut from the base. The ideal stage for cutting spikes is when first pair of flower is fully open. Immediately after harvest the spikes are kept in water. Picking of flowers is to be done during cool hours of the morning or evening. An average of 5-6 tonnes of loose flowers per acre can be obtained while 40000 to 50000 spikes in double flowered varieties may be obtained from an acre.

Post harvest handling and packaging of flowers

The stem portion of the bundles is wrapped with wet newspaper. Soft tissue paper or thin polythene sheets can be used for packing the entire bundle in order to avoid damage including the flowers and buds. The flower spikes are graded according to length and quality of flowers in bundles of 100 spikes each. Later these bundles are packed in cardboard boxes or cartons and transported. Bamboo baskets with a capacity of 10-15 kg are used for packing loose flowers.

Bulb harvesting and storage

The bulbs attain maturity when plant growth ceases and flowering is over. Irrigation is withheld at this stage. The leaves are cut at the ground level and bulbs are taken out of the soil. The clumps are cleaned and bulblets are separated before grading the bulbs based on their size. They are stored separately in cool, dry and shady place or in cold storage at 10°C for 30 days. During storage care should be taken to check the spread of mould or rot. Storage of 4-6 weeks is necessary before the plants are ready for planting. A bulb yield of 8-10 tons per acre can be obtained.

Diseases

Stem rot caused by Sclerotium is an important disease in tuberose cultivation. Mycelial growth on leaves at or near the soil is the initial symptom. The infected spots loose green color due to rotting which extend and cover the whole leaf. Further the leaves get detached and fall on the soil surface. As a result the plant becomes weak and flower production gets affected. Drenching with carbendazim (1g/l) will control the disease.

Insect-pests

Aphids, thrips and sometimes caterpillars damage tuberose. While aphids damage the flower buds and growing points, the thrips feed on leaves, flower stalk and flower by sucking the sap. In severe infestations inflorescene is malformed. Spraying with dimethoate (0.1%) or with 0.1% malathion is effective. Nematodes affect the crop resulting in poor growth and flower production. Use of healthy planting material, neem cake and carbofuran will check the damage caused by nematodes.

Crossandra

Introduction

Crossandra is an important traditional flower crop of south India. It is small evergreen hardy shrub and produce beautiful flowers with a range of colours almost throughout the year. This has made it extremely popular in southern states of India, where the flowers are widely used in temple offerings and for making gajras and venis to use by ladies as hair adornments.

Owing to their attractive colours, light weight and good keeping quality, crossandra flowers are ideal for garland making either alone or along with jasmine to produce charming colour contrasts. Steady market demand and assured and regular income have made crossandra a profitable crop for farmers in South India. .

Species

The genus Crossandra belongs to family Acanthaceae. This genus consists of 40 –50 species of tropical plants with only a few cultivated species such as *Crossandra undulaefolia*, *C. mucronata*, *C. guineensis* and *C. sebacaulis*. *C. undulaefolia* is an important species which is grown commercially across the world. The description of a few important species is as under:

C. guineensis: It is free flowering and reaches about 30 –60 cm in height. The bracts lack spines and are pale lilac in colour. Suitable for both sunny as well as partial shade situations.

C. nilotica Oliver: The plants are about 60 cm in height. Stem is semiwoody. Leaves are elliptic, dark green and glossy, borne in whorls of 4 and are about 5-7 cm in length. Brick red flowers are borne in dense spikes. Bracts are green, about 2 cm long, veined and hairy. Each flower consists of a long slender corolla tube. Lower 3 petals form a lip. Grows well under partial shade.

C. flava: It is an unbranched shrub about 15 –20 cm tall with erect and green stem. Leaves are glabrous, opposite, obovate to lanceolate in shape and dark green in colour. Lower leaves are stalked, while-upper ones are sessile. Spike is 4 sided with spiny yellow green bracts. Flowers are bright yellow and the corolla tube is much exerted and jointed.

***C. undulaefolia* Salis** (*C. infundibulijormis* Nees): It is also known as Fire Cracker plant due to the cracking sound of the splitting of the seed pod.It is a short branching perennial shrub about 30–90 cm in height. Dark green, lustrous and pointed Leaves arc about 5 –12 cm long. Leaves are generally opposite, narrowly oval having more or less wavy margin. Flowers are showy, bright orange, salmon to scarlet in colour. They are borne in a large densely packed spike. The spikes are 7.5 to 12.5 cm long and 4 sided. Individual flowers consist of long slender tube about 2.5 cm long which terminates in a 5 lobed limb. Flowers have prominent spiny bract. The calyx is also 5 lobed and irregular and there are about 4 stamens.

***C. pungens* Lindav**: It is dense subshrub upto 60 cm tall. Leaves oblong, dull green, traced with pearly veins. Spikes are conjested, bracts broadly ovate, margins bristly to spiny with orange colour of flowers.

Origin and Distribution

The word crossandra is derived from Greek words 'krossoi'meaning fringe and 'aner'meaning male, thus word crossandra means fringed stamens. *Crossandra* genus has about 50 species which are shrubs or herbs. *Crossandra undulaefolia* Salisb. is a native of East Indies. However, it is also considered to be native of south India and Ceylon. *C. guinensis* is native of east Africa while *C. nilotica* has its origin in east Africa and Mozambigue. *C. flava* is a native of tropical West Africa.

The crossandra plants are distributed in tropical as well as subtropical regions of the world such as south Asia, South America, South Africa and Madagascar. In India, it is commercially cultivated in southern states of Karnataka, Tamil Nadu, Kerala and Andhra Pradesh.

Varieties

Crossandra shows a remarkable range of colours, varying from orange, pink, red, yellow and double coloured blue types with white throat. Recently white, light green and violet types have become popular. Important varieties of crossandra are as follows.

Orange Crossandra: Orange Crossandra is tetraploid (2n = 40) and produces bright orange flowers in profuse number.

Delhi Crossandra: Delhi Crossandra is triploid (2n = 30) and bears deep orange flowers. It does not set seeds and therefore propagated by cuttings.

Lutea Yellow: It is a tetraploid variety bearing orange yellow colour flowers. It is good for planting in pots, front line of shrubbery and hanging baskets. It provides an impressive view when mass planted in pockets.

Sebacaulis Red: It is a tetraploid variety and is tolerant to nematodes.

Mona Wall Head: Plant is compact, upto 50 cm tall. Leaves are lustrous and vivid green while flowers are deep salmon to pink.

Lakshmi: It is a high yielding mutant of Delhi Crossandra and is developed through gamma irradiation technique. This variety has a potential to yield upto 75 kg flowers per hectare in a day, starting from second year of planting. The improved variety is same as parent in appearance but it is differentiated by pleasing orange colour large flowers.

Dr. A. P. J. Abdul Kalam: It is a deep red colour mutant of Delhi Crossandra developed by using gamma irradiation techniques. Flowers of this variety have gentle crater-like patterns and a channel like formation at centre. These patterns and its long shelf life have made it a promising variety. It bears deep red flowers which are larger than Delhi Crossandra flowers. The bold flowers make it suitable for long transportation. It can flower throughout the year and is resistant to major pests.

Maruvur Arasi: It is a red colour mutant of Delhi Crossandra which has been developed by gamma irradiation techniques. It shows bright deep red colouration on flowers. The flower stalks are longer and bears over 75 flowers.

Hybrid: Tamil Nadu Agricultural University has developed a resistant hybrid by crossing 2 local types from Palani and Marudhamalai hills. This hybrid is characterized by large attractive yellowish orange flowers showing higher degree of tolerance to nematode-fungal complex.

Climate

It requires warm humid areas for its cultivation with average temperature ranging from 20 to 32 °C and it cannot tolerate temperature below 13 °C and dies under frost conditions. Luxuriant growth occurs at 30 °C. Seeds require 25 –28 °C temperature for germination however for cotyledon emergence and true leaf expansion, 21 –22 °C temperature is ideal. The optimum minimum night temperature is 18 –24 °C, while minimum day temperature requirement is 21-28 °C. Plants can be grown under sunlight as well as partial shade. Photoperiod does not affect flower induction.

Soil

Although crossandra can be cultivated in a wide range of soils, fertile, red loam soil with pH 6 to 7 are ideal. Soil should be well drained, rich in organic matter with an EC range of 0.75 to 1.0 mmhos/ cm. In alkaline soils, due to nutrient deficiency, plants develop chlorosis and poor flower production. Soil should be free from nematodes.

Propagation

Propagation is done by seed, stem cutting or layering. In triploid varieties propagation is by asexual means only.

Seed: Approximate seed count of Crossandra is 200–250 seeds/g. Good quality seeds are obtained when crop is raised at 50 cm x 50 cm spacing. Seeds attain physiological maturity at 55–60 days after flowering and can be stored for 6 months after treating with captan at 2 g/kg of seed. Seeds are sown in June – July on raised beds or in polythene bags at the rate of 2.25kg seeds/ha.

The 10 cm raised beds for seed sowing should be 1 m wide. The seedlings that are raised in polythene bags can withstand transplanting and for this purpose perforated plastic bags of 22.5 x 15 cm size with 100 gauge thickness are used. The bags are filled with mixture of FYM and soil in 1:3 ratio and 2 to 3 seeds are sown in one bag though only one is retained after germination. Seed viability is lost in 3 months, hence only freshly extracted seeds should be used. Care should be taken to protect the germinating seedlings from cut worm by dusting carbaryl 5% on the beds.

Layering: Triploid crossandra can be propagated by ground layering. Well conditioned layers can be planted from June to July in pits of 15 cm x 15 cm x 15 cm at a spacing of 90 x 90 cm. About 12,500 layers are needed to cover one hectare area. Plants starts flowering within a month after planting.

Cutting: Softwood cuttings of about 5-8 cm long are put for rooting under mist from March to June. Cuttings should be transplanted when sufficient roots have developed. Cuttings are rooted in equal parts of peat and sand. For better results rooting hormone IBA upto 3000 ppm can be used. Bottom heat enhances root development, therefore, media temperature should be maintained at 21 to 24°C.

Micropropagation: Triploid crossandra do not set seed and are commercially propagated by cutting. Cuttings have low rate of multiplication, hence micropropagation can be used to multiply triploid crossandra. The best response to multiple shoot regeneration from shoot tips was recorded on full strength MS medium fortified with BAP at 1 mg/1. Callus induction from leaf bits was recorded best on full MS medium fortified-with 2, 4-D 3 mg/l + BAP 0.5 mg/l.

Land Preparation

The land is thoroughly ploughed 2-3 times and about 25 tonnes of FYM is incorporated in the soil at the time of last ploughing. After proper levelling, beds are prepared about 1 m wide and of required length. Beds should be kept 10 cm high from the base to facilitate drainage.

Manure and Fertilisers

High fertility soil is needed for better flower production and FYM at 25 t/ha is incorporated in the soil at the time of last ploughing. For obtaining better yield, a mixture of ammonium sulphate and groundnut cake at 1:2 ratio is applied in two doses, once in first month after planting and again six months after planting. The recommended fertilizer dose is 33.3, 60, 60 kg/ha of N, P and K, respectively applied as basal and 33.1 kg N as top dressing 3 months after transplanting. In another recommendation, fertilizers should be applied at 55-60 days after planting at the rate of 50 kg of urea, 100 kg of single super phosphate and 60 kg of muriate of potash/ha and again at interval of 6 months. Constant liquid feed at 200-250 ppm with a balanced liquid fertilizer and supplemental magnesium at 30-50 ppm is beneficial. Iron and manganese supplied in trace amount is good for yield and quality.

Planting

Seedlings can be transplanted at 4-5 leaf stage. Planting is generally done in June –July and October –November. Seedlings are transplanted at 30 cm x 50cm spacing. At this spacing 1,25,000 seedlings per hectare can be accommodated.

Cultural Operations

Immediately after transplanting, the crop is irrigated twice and later once a week. When new growth appears, watering is done more freely.

Weed control: Weeding, application of manures, fertilizers and earthing-up are combined for easy upkeep of crop and saves labour. Metachlor @1.25 kg/ha decreased weed population and found favourable for plant growth, flower yield and quality of crossandra flowers.

Pruning: It is recommended to shorten all flowering growth by 1/2 after flowering during late winter.

Mulching: The quality and flowering attributes are significantly influenced by both organic and inorganic mulches. Black polyethylene mulch resulted in maximum length of spike and more flower yield per hectare.

Pot culture: Crossandra plants can be easily grown in pots having potting mixture of red soil, sand and compost in 1:2:1 ratio. An estimated yield of 91.88 g of flowers can be obtained per pot.

Plant Growth Regulators

The yield can be increased by using GA_3@200 ppm alone or GA_3 + 1 % urea applied one month after planting. SADH spray decreases internode elongation.

Diseases

Wilt: Wilt is cause by *Fusarium solani*. The infected leaf margins show pinkish brown discolouration and the stem shriveled. The leaves become pale and droop. On the root, dark lesions are seen which extends upto the collar region. The disease is observed in the field within short time after transplanting in scattered patches and ultimately the plants dry up completely. Wet Ceresan (0.1 %) along with phorate at 1 g/plant is effective in checking the disease.

Alternaria leaf spot: Small circular or irregular yellow spots on the upper surface of the leaves caused by *Alternaria amaranthi* var. *crossandrae* are characteristic symptom of this disease. The spots enlarge and develop to brown concentric rings. Premature droppings of leaves are observed in disease infected plants. Application of indofil M-45 (0.3%), or carbendazim (0.2%) effectively controls the disease.

Stem rot: This disease is caused by *Rhizoctonia solani* and is characterized by the brown to black lesions on stem just above the soil level resulting in girdling of stem, ultimately the disease extends upto the upper part. Roots also develop rotting in severe infection. Badly infected plant should be removed. Fortnightly spray of benomyl at 0.2 % and captan at 0.25 % in lightly infected plants controls the disease.

Root and crown rot: The disease causes violet discolouration of leaves and foot region of plants rot and is caused by *Phytophthora nicotianse*. The causal fungus is soil inhabiting and infection arises through damaged tissues. It is better to destroy the affected plants and replace with the fresh stock at a site far from original, To control the disease drenching of fosetyl-Al (0.2%) is effective.

Leaf blight: Leaf blight is caused by *Collectorichum crossandrae* and is characterized by development of brownish and depressed necrotic areas in leaves. The spots become darker as they expand. Infected leaves ultimately shrivel and drop off, leaving a barren stem with a whorl of young leaves at the top. At the first sign of the symptoms affected parts are destroyed and the plants are sprayed with a sulphur suspension and repeated at intervals of a week or ten days.

Insect Pests

Scales: Colonies of nymphs and mature females of scales (*Saissetia nigru, Saissetia viridis*) infest leaves, stems and flowers. They feed on sap and excrete honeydew, which makes plants sticky and encourages growth of sooty moulds, persistent infestations weaken growth. Remove and burn infested portion of plant to prevent further spread. Chemical treatments are only effective when applied on young crawlers before they started to form protective scale. Apply Carbofuron 3 G @ 1.0 kg a.i./ha. Spray with Ethion 50 EC @ 1 ml/l of Chlorpyriphos 20 EC @ 2.5 ml/l or dimethoate 30 EC @ 2 ml/l followed by pongamia oil 1 %.

Spike borer: Larvae of *Helicoverpa armigera* food on spikelets, flowers and developing seeds. Early instar larvae bit holes into spikelets whereas full grown larvae bore into mature spikes. Spray with Methyl parathion 50 EC @ 1 ml/l in combination with Diflubenzuron 25 WP @ 2 g/l for 2 to 3 times at 15 days interval.

Crossandra bug: *Gynencia affinis* is very specific on this crop. Nymphs and adults sick sap resulting in brown lesions. Severly infested plants fail to produce flowers. Spray with Acephate 75 SP @ 1.5 g/l. Carbaryl 50 WP @ 2 g/l. Subsiquent spray with neem oil 1 % at 10 days interval will manage the pest.

Nematodes

The roots exhibit brown to black lesions and prominent galls due to infestation of *Pratylenchus delattrei* and *Meloidegyne incognita*. It causes chlorotic leaves which ultimately turn white at the advanced stage of infestation. In case of severe attack, plants become stunted and defoliated without any side shoots. Nursery bed treatment with *Trichoderma harzianum* @ 50 g/sq.m. application of 2 tons of farm yard manure enriched with *T. harzianum* and *Paecilomyces lilacinus* per acre before transplanting, along with 100 –200 kg of neem or pongamia cake will manage the nematodes.

Harvesting

The crossandra plant generally flowers 2-3 months after planting. Flowers are borne round the year except rainy season. Flower opening starts from the base of the spike. Two diagonally opposite flowers will open together. Complete opening of flower takes 2 days, while complete flowering occurs in 15-25 days. The flowers are harvested on alternate days in the early morning by pulling corolla out of the calyx. Harvesting generally starts 3 month after planting. After flowering is over, dried spikes should be removed.

Postharvest Management

While fully opened flowers may remain fresh for about 3 days on the plants, the harvested flowers will fade away in 36-48 hours. Hence, freshly harvested flowers should be packed in cloth or polythene bags for local market. Shelf life of flowers is 2-3 days and soaking flowers in 0.5% boric acid or 1 % aluminium sulphate can extend it for 3 hours.

Yield

Flower yield depends on fertility status of soil, cultural operations, variety and climate of the area. There is a drop in production during rainy season. With a plant population of 40,000 plants per hectare by following a spacing of 50 cm x 50 cm one can obtain an estimated yield of 4.5 t/ha. However, the high yielding cultivars may yield upto 7-9 tonnes per hectare. One kilogram flower weight consists of about 15000 flowers. The crop can be retained in the field for about 3 years after which it has to be removed as it is not economical.

China Aster

Introduction

China aster (*Callistephus chinensis* L. Nees) is one of the most popular annual flowering plant grown throughout the world. It can easily be grown in the open fields and lath houses for the production of cut flowers. Among annuals it ranks next to chrysanthemum and marigold. Cut asters has longer post harvest life and are used in vases and floral decoration. The prominent colours of flowers are blue, violet, lavender, rose, white, etc. Some strains are used as pot plants, while dwarf cultivars are suitable for edging and window boxes. When asters are grown in large mass, they make very showy bedding plants and are useful in filling up the gaps in mixed herbaceous borders. In tropical and subtropical countries aster is extensively grown for bedding and cut blooms during winter months and spring. China aster is one of the most important traditional flowers in India and its commercial importance is increasing especially in Karnataka, Tamil Nadu, West Bengal and Maharashtra.

Description

China aster (*Callistephus chinensis* L. Nees) belongs to family Asteraceae. The genus *Callistephus* has only a single species *i.e.*, *C. chinensis*. It is a half hardy diploid annual (2n = 18). Plants are erect having hispid hairy branches bearing alternate, broadly ovate or triangular ovate, deeply and irregularly toothed leaves. Flower consists of many small flowers usually called florets, which combine together and make a single inflorescence called as head. The flowers are of two distinct types. Some produce both pollen and ovules and are termed as perfect flowers, while the other types lacks anther for pollen production but have pistils for ovules production and are called pistillate flower. Flowers are solitary and blue, lavender rose and white are the prominent colours.

Origin and Distribution

Two Greek words, Kalistos meaning most beautiful and stephos means a crown referring to the flower, makes the genus *Callistephus*. Cassini described the China aster as *C. hortensis*. Previously it was named by Linnaeus as *Aster chinensis* Nees and later on renamed to *Callistephus chinensis* L. Nees.

The present day aster has been developed from single form of wild species, *C. chinensis*, a plant native to China. The plant was introduced in Europe in 1731 by Jesuit Missionary R. P. D Incarville. The original plant had single flowers with two or four rows of blue, violet or White ray florets. The prolongation or development of central florets which lead to quilled flowers was the first novelty developed. The French were the pioneer contributors in aster improvement and later on the British joined in the breeding efforts. However the credit for recent improvement in the crop goes to American breeders. French florists contributed much in development of double forms. German scientists also evolved many doubled cultivars. These asters were known as German asters because the first great advances in the evolution of the plant were made in Germany. The comet types introduced in 1886 or 1887 subsequently replaced the quilled types. With the introduction of new branching types by James Vick in 1893, the USA became the main centre of development of this plant as well as for the production of seeds. Since its introduction into Europe, it spread to almost all the countries of the globe as a garden annual. A considerable amount of work on China aster on improvement and agronomy has also been done in India.

Varieties

Annual aster has many types and cultivars varying in growth habits, flower shape, size and appearance of florets. Varied colours are also found in a particular type. Double flowered asters are classified broadly as Tall, Medium and Dwarf types. The plant height varies from 70-90 cm in tall group and important varieties in this group are American Branching, Bouquet Powder Puffs, Princess, Giant Princess, Giant California, Totem Pole, Chikuma strain and Giant Massagno. Plant height varies from 40-60 cm in medium group and varieties belonging to this group are Ostrich Feather, Giant Comet, Giant Grego, Early Burpeeana, Pompon, Rubens, Liliput and Unicum. In dwarf types plant height varies from 20-40 cm and important varieties are Pinocchio, Colour Carpet, and Dwarf Chrysanthemum.

Some other important varieties of China aster are Queen of the Market, Early Wonder, Early Express, King, American Beauty, Heart of France, Asterix and Sunshine. In India, Local Pink, Local Violet and Local White varieties were very popular. The description of four promising varieties of China aster released from Indian Institute of Horticultural Research, Bangalore are given.

Kamini: A cross of two pure lines AST -6 x AST -36 and developed through pedigree method of breeding. Deep pink flower of this variety is more attractive than cv. Local Pink. It grows to about 60 cm and takes' about 138 days to flower. Diameter and weight of flower is 6 cm and 2g respectively. It produces about 50 flowers. It has 30 cm stalk length and vase life of flower is 8 days.

Poornima: Advanced pedigree selection of the cross between AST-29 and AST-3 bearing pure white flowers of powdery puff type. It takes 105 days to flower and grows to about 50 cm. Flower diameter is 5 cm. The plant produces 25 flowers and individual flower weight is 3.5 g. It has a stalk length of 25 cm and vase life of 7 days.

Shashank: A cross of a local variety (Local Pink) and a pure line variety (AST-2). Its flowers are also. It bears powdery puff and creamy white flowers in 124 days after sowing. Plant height is 55 cm. Flower size is 6 cm and weight of flower is 2.5 g. It produces 45 flowers which is more than double of Local white variety. It has 25 cm stalk length and a vase life of 9 days.

Violet Cushion: Developed through pedigree selection from a cross between Local Pink and AST -2. It is a floriferous variety bearing pompom type flowers of violet colour. Its flower are having 4 to 5 row of ray florets with tubular disc florets. It produces flowers in 130 days. The flowers are 4.5 cm in diameter and weigh 2 g each. It bears 70 flowers. It has a stalk length of 20 cm and vase life of 8 days.

The NARP research station of MPKV, Ganeshkhind, Pune also released following four improved varieties.

Phule Ganesh White: It is a medium tall, erect, late growing variety. It produces snow white double chrysanthemum type flowers. Yield potential is 47.07lakh flowers/ha and is suitable for cut flowers as well as for garlands.

Phule Ganesh Violet: It is an early, semi-erect variety producing medium sized violet flowers of double chrysanthemum type. Yield potential is 60.47lakh flowers/ha.

Phule Ganesh Pink: It is early flowering, semi-erect variety. Produces superior quality double chrysanthemum flowers of rosy pink colour. Produces 43.03lakh flowers/ha.

Phule Ganesh Purple: A semi-erect variety with a yield potential of 46.82 lakh flowers/ha. It produces bright purple flowers suitable for artistic floral arrangement and pasting works.

Climate

The China aster is a winter grown annual but it can be grown in summer months. The summer crop is commonly grown in shade houses. Although the seeds germinate in the dark, light increases germination percentage. The seedlings should receive a minimum of 12 hours of light per day. Seeds germinate well when soil temperature is 21°C. The duration of light has significant influence on growth and development of this crop. The rosetting of leaves occur in shorter day length (<14 hours). Supplementary lighting should be provided from the seedling stage until flowering in short day conditions. Though a winter season flowering annual, China aster is quite susceptible to frost injury, therefore, sowing time should be adjusted to avoid freezing temperature. At higher temperature, plant grows well but will have weaker stems and produce poor quality flowers. Flower express better colour in the temperature range of 20–30 °C during day and 15–17 °C during night with a relative humidity of 50–60%. Annual rainfall requirement need is approximately 500–700 mm spreading over June to September.

Soil

China aster can be grown in a wide range of soils. It grows well in rich, well drained loamy soil however sandy loam soil with sufficient amount of organic matter is preferable. Soil pH should be 6.8–7.0. Nursery soil should not be treated with methyl bromide because it causes poor germination. High calcium content in soil is not preferred for China aster cultivation. The crop is highly susceptible to water logging therefore, heavy loam soil is not suitable.

Propagation

Asters are propagated by seeds, which are tan to dark brown in colour. There are approximately 450 seeds per g. The seeds are about 4 mm long, oblong and flat, with one end pointed and other rounded. Seeds should be collected from the healthy and disease free plants. It will germinate in a week. Seeds of aster have no dormancy and freshly harvested seeds have a narrower temperature requirement for germination than older seeds. The seed cannot be stored for a long period in ordinary condition and lose their viability. The seeds can be stored in polythene bags and kept in the cold store at 2 °C upto 24 months. For commercial cultivation seeds are sown on raised beds made up of river sand and FYM. Four seed beds of size 7.5 m long, 1.2 m wide and 20 cm high are sufficient for raising nursery required for one hectare planting. Beds should be mixed with 10 per cent carbaryl dust and later drenched by captan or mencozeb solution (0.25%) before sowing of seeds. The seeds are sown thinly at a spacing of 10-12 cm apart and covered with a mixture of soil and farmyard and irrigated daily with water can having fine nozzle. After one month seedlings are ready for transplanting. Transplanting should be done during early morning or evening hours to avoid high temperature. After transplanting the soil around seedlings should be firmed and watered thoroughly. Treating the seedlings with 0.3 per cent solution of cooper oxy chloride at planting stage helps in avoiding the incidence of Wilt disease. The seed rate requirement is 2.5-3.0 kg/ha and sowing should be staggered by 10 to 15 days to ensure regular supply of flowers and avoid glut in the market. If seedling are held for an extended period before transplanting, its feeding once with a light application of a well balanced fertilizer at 50 to 100 ppm is advisable.

Land Preparation

The field should be ploughed thoroughly and brought to a fine tilth. About 10–15 tonnes of well decomposed farmyard manure may be incorporated in the soil at the time of soil preparation. In area receiving high rainfall the seedlings should be cross-planted on ridges to avoid chances of *Fusarium* wilt.

In China aster normally recommended spacing is 30 cm between plant to plant and row to row. In light soil 45 cm x 20 cm and in medium soil 60 cm x 20 cm planting distance should be maintained by following ridges and furrows method.

Manure and Fertilizers

Proper manuring and fertilization are important for aster cultivation. Deficiency of nutrients results in poor growth and flowering. Nitrogen deficiency cause stunted growth and production of dull coloured flowers. Phosphorus deficiency affects vegetative growth and delays flowering. China aster responds well to micronutrient application of Zn, B, Cu and Mn and improves quality of flowers. Application of a complete fertilizer improves growth, flowering and seed production and wilt resistance in áster. Application of decayed manure as 5-6 kg/ m^2, followed by 45-60 g sulphate of ammonia, 45-60 g super phosphate and 15 g potassium chloride/m^2 increased plant height and flower yield. Top dressing with 120 kg N, 80 kg P_2O_5 and 120 Kg K_2O/ha at the 10-12 leaf stage resulted in early bud formation and flowering. In Bangalore, recommended dose of N, P, K *i.e.*, 90, 60, 60 kg/ha, respectively, at the time of preparation of land and a top dressing of nitrogen 90 kg/ha, 40 days after transplanting proved beneficial. Nutritional trials conducted at Kalyani revealed that application of 400 kg N along with 200 kg P_2O_5 gave highest yield of flower. In Pune conditions application of 300 kg/ha N and 200 kg/ha P improved yield.

Planting

Seedlings are ready for transplanting within 30 to 45 days, when they have developed about three to four leaves. The seedlings are hardened sufficiently before planting. The transplanting should be done preferably during early morning or evening to avoid bright sunshine. After planting, the soil around the seedling is to be firmed and watered thoroughly.

Transplanting time is crucial in the success and failure of China aster crop. In India, it is grown in different months in its major growing areas. In and around Bangalore the crop can be grown throughout the year but normal growing season are May –June and September –October. In and around Delhi or places having similar type of weather conditions it is not advisable to grow aster during winter season, because frost occur during this period. Therefore, sowing and transplanting can be taken up from August –September. Similar to Delhi climate, planting may be advanced to July, where rainfall is low. In West Bengal, the best time for transplanting of aster is during first week of October. It is also observed that November planting of China aster seedlings produce flowers with larger diameter, whereas planting in the early season (September) results in early flowering with longer flowering period.

Irrigation

China aster is a shallow rooted crop and therefore, requires continuous soil moisture. Irrigation requirement depends upon the weather, type of soil and season of the crop grown. In general, irrigation at intervals of 7 to 10 days is recommended. In a study, three soil moisture regimes (pF 2.0, 2.5, 2.8) was compared and observed that plants grew best with medium regime (pF 2.5). Supplementary irrigation of aster field was found highly beneficial.

Cultural Operations

Weed control: Regular hand weeding is required in early stages of crop growth but it is very expensive and may not be economic for large scale cultivation. Therefore, use of some suitable herbicide is advisable. Application of CDEC and chlorprophan and dectahl (dimethyl 2,3,5,6-tetra chloroterephthalate) at 9-13 Kg/ha after 4-6 weeks of transplanting was found very effective in controlling weeds. Application of veenolate at 8.5 kg/ha as preplant-treatment showed excellent weed control in aster. Spraying of oxadiation at 1.25 Kg/ha at three days after planting of aster gave best results in long term weed control.

Pinching: Pinching is found beneficial in some aster varieties. Pinching of main shoot at one-month after transplanting in variety Ostrich Plume Purple had significantly increased number of nodes, branches, flowers per pant, flower yield/ area and delayed flowering by 8-12 days.

Plant Growth Regulators

There are different types of plant growth regulators which influence growth and flowering of China aster. The number of flowers per plant and the duration of flowering increased in China aster by spraying of GA_3 at 200 or 300 ppm. Application of MH delays the flowering while NAA had little effect on growth and flowering.

Foliar application of GA_3 at 500 ppm produced more number of branches per plant than control and anthesis got enhanced by GA_3 at 250 ppm. Spraying of GA_3 improved the flower quality but did not influence flower yield. It was observed that cycocel spraying and pinching the plants at 30 days after transplanting significantly reduced the plant height and produced more number of flowers. Foliar application of SADH when the stem began to elongate results in compact growth of plants.

Diseases

Wilt: This disease is caused by *Fusarium oxysporum* f. *callistephi* but it has also been observed that *Acrostalagmus vilmorinii* and *Verticillium albo-atrum,* together with *F. oxysporum* f. *callistephi* also cause this disease. Vascular wilt disease of aster caused by *Phialophora fastigiata* was first reported in New Zealand and cause stunted plant growth. Soaking of seeds for 30 minutes in 0.1 per cent mercuric chloride solution and steam sterilization of soil are effective for preventing wilt occurrence.

Collar rot: It is caused by *Phytopthora cryptogea* usually under high moisture conditions in soil. Symptoms are sudden wilting of the plant and rotting of stem at the ground level and root rot. To avoid this disease planting should not be done in the field which has shown the disease in previous years. Soil should be sterilized and ridge planting should be practiced.

Grey mould: It is caused by *Botrytis cinerea*. During cool and humid weather this disease spreads rapidly. Incidence of this disease can be reduced by proper drainage and removal of residues and weeds. Spraying of zineb or indofil M-45 at 1.5 g/l water is found effective to control this disease.

Rust: Caused by fungus *Coleosporium solidaginis*. Bright yellowish orange spots appears on the lower surface of the leaves. Its appearance is more prominent on young plants. This disease can be controlled effectively by spraying of calixin or benlate (0.1%) during the grown season.

Leaf spot: Various types of fungus such as *Ascochyta asteris*, *Stemphylium callistephi* and *Septoria callistephi* are associated with this disease. Yellowish spots first appear in this disease and then become dark brown and black. Size of spot increases gradually. In this disease, lower portion of leaves is infected first. Spraying with Bavistin (0.1%) or blitox (0.3%) control this disease.

Stem rot: It is caused by *Pellicularia flamentosa*. Plant dies due to rotting of stem. Use clean soil and treatment with PCNB (terraclor) are effective control measures.

Canker: *Phomopsis callistephi* causes canker. It appears on lower portion of stem. The distal part is collapsed in this disease but roots of the plants are not affected.

Yellows: This disease is caused by the virus *Chlorogenus callistephi* which is transmitted by a leafhopper, *Macrosteles fascifrons*. The disease is characterized by large number of adventitious shoots and pale leaves. In extreme conditions plant becomes dwarf and flowers are off colour. Destruction of diseased plants immediately after the appearance of the symptoms and control of leafhopper with spraying of Dimethoate 30 EC or Methyl Demeton 25 EC at the rate of 2ml/l of water helps in minimizing the spread of the disease. Disease development can be suppressed by the use of antibiotics like chlorotetracycline and tetracycline. This disease can also be controlled by root treatment with oxytetracycline.

Chrysanthemum mosaic virus: It is a seed borne disease resulting in the production of distorted blooms. Disease free seed can be used to control it.

Phylloidy: It is caused by mycoplasma. Two sprays of systemic insecticides like monocroptophos 36 SL @ 1.5 ml/l can effectively control it.

Insect Pests

Leaf beetle: Tiny beetle (*Aulacophora foveicollis*) attacks newly planted crop and makes holes on leaves and cut tender shoots of China aster. Soil application of phorate 10G @ 1.0 kg a.i./ha is recommended to control it.

Semilooper: Semilooper (*Ctenoplusia albostriata*) is severe during June. The loopers feed on leaves. These insects can be controlled effectively by spraying of quinalphos 25 EC @ 0.5 ml/l or carbaryl @ 1.0 ml /l.

Stem borer: Larvae of stem borer (*Platyptilia molopias*) bore into stem and side branches and results in hollowing and wilting of affected shoots. Collect and destroy the infected shoots. Spray with profenofos 50 EC @ 1 ml/l of ethofenprox 10 EC @ 1 ml/l.

Flower eating caterpillar: Flowers of China aster are mainly damaged by *Helicoverpa armigera* and *Phycita* sp. These insects feed on flower head by making groves in the receptacles. They feed on the ovaries and stamens of the flower buds. Spray with neem seed kernel extract 5 per cent will manage the insect damage. If the incidence is severe, spray with phenthoate 50 EC @1 ml/l or fenvalerate 20 EC @ 0.5 ml/l or monocrotophos 36 SL @ 1.5 ml/l water is effective to control flower eating caterpillars.

Leaf hopper: Leaf hopper (*Macrosteles fascifrons*), feed on plants and also transmits yellows virus. Treatment with dimethoate 30 EC @ 2 ml /l water control this insect effectively.

Black blister beetle: The black blister beetle (*Epicauta pennsylvanica*) is also known as aster beetle. This insect attacks on plant and damage leaves and flower heads. Infestation by this insect can be effectively controlled by spraying of malathion @ 2 ml/l water.

Aphids: Aphis gossypii mainly infest lower side of the flowers and base of petals. Nymphs and adults suck sap from the flowers causing discoloration, deformation and withering. Spray with oxydemeton methyl 25 EC @ 2 ml/l or dimethoate 30 EC @ 2 ml/l. Repeat the spray with neem or pongamia oil 1% after 10 days.

Potato aphid: This aphid (*Macrosiphum solanifolii*) attacks on all the tender parts of the plant. It damages foliage, stems and flowers. This insect can be controlled by spraying dimethoate 30 EC @ 1.0 ml/l or malathion @ 1.0 ml /l.

Spider mite: Spider mites (*Tetranychus telarius*) are very small non-insect arthropod pests. Damaged leaves become dull in colour. Spray of kelthane 50 WSP (0.2%) or metasystox 25 EC (0.1 %) is found effective to control this pest.

Leaf miner: This insect (*Liriomyza compositella*) feeds on foliage and leaves become disfigured. Adults lay eggs on leaves and maggots feed inside leaves and flower calyx, leaving characteristic mines. Spraying of monocrotophos 36 SL @ 0.5 ml/l is effective for its control.

Tarnished plant bug: This bug (*Lygus lineolaris*) is light brown and deeply spotted insect. This is very destructive pest and punctures the terminal shoot below the flower buds which causes drooping of flower and ultimately drying. Effective control measure is spraying of metasystox 25 EC (0.1%).

Nematodes

China aster is also affected by nematode. Infestations are caused by *Meloidogyne incognita* and *Aphelenchoides ritzema-bosi*. The leaves of infested plants become dwarf, malformed and growth is checked. Nursery bed treatment with *Trichoderma harzianum* @ 50 g/sq.m. application of 2 tons of farm yard manure enriched with *T. harzianum* and *Paecilomyces lilacinus* per acre before transplanting, along with 100 –200 kg of neem or pongamia cake will manage the nematodes.

Harvesting

Right stage of harvesting plays an important role in determining the flower quality. China aster is harvested in two different ways. Loose individual flowers are harvested for decoration, making garlands and worship purposes, whereas cut flower along with stalk or the whole plant just above the ground are cut (when they develop their original colour) for cut flower purpose. The harvesting should preferably be done in the early morning or evening hours. Stems are tied in bundles each consisting of 20 flowers. In case of loose flower producing cultivars, harvesting is completed in 8 to 10 pickings in early and 15–20 pickings in late varieties. The loose flowers are kept under shade after harvesting.

Postharvest Management

Cut flowers with stalk are put vertically in a container having clean water immediately after harvest. Lower leaves attached to the stalk should be removed to minimize transpiration. Grading is done according to stalk length, size and shape of flowers, freshness and other quality. The flowers are graded and cleaned by removing unwanted lower leaves and bunches are prepared. A bunch accommodates 3–5 plants and 25–30 bunches are kept together to make one bundle. While packing, care should be taken so that there is no direct contact between the flower and packing material. Clear plastic films improve the attractiveness of flower and also provide protection to the flowers during transportation. The post-harvest life can be extended by addition of floral preservatives. A specific odour, which is usually generated by China aster stems kept for a long time in vase water can be reduced by addition of few drops of formaldehyde in the water. For prolonging vase life of asters, holding solution of 0.2% $Al_2(SO_4)$ and 2 % sucrose is advisable.

Yield

The yield of China aster flower varies with cultivar and cultural practices. About 10-12 tonnes/ha flower is produced under normal conditions, whereas, with improved package of practices a yield of 18-20 tonnes flower/ha can be obtained. About 400-500 big bundles of flower containing 30 bunches in each bundle can be produced from one ha area.

References

Bose, T.K. and L.P. Yadav (1989). Commercial Flowers. Naya Prokash, Calcutta, India. Pp 874.

Chadha, K.L. (2006). Handbook of Horticulture. Indian Council of Agricultural Research, New Delhi. Pp 1031.

Kanwar Pal Singh, Namita, Prabhat Kumar and Sapna Panwar (2014). Avenues in Floriculture For Livelihood. Division of Floriculture and Landscaping, Indian Agricultural Research Institute, New Delhi. Pp. 160.

Randhawa, G.S. and A. Mukhopadhyay (2007). Floriculture In India. Allied Publishers Pvt. Ltd. New Delhi. Pp. 656.

Sindhu, S.S., Kanwar Pal Singh, Gunjeet Kumar (2008). Alankrit Bagwani. Division of Floriculture and Landscaping, Indian Agricultural Research Institute, New Delhi. Pp. 130.

Ornamental Horticulture, pp 389-403

New India Publishing Agency, New Delhi - 110034, India

17

Dried Ornamentals: An Option for Commercial Floricultural Enterprises

Desh Raj

Flowers and plants are associated with mankind from the dawn of civilization. In the modern era these have become an integral part of human life, as love for flowers and ornamental plants is a common natural instinct in human beings. The catchy flowers glowing and glowing in all their grandeurs with simplicity always make them tempting to wish the season to continue for never ending, but it may be beyond reach of human beings. However, the answer for it could be by producing and preserving ever-lasting or long-lasting flowers and other attractive ornamental plant parts. These are now becoming favorites especially with the adventurous flower lovers across the globe. The beauty and fresh look of cut flowers and cut foliage is generally lost due to microbial activities and biochemical changes thus these are retained only for a few days even by using the best techniques of post harvest management (Datta, 1999). The charm of dried ornamental plant parts can be maintained from a few months to years with lesser cost if protected from the damage of high humidity as in dried flowers, the microbial activities in the aging process come to stand still.

The main characteristics of dried ornamentals are novelty, longevity, aesthetics, flexibility and year-round availability. The dehydrated or dried

Department of Vegetable Science and Floriculture, College of Agriculture, Palampur, Himachal Pradesh

ornamentals are generally inexpensive and are sought for their everlasting and attractive appearance globally (Smith, 2000). Being largely focused on the qualities of attractive flowers, living plants and plant parts, floriculturists generally pay little attention to dried and preserved flowers and plants. Therefore, these products remained overlooked and underrated throughout the world. These are overlooked in the sense that their considerable economic importance is generally unrecognized and these are underrated in the sense that their aesthetic contribution is largely discounted. In contrast to other areas of floriculture including post-harvest management, relatively very little public funded research and development projects were undertaken on the dried flower industry across the globe (Joyce, 1998). This has slowed the pace in optimizing the dehydration technology and quality of the produces. Privately funded research and development work has sustained, but such findings are usually kept secret and limited information are available to the common man. Thus, this component of floriculture industry has remained ill defined with respect to the processes, products and organizations in the developed and developing countries. As we see in major cut flower crops, there are defined standards available for judging the quality, however these are lacking in dried ornamentals. There is an urgent need to start war- footing efforts and awareness on the dried ornamentals to attract greater public support to conduct concentrated research and development with more professionalism in the trade of these products.

Status of Dried Ornamentals

The demand for dried ornamentals, dried floral arrangements and floral craft items has increased manifold during the last two decades. The world import of dried flowers and plants has increased from 14.88 million US$ in 1996 to 230.73 million US$ in 2008 and decreased to 167.34 million US$ in 2009 (*Source*: UNCOMTRADE). The United Kingdom is the biggest market for dried flowers. The leading dried ornamental products exporting countries

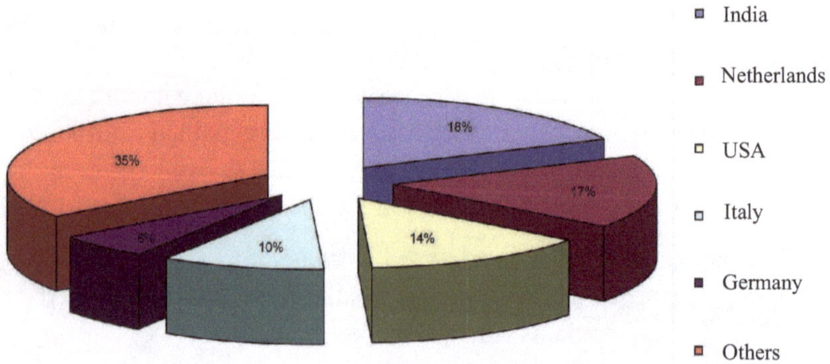

Export of dried ornamentals in the world during 2009

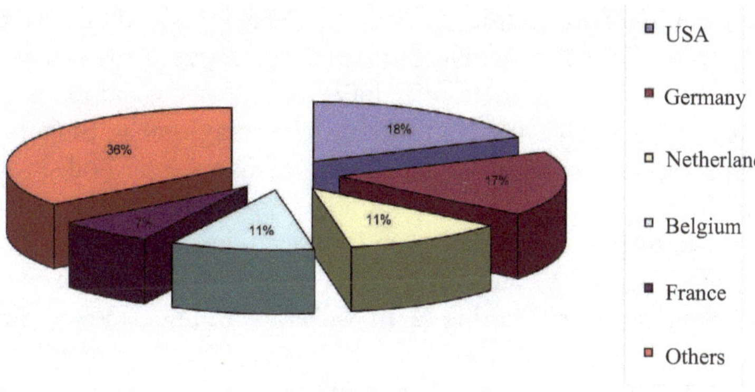

Import of dried ornamentals in the world during 2009

during 2009 in the world were India (18%) followed by the Netherlands (17%), USA (14%), Italy (10%) and Germany (6%). The leading dried ornamental products importing countries during 2009 in the world were USA (18%) followed by Germany (17%), the Netherlands (11%), Belgium (11%) and France (7%).

The dried ornamentals market is the most profitable in terms of total sales and unit value among all the floricultural products. The top ten dried flower genera as ever-lasting cultivated are *Helichrysum, Helipterum, Limonium,*

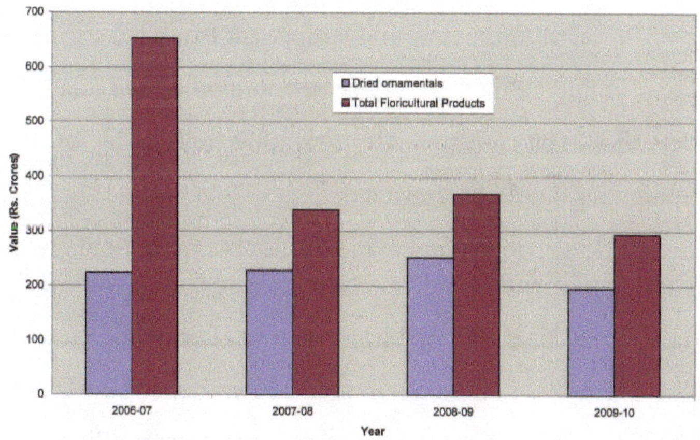

Nigella, Gypsophila, Delphinium, Amaranthus, Papaver, Carthamus and *Rosa* (Anonymous, 1989).

In India flower export has sharply increased after liberalized EXIM policy from Rs. 145 million in 1991-92 to 6527 million in 2006-07. However, total export of floricultural products has now decreased to 2945 million during 2009-10. The contribution in total export of dried flowers and plants is over 60 per cent and at present the export from India is over 20 thousand tones of dried ornamentals to USA, UK, the Netherlands, Germany, Japan, Singapore and other European countries. The main exporting items are lotus pods in addition to dried flowers of camellia, dahlia, bell cups, marigold, jute flowers, wood rose, wild lilies, paper flower and naturally dried plant parts from Himalayan region. The dried flower business is confined to about sixty private companies operating from Kolkata, Tuticorin (Tamil Nadu) and Mumbai close to sea ports, as the material being bulky is usually shipped through the sea routes (Raghava, 2001). Ramesh Flower Limited at Tuticorin has about 50 per cent share of the total dried flower export from India.

Ornamental Plant Parts for Drying

The range of dried flowers and other attractive plant parts is extensive *viz.*, stems, roots, shoots, buds, flowers, inflorescence, fruits, fruiting shoots, cones, seeds, foliage, bracts, thorns, bark, lichens, fleshy fungi, mosses, sellaginellas, etc. (Desh Raj, 2001).

Precautions During Dehydration

For insuring better results and quality products ,the following points should be considered:

i. Collect material after the dew and surface moisture have evaporated.

ii. Collection should be made in dry season and on a sunny day.

iii. Materials should be collected after irrigating fields a day or two.

iv. Fresh materials should be collected.

v. All stages of flower development in an inflorescences should be collected, which have sufficiently hardened as immature shrivel very fast.

vi. Materials should be embedded immediately after plucking.

vii. One type of flowers and foliages should be embedded at one time.

viii. Undesirable plant portions should be cut before embedding.

ix. Spread uniformly all plant parts/ petals in herbarium press.

x. The plant parts should be harvested when the highest possible quality product can be harnessed.

Used of Dried Ornamental Plant Parts

These dried items may be used with fresh flowers or alone as floral arrangements, bouquets, gift boxes/packs, festive decorations, collages, flower pitchers, floral balls, pomanders, wall sceneries and sweet smelling pot-porris which have become very popular among the flower lovers especially in European countries.

Pot-Porris

Pot-porris, a special dried floral arrangement, is the mixture of sweet smelling leaves, spices, seeds, roots and distilled essential oil which are filled in pillows or sachets. The important dried ornamental plants used for making pot-porris are roses, lilacs, lavender, pinks, hyacinths, lilies, violets, wall flowers, marigold and many others which are associated with perfumes viz., geranium, daisy bush, bergamot, sage, savory, thyme, angelica and sweet cicely. In addition, foliage of ferns, pine- cones, lily pods and driftwood and other items from natures' treasure of exotic bloom are collected from forests and gardens. The materials used should be resistant to mold, non-toxic, free from noxious odors and sufficiently robust to withstand mechanical bending. Pot-porris is normally designed in a transparent glass bowl or a ceramic jar or stored in a colorful salin or muslin sachet, while other arrangements are designed in a beautifully colored or specially shaped pots (Anonymous, 1999 and Datta, 1999).

Dehydration Technology

Dehydration technology means to dry something under artificially produced heat through controlled temperature, humidity and air- flow. In dehydration process, moisture is removed from flowers and foliage. In nature the plant materials tend to discolor during drying largely due to oxidative reactions associated with loss of compartmentalisation within the plant cells during desiccation of the plant tissues. Such plant- material turns brown and is particularly used as filler in various floral arrangements. Artificially there are many varied approaches/ methods to dehydrate or dry flowers and ornamental plant parts as described by many workers (Kher and Bhutani, 1979; Dubois and Joyce, 1989; Westland, 1993, Westland, 1995 and Bhutani, 1995). The principle involved is common to all processes, as per which the plant material is exposed to a vapor pressure deficit (v.p.d. = $\text{v.p.}_{\text{source}} - \text{v.p.}_{\text{sink}}$), which induces water vapor to move by transpiration/ evaporation from the plant material (source) into the surrounding environment (sink). The flux of water vapor (J) is proportional to the vapor pressure deficit: *viz*., J = k x (v.p.d.); where k is a constant dependent upon the water vapor transfer properties of the particular product (Joyce, 1998).

1. Natural Drying

In this method, ornamental plant parts are allowed to dry naturally on the plant itself. These are collected as and when completely dried during excursion trips to forests, countryside, gardens or cultivated fields. The important naturally dried plants were identified by Desh Raj and Gupta, (2002) in the outer Himalayan region as follows:

The plants viz., *Aegle marmelos, Bambusa* spp., *Bauhinia retusa, Cassia fistula, Caesalpinia sepiaria, Clematis grata, Dioscorea deltoides, Mallotus philippensis, Rosa moschata, Oroxylum indicum, Pinus roxburghii, Picea smithiana, Sapindus mukorossii*, etc. have beautiful fruiting shoots, whereas plants like *Abrus precatorius, Aesculus indica* and *Sapium sebiferum* have beautiful seeds.

2. Vertical Hanging for Drying in Air

This is the most common process for drying ornamentals. In this process only rope/ wire and newspaper/ blotting sheets are required. Plant material is simply kept hanging down under cover on racks or from rafters in dark or kept on newspaper/ blotting sheets. Drying in the sun result in discoloration and browning due to oxidation. Improved control over the rate of drying can be achieved by ensuring air movement around each and every stem and by lowering the relative humidity of the air. Good air circulation assists in sweeping away moisture vapors, thereby decreasing the thickness of the unstirred boundary layer of the water vapor with fresh plant material in surrounding. Relative humidity is lowered by adequate ventilation, heating, dehumidifying and pressure reduction. Ventilation with outside air will flush away the moisture vapors released during the process of drying the plant material. Heating will lower the relative humidity because warm air can hold more water vapors per unit volume than can cool air. Kiln drying is the common method of flower drying in India followed by most of the export oriented units in South India. Another energy intensive system is air tunnel, which operates on a convection air system at a minimum 50°C temperature. Dehumidification can be achieved with moisture scrubbers (e.g. silica gel- based or by frosting water vapor out into cold coils. Pressure reduction may be achieved by large exhaust fans fitted to rooms with a restriction imposed on the supply of the air flow or with vacuum pumps. Thus, the entire process of air-drying depends upon the relative humidity, air velocity, pressure, moisture in plant material and type and shape of flowers.

This is the most simple and cheep process of dehydration/drying. The main drawback is its weather dependence and petals/foliage shrinkage, thus lowering the aesthetic value of plant material. Hanging drying of small quantity of plant material with reasonable success can be achieved by drying in microwave oven, hot air oven or solar dryer within short span in comparison to air-drying. Flowers with crisp texture viz., *Acroclinum, Anaphalis, Helichrysum, Delphinium*, Lemon mint, *Oregano, Limonium, Rumex, Holmskioldia*, etc. are more suitable for vertical hanging drying in microwave/ hot air oven/ solar dryer.

3. Embedded Drying

It is rather difficult to avoid shrinkage and changes in morphology of dehydrated ornamental plant material during vertical hanging drying in air/ microwave/hot air oven/ solar dryer, due to the loss of moisture/ water vapors from the cells. The petals/foliage generally droop- down and as a result aesthetic value of that material reduces. In embedding the drying medium/desiccant supports the flower/ foliage from all around and maintains perfectly the original shape.

Characteristics of Desiccants

The material which removes the moisture quickly from the ornamental plant parts embedded without reacting with the water vapors released during drying or bleaching of petals and foliage, *etc*. The drying material/ desiccant should normally have the ideal size of 0.02-0.2 mm or 20-200 mesh. It should be heavy to keep the plant parts perfectly in its original shape. It should not stick to the plant parts during drying. It should also not be very costly as the market value of most of the dried ornamental floral arrangements and craft is low.

Commonly used desiccants

Silica gel (white and self indicating blue), borax, boric acid, river sand, alum powder, aluminum sulfate, saw dust and corn granules are commonly used desiccants in purely single form or in combinations of two or more desiccants so as to be more effective and economical in drying ornamental plant parts.

Embedding process

Various types of containers are used for embedding different ornamental plant parts which suit to the plant materials and the process of drying in room, sun, solar dryer, hot air oven or microwave oven. The containers should uniformly get heated so that the material evenly dry out from all sides. Aluminum or tin containers are normally used which may be cylinder or tray type. In microwave nonmetallic containers like glass and china clay, heat tolerant plastics are used. Spread the layer of the desiccant on the bottom of the container. Arrange the plant parts according to type, well spaced and with no two items touching each other or the sides of the containers. Pour the desiccant gently, gradually and completely from all the sides until it forms a top layer for about 1.5 cm above the plant parts. In flowers where pedicels/ inflorescence are very delicate or hollow, insert a thin wire before embedding so as to maintain the original shape. After embedding ,the containers are left in the room for drying. But it takes longer time to dry. The containers for quick drying are kept in sun, solar dryer, hot air oven or microwave oven, in which drying is faster with superior quality of the dried plant parts. These are described as follows:

i. Drying in Sun

After embedding the plant materials, the containers are daily exposed to sun, resulting in rapid dehydration. Containers are shifted under a room in the evening and again on the next morning these are kept under the sun light. The efficiency of drying depends upon the prevailing weather in the region.

ii. Drying in Solar Dryer

The containers are kept in solar dryer in which movable tray at the bottom is fitted to regulate the temperature. This is a quicker method of drying and drying is done without use of energy. The results are at par with hot air oven or microwave- oven thus is a more economical method.

iii. Drying in Hot Air Oven

The containers are kept in the hot air oven at 45 to 60°C for a few hours to three days depending upon the moisture content in the plant material to be dried. Drying technology for number of cultivated ornamental plants has been standardized at NBRI Lucknow by Kher and Bhutani (1979).

iv. Drying in Microwave Oven

The plant material in microwave oven is dried with the help of electronically produced microwaves. Non-metallic containers in which various plant parts are embedded in desiccants are kept in microwave oven for 2-5 minutes. These are then kept in the room for 2-5 hours known as setting time before taking out from the desiccant. In microwave oven ,the mostly plant material is dried at a range of 300-700 Hz., which may be given once or in splits. The results in split drying are comparatively better.

v. Vacuum Drying

The embedded plant material may be dried/ dehydrated under vacuum. The vacuum is created under thick-walled chamber fitted with a heating device, a vacuum pump and condenser. Low temperature is very effective and the quality of the product is excellent.

Taking Out Embedded Plant Material

The embedded plant material after drying/dehydration is taken out by gently tilting the container to remove the desiccant over and around the plant material. The dried material is either picked up with the help of fingers or tweezers. It is cleaned by inverting and tapping with fingers slowly and gently. Any remaining desiccant is finely removed with the help of a fine camel hair painting brush.

4. Drying in Freezer

Freeze-drying is now becoming very popular although a more expensive method of producing dried flowers and plant parts throughout the world. The worldwide market of freeze-dried products during 1997 was about $18-20 million. The necessary equipment for freeze-drying is more expensive than vacuum drying, but the quality of the finished floral product is almost double and this process is relatively simple. The flowers are arranged in the specimen chamber. Then, these are frozen unto -35° C. Any frozen ice crystals on the flowers are sublimed or vaporized with the application of heat. After transforming into condensation, the vapor is captured into a separate chamber as ice. By eliminating the water, the flowers dry up with life freshness and retain better integrity and more durability. Thus, rapid dehydration particularly at low or ambient temperature tends to result in better-colored products (Anonymous, 1997 and Chen *et al.*, 2000).

5. Drying in Water

The flowers and foliage are kept in water and allowed to dry. Water drying, which usually seem to be a contradiction in terminology, but gives fairly good results with flowers like hydrangea, corn flower, babys' breath and fever few. In water drying first the lower leaves and flowers from the stem are stripped and are kept in standing cold water with about 5 cm depth. The material is kept undisturbed in a dry, well-aerated room until both the holder and flowers are dried. This takes 6-7 days depending upon the water content in plant materials and water taken in container/holder.

6. Press Drying

The flowers and foliage are placed between the pages of any old book or magazine or between the folds of newspapers or blotting sheets by giving some space between the flowers/foliage. These sheets are kept one above the other and corrugated cardboard sheets of same size are placed between the folded sheets so as to allow the water vapors to escape. The whole bundle is then kept in the room. For quicker drying herbarium press can be kept in hot air oven at 45°C for 12 to 24 hours depending upon the moisture in the flowers or foliage to be dried. Take care that too much material is not pressed simultaneously. Materials of varied thickness should also be avoided in the same folders. When press drying is practiced in room, one must over-turn the flowers/foliage daily which will help in maintaining the original color and prevent the plant material from discoloration. This method is laborious and time taken for drying is more than drying in solar dryer/ hot air oven.

The press-dried material can be stored in cardboard boxes. These boxes are lined with blotting sheet/ paper, which keep flowers/foliage dry for longer duration. These boxes are better stored in herbarium vasculums fitted with cork sheet. The press-dried material is used for preparation of various kinds of greeting cards, wedding cards, wall plates, sceneries, table- tops, interior designing and other festive decorations.

7. Polyset Preservation

Polyset is a polymer preservation method and is applied to the flowers/ foliages about 45 minutes before drying. It is a chemical pre-treatment application, which is used before air drying to improve the quality of the dried flowers. This method lessens the drying time and improves the intensity of flower color. It also minimizes shattering and wrinkling of the petals, which may occur during drying process. Polyset treatment is applied to the freeze-dried flowers to hold them together or to help the flowers in retaining their color for longer duration.

8. Humectants Preserving

Ornamental plant parts can be preserved (plasticized) by treatment with humectants. Humectants are hygroscopic chemicals, which help to maintain the suppleness of dried plant material by attracting water vapor from the surrounding atmosphere. Therefore, the preserved plant material is less brittle than dried material, making it less prone to shattering and mechanical damage. This material is more life like in appearance and touch. Humectants used for preserving plant tissues are drawn from four general groups of chemicals *viz.*,

Diversity of Dry Flower Products

sugar and sugar alcohols (e.g. Glycerol), polyols (e.g. Polyethylene glycol), salts (e.g. KCl) and quaternary ammonium compounds (e.g. Comfort). These humectants have different moisture sorption isotherm characteristics. These can be applied by uptake or immersion, where the former method is not used commercially. Relationships between specific plant characteristics and humectants uptake are poorly defined, despite the common perception that some species preserve easier and better than others. Although relatively more volume of solution is required for immersion method of preservation but this offers some unique advantages as dried and/ or bleached plant parts can be treated. There is more uniform penetration and distribution of the humectant. Waxy plant cuticles are removed by NaOH or organic dip treatment prior to preservation in order to facilitate penetration of aqueous humectant solution. The risk of sweating can be reduced by using any humectant with a relatively low moisture attracting capacity and/ or by reducing the amount of humectant taken up i.e. shorter treatment time at low concentration of humectant (Joyce, 1998).

9. Bleaching and Sulfuring

Otherwise aesthetically attractive plant material, which is inherently colored by unwanted pigments or brown or discolored due to oxidative browning can be bleached or sulfured for their use in various floral arrangements. In bleaching, both the oxidative or reductive bleaching agents can be used on the plant material, where former tends to break down colored compounds and the latter tends to modify into colorless compounds (Dubois and Joyce, 1988). Multiple bleaching specially with hypochlorite- chlorite peroxide as used for lilen offers many advantages in terms of reduced brittleness, cost and yellowing. In sulfuring, sulfur dioxide acts to bleach colored (e.g. green) plant material and below certain concentration, to fix red colors in some flowers. Color fixation is associated with acidification of the tissues. Sulfuring of fresh material is achieved by burning sublimed sulfur powder (e.g. $10 g/m^3$) or by injection of sulfur dioxide gas (1-3%) into a sealed chamber. The plant material is usually treated over night prior to ventilation of the chamber and subsequent completion of the drying process. Both bleaching and sulfuring are dangerous processes by virtue of the use of chemicals, thus should be taken up by trained persons with all safety equipment.

10. Colouring

Colours or dyes can be applied to fresh, dried or bleached ornamental plant parts to increase their aesthetic value as per the changing seasons and fashions. Dyes are usually applied by immersion to dried and bleached plant material. In Australian plants *viz*., *Banksia hookeriana* and *B. prionotes,* inflorescence is commonly marketed after coloring. Dyeing of ornamental plant material is considered much as an art. Accordingly for determining the optimum treatment protocol for a particular product is often a matter of systemic trial and error. Immersion dyeing is improved by removing the waxy cuticle with NaOH from dried plant materials and by adding surfactants to improve the contact between the dye bath solution and the plant material. In addition to dyeing, which is properly called staining the plant material can also be coloured by painting and commonly silver or gold paints are used.

Packaging, Handling and Storage of Dried Ornamentals

The major benefit associated with dehydrated/dried plant material is its relatively non-perishable nature as compared to fresh parent plant material. As a result ,packaging and handling should theoretically offer no real problem, but due to less moisture, dried plant material is very brittle. Packaging for delicate dried plant material should serve to contain and protect the product both. It is advisable to use superior grade or standard cartons/ boxes for packing dried plant material. Dried plant materials should not be roughly handled during transportation and distribution. It can be done through proper education by creating awareness of the product characteristics. Dried plant materials should be protected from moisture throughout the marketing channel. Dried plant materials absorb atmospheric moisture and loose its shape, therefore, should be stored immediately in moisture proof containers. Different containers like glass desiccators, tin boxes, cartons, wrapped with plastic sheets or wax paper and herbarium vasculums fitted with cork sheet are used for storage of dried ornamental plant materials. Small quantity of silica gel sachets may be placed at the bottom and sides to absorb moisture. Protect the materials from direct sunlight or more light intensity espcecially from incandescent lamps. The storage should be dust free and keep the cartons/ boxes clean from time to time. Store separately different types of dried ornamental plant material and fix proper label outside with complete description of the product and materials kept inside, which will be very helpful during handling, transportation and marketing.

Insect-Pests and Diseases

Insect-pests can badly affect all types of dehydrated plant materials. These can be controlled by application of various insecticides applied in solid pest strips (dichlorvos), liquid (synthetic pyrethroids, ethyl parathion- 0.01%) or gas (methyl bromide, phenyl tablets). Book lice, silver fish and mice are the major pests infesting dried plant materials. Microbes notably fungi can be controlled by regulating the relative humidity to keep the water activity of the dried plant material below the level, which does not allow spore germination, mycelial growth and sporulation. As a general principle, the development of fungi is negligible below 65% relative humidity (Wilson and Abramson, 1992). The dried plant materials should be packed at low relative humidity and the containers should be properly sealed which preclude fungal growth and prevent suppleness during handling later. The main genera of fungi viz., *Aspergillus, Penicillium* and *Rhyzopus* infest dried plant material. The plant material before collection should be treated with Dithane-M. 45 (0.2%). Sulfur burning or sulfur dioxide fumigation also controls these fungi during storage.

Marketing of Dried Ornamentals

In comparison to fresh ornamental plants ,the profitable marketing of dried ornamental plant material requires understanding of various aspects viz., economics, psychological attraction for such products and physical considerations like ware- housing, distribution and sale facilities. Being delicate materials, losses are more and little attention is given to maintain the quality. This sector could get the least attention; therefore, both consumers and producer s are at loss, whereas middleman is earning the huge profit. Bhutani (1993) carried out the feasibility study for export, inland marketing and commercial exploitation of dried ornamental plant materials and floral crafts. It is possible to start a home scale establishment with a meager amount of Rs 15,000 and one may supplement his/her income by employing other members of the family. If properly explored, it can develop into a successful cottage industry in our country especially in rural areas. As industry, it is poorly defined in terms of producers, products, handling systems, marketing and consumers/ customers. It can earn huge financial gains if professional profile is enhanced as a discrete entity. Further quality assurance has to be given and followed in respect of processing conditions, defined temperature and relative humidity limits, proper description of the product, performance characteristics, longevity, uniformity in grading, packing and overall quality of the product. This warranted more scientific investigations on dehydrated/dried ornamental plants to develop it into a full-grown separate industry.

References

Anonymous, (1989). Everlasting flowers from Holland: a new product line in the ornamental range. *Agri-Holland*, 1: 1-5.

Anonymous, (1997). A report on Indian dry flower Industry. APEDA, New Delhi, 65p.

Anonymous, (1999). Making Pot-porris. *Floriculture Today*, 4 (6): 44-45.

Bhutani, J. C. (1993). Economic potential of dried flowers. *Agrl. Marketing*, 36 (1): 43-46.

Bhutani, J. C. (1995). Drying of flowers and floral craft. *In:* Advances in Horticulture-Ornamental Horticulture Vol. 12 (Eds K. L. Chadha and S. K. Bhattacharjee), Malhotra Publishing House, New Delhi, 1053-1058pp.

Chen, W., Gast, K. L. B., Smithy, S. and Chen, W. (2000). The effect of different freeze drying processes on the moisture content, colour and physical strength of roses and carnations. *Scientia Hortic.*, 84 (3&4): 321-332.

Datta, S. K. (1999). Dehydrated flowers and foliage and floral crafts. In: *Floriculture and Landscaping* (Eds T. K. Bose, R. G. Maitti, R. S. Dua and P. Das), Naya Prokash, Kolkata, 696-703pp.

Desh Raj and Gupta, P. (2002). Naturally dried ornamentals native to outer Himalayas. In: *National Symposium on Indian Floriculture in the New Millenium* held at Bangalore w.e.f. 25-27, February 2002 (Abstract).

Desh Raj, (2001). Making floral crafts from the forest products of the Himalayas. *Indian Hort.*, 46 (3): 26-27,36.

Dubois, P and Joyce, D. (1989). Drying cut flowers and foliage. Farm-note No. 10/89. Department of Agriculture, Western Australia, 3p.

Dubois, P. and Joyce, D. (1988). Bleaching plant foliage. Farm-note No. 87/88. Department of Agriculture, Western Australia, 4p.

Joyce, D. C. (1998). Dried and preserved ornamental plant material: not new but often overlooked and underrated. *Acta Hort.*, 454: 133-145.

Kher, M.A. and Bhutani, J.C. (1979). Dehydration of flowers and foliage. Extn. Bulletin, EBIS, NBRI, Lucknow, 20p.

Raghava, S.P.S, (2001). Floriculture- A viable diversification option in agri-business. *Indian Hort.*, 45 (4): 41, 44-45.

Smith, A. (2000). Even better than the real thing. Flower Business International, September: 25-28.

Spriggs, P. and Webb, M. (1994). A study of the western Australian processed wild flower industry. Department of Agriculture, Western Australia, *Miscellaneous Publication* No. 44/94, 25p.

Westland, P. (1993). Dried Flowers. Annerss Publishing Limited, London, 96p.

Westland, P. (1995) Dried Flower Arrangements. Cassell, Wellington House, London, 144p.

Wilson, D. M. and Abramson, D. (1992). Mycotoxins. In: *Storage of Cereals and their Products* (Ed. D. Sauer) American Assn. Cereal chemists, St. Paul, 341-91pp.

Ornamental Horticulture, pp 405-418

New India Publishing Agency, New Delhi - 110034, India

18

Design of Protected Structures for Indian Conditions

M. Hasan

A greenhouse with all measures of climate control is one important component for integrated greenhouse crop production and protection. However, the growing period is very often limited to few months in the year, because of insufficient ventilation and cooling in summer as well as heating in winter. Structures and shapes of greenhouses are adapted insufficiently to climate conditions. Climate control and greenhouse structures have to be adapted and optimized with regard to outside climate, and growers must be trained in these adapted technologies. A greenhouse structure with light transmittance (through framework and cladding material), ventilation, heating, cooling, and protection from insect pests by screening, as well as all influences of management, irrigation, fertilization, water quality, physical and biological plant protection is an integrated system. Greenhouse structures should be designed according to the climate conditions and to the general design criteria, as well as to the locally available and cost-effective materials. The import of greenhouses from countries with a different climate fails very often, because they are not adapted to the local conditions. Even an appropriate greenhouse structure can fail if mounting and continuous maintenance are not guaranteed. Temperatures and global radiation allow open-air crop production throughout the year in many subtropical and tropical countries.

[1] Centre for Protected Cultivation Technology (CPCT), IARI, New Delhi

Vegetable and flower production are significantly influenced by the seasonality and weather conditions. The extent of their production causes considerable fluctuations in the prices and quality of vegetables. Striking a balance between all-season availability of vegetables and flowers with minimum environmental impact, and still to remain competitive, is a major challenge for the implementation of modern technology of crop production.

The crop productivity is influenced by the genetic characteristics of the cultivar, growing environment and management practices. The plants' environment can be specified by five basic factors, namely light, temperature, relative humidity, carbon dioxide and nutrients. The main purpose of protected cultivation is to create a favorable environment for the sustained growth of plants so as to realize its maximum potential even in adverse climatic conditions. Greenhouses, rain shelters, plastic tunnels, mulches, insect-proof net houses, shade nets etc. are used as protective structures and means depending on the requirements and cost-effectiveness. Besides modifying the plants' environment, these protective structures provide protection against wind, rain and insects.

Protected cultivation is relevant to growers in India who have marginal and small land holdings, which help them to produce more crops each year from their land, particularly during off-season when prices are higher. However, growing vegetables and flowers under protected conditions require comparatively high input cost and good management practices, which have direct bearing on the economic viability of production system. Even if the protective structures are cost effective, proper planning, management and attention to details are needed to achieve maximum benefits.

Concept of Protected Cultivation: Protected cultivation technology is based on greenhouse effect. Greenhouse effect refers to the absorption of infrared energy by the atmosphere and the earth, which maintains the optimum temperature range on the earth suitable for life. The earth would be a frozen planet without greenhouse effect with an average temperature of about -18°C. Greenhouse gases like carbon dioxide, water vapour, nitrous oxide, methane *etc* allow incoming short wavelength (0.3-2.3 μm) solar radiation to reach the earth surface but restrict the outward flow of long wavelength (>2.3 μm) (Cengel, 1998). They absorb and reradiate the outgoing radiation after storing some heat in the atmosphere, which results in the warming of the earth surface through greenhouse effect. Greenhouse is an inflated structure made with G.I or steel pipes covered with plastics and nets, which can be used for crop production under controlled environment conditions. Micro climate inside greenhouse is created and maintained for high quality crop production mainly of vegetables and flowers round the year.

Characteristics of Protected Agriculture: The major characteristics of protected agriculture are as follows.

- Round the year production of vegetables and flowers
- High yield and good quality production
- Optimum use of water and nutrients
- Minimum use of insecticides and pesticides
- Suitable for small land holdings
- Suitable for water harvesting

Plant Environment-Interactions for Protected Agriculture: An understanding of plant-environment interactions is essential for maximizing plant productivity under a set of operating constraints. The growth of plant could be modified by changing the parameters of its surrounding environment, that is, its microclimate. Logically, for a given plant, it is possible to arrive at an optimized set of microclimatic parameters to maximize the plant growth. A greenhouse is, essentially, a practical means to achieve maximization of plant productivity by maintaining the optimum microclimate.

The microclimate of a plant is specified in terms of temperature, light, air composition and nature of root medium. Air and soil temperatures are generally the parameters considered in studying the plant environment interactions, but some of the recent studies show that leaf and root temperatures are more appropriate. Temperatures of other objects in the surroundings affect the plant temperature and, therefore, are important. The parameters of light which influence plant growth are sources of light, total energy content of the light, spectral quality, photosynthetically active radiation (PAR) and albedo of the surroundings. Crop growth has also recently been correlated with the U.V. content of light and the ratio of energies in the far-red and red portions of the light. Gaseous composition of the air surrounding the plant is important. Contents of carbon dioxide, oxygen, nitrogen, water vapour and other trace gases influence the plant metabolism immensely. Physical, chemical and microbial compositions of soil define the nature of root medium in soil culture. However, it is no more necessary to depend on soil for development of plant roots. Use of synthetic root media, such as Rockwool, or continuous/intermittent flowing nutrient solutions are becoming commercial realities for root development. Complete specifications of the root medium must be known to interpret the plant growth data and to execute accurate watering and nutrition programmes.

A number of studies were conducted to reveal their interactions and relationships among plant growth, both vegetative and reproductive, and the microclimate. While these studies could be used to explain some of the crop productivity results, a more comprehensive utility is to develop a crop-environment model to be used as a management tool. Several such efforts have already been made and the scope is unlimited in terms of the types of models and the crops for which such models could be usefully developed. Light, temperature and relative humidity are the most important environmental parameters effecting the greenhouse environment.

Light

Light is the most important component of greenhouse environment because it is essential for plant growth. Sunlight is the only resource as far as the open field cultivation is concerned. Electrical light has, however been utilized in greenhouse cultivation whenever needed. Availability of sunlight at a place is time dependent mainly, due to the season, latitude, atmospheric turbidity and cloud cover. Early greenhouse developments took place in cold and cloudy regions where greenhouse structures were designed to admit as much sunlight as possible for better crop productivity. Provisions were made for artificial lighting of greenhouse crops under cloudy or low radiation periods. Sometimes, white or reflective mulch films have also been laid in rows to reflect the sunlight back to the plants. However, in tropical regions, high air temperatures often accompany the bright light. In these situations, shading is generally practiced to control the temperatures in the crop zone. In fact, most plants of economic value are observed to saturate at 1/3 to 1/4 of the available sunlight on a clear day. Therefore, shading does not decrease the net photosynthesis. In recent times, artificial lights are used for extending the day length for supplementing the natural light in commercial greenhouses in the developed world. In Northern Europe, seedlings are economically grown using artificial lights during winters. The use of artificial lights was also found economically viable for timing and maturity of flower crops. In tropical situations, artificial lighting has seldom been used for supplementing the natural light.

Temperature

A plant has a temperature range in which it can be grown. The lowest temperature is the one at which ice forms within the tissue tying up water necessary for life processes. At this temperature the life stops. At the other end of the temperature scale, enzymes become inactive and again the life processes stop. In most cases, the lowest possible temperature for keeping the plant alive would be a couple of degrees higher than the freezing point. However, the highest temperature at which a plant could survive depends upon the relative humidity. Plants are noticed to survive even beyond 50 ^{0}C provided the air humidity is kept close to saturation. Generally, day temperatures are higher than the night temperatures and, thus, the photosynthetic rates exceed the respiration rates to achieve positive plant growth. It can be expected that if the nights are cooler, the plant growth will be higher because the loss due to respiration would be less. In addition, if the air is enriched with carbon dioxide the permissible day temperatures could be increased by 2 to 3 ^{0}C.

Greenhouse use has largely been promoted by the need to control the crop temperature so that summer crops could be grown in winters and *vice-versa.* Besides, if the ambient conditions are too hot or cold for a given crop, greenhouses permit the maintenance of temperatures in the favourable range. The temperature control in the cold climatic conditions has been partially achieved through ģreenhouse effect' Heating systems based on coal, gas, liquid fuels and electricity have been developed. Greenhouses are redesigned to improve the energy conservation properties and, thereby, reduce the energy intensiveness of the greenhouse temperature control processes. Natural ventilation systems have been devised. Uniformity of temperature control over the crop zone is essential for high quality production and, therefore, the heat distribution systems need special attention. Needless to say that greenhouse heating has been studied in greater detail then greenhouse cooling.

Use of greenhouses in warmer climatic conditions has been increasing due to a number of reasons. Evaporative cooling is the only practical method for lowering temperature in commercial greenhouses at present. Sometimes variations of evaporative cooling, e.g., fan pad systems, fogging and misting have been devised. The theoretical limit of single stage evaporative cooling does not permit the greenhouse air temperature to go below the wet bulb temperature for that place and time. Under high humidity ambient conditions, this is a serious constraint. There are some efforts to develop alternate cooling methods which could be commercially viable, such as heat pipes, thermal storage systems (both sensible and latent), earth-air heat exchangers, etc. These systems have, in general, not become commercially acceptable yet.

Relative humidity

Humidity of the greenhouse air is normally higher than that of ambient because of a lack of mixing of the greenhouse air with the ambient air. A complete moisture balance on the greenhouse air including evapo-transpiration, condensation, infiltration-exhilaration and moisture storage in the air determines the instantaneous humidity of the greenhouse air. However, evapo-transpiration is the major parameter influencing the greenhouse air humidity during sunshine hours. Control of humidity is generally possible through ventilation, heating and misting.

The Centre for Protected Cultivation Technology (CPCT) at IARI New Delhi has done very good work in the area of protected cultivation technology. The centre has demonstrated and standardized many protected cultivation technologies and has given training to the farmers, officials and entrepreneurs and carried out active research and post graduate studies related to different aspects of protected cultivation.

Protective Structures

The kinds of protective structures for crop production range from simple provisions such as rain shelters, shade houses, mulches, row covers, low tunnels and cloches to greenhouse structures with passive or active climate control.

Greenhouses

Greenhouse refers to a specialized structure, providing suitable conditions for optimum plant growth and round the year crop production (Tiwari, G.N 2003).

A greenhouse is a quasi-permanent structure, covered with a transparent or translucent material, ranging from simple homemade designs to sophisticated pre-fabricated structures, wherein the environment could be modified suitably for propagation or growing of plants. Materials used to construct a greenhouse frame may be wood, bamboo and steel or even aluminum. Coverings can be glass or various rigid or flexible plastic materials.

Plant Environment and Greenhouse Climate

A plant grows the best when exposed to an environment optimal for that particular plant species. The aerial environment for the plant growth can be specified by the following four factors:

- Heat or temperature
- Light
- Relative humidity
- Carbon dioxide

Materials for Greenhouses

As mentioned earlier, the purpose of a greenhouse covering is to allow sunlight to pass through it so that the energy is retained inside. Glass was the main covering material in early greenhouses. With the introduction of plastic materials, there are now several alternatives available for greenhouse coverings. A brief description of greenhouse covering materials is given below:

- **Glass:** A clean, transparent glass provides the maximum light transmittance to the extent of 90%. However, being heavier in weight, it requires elaborate structure for adequate support. It is brittle and can break with minimum shock or vibrations resulting in high maintenance costs.

- **Acrylic:** This material has long service life, good light transmittance (80%), moderate impact resistance, but prone to scratches. It has a high coefficient of expansion and contraction. Being inflammable and costly, it is not a preferred material.

- **Corrugated / multi wall polycarbonate sheet:** It is available in single or double wall sheets of different thickness. A new polycarbonate sheet has good light transmittance of about 78%, but reduces with age. It has excellent impact resistance and low inflammability. High cost limits its use on large scale.

- **Fiberglass reinforced plastic panels (FRP):** These plastics consist of polyester resins, glass fiber stabilizers etc. It has an initial light transmittance of about 80% and has high impact resistance with a service life ranging 6-12 years. Good quality FRP materials for greenhouse coverings are not quite assured.

- **Polyethylene film:** A clear new polyethylene sheet has about 88% light transmittance. Its higher strength and low cost have made it the most popular replacement to glass. An ultra-violet (UV) stabilized plastic sheet can have a service life of 3 years. These sheets are generally available in 7 and 9 meter widths with 200 micron (0.2 mm) thickness.

- **Thermal and shading net:** These nets are used mainly for creating shades of different intensity. It is made of aluminium profile or nylon. It should be UV stable and give shading in the range of 50-80 %. Its life is 3-5 years depending mainly upon UV stability parameters. Shading nets are now available in black, green and red color. Thermal net is usually white in color made of aluminium profile. It is costly but effective and used mainly inside greenhouses. Thermal net is effective in both summer and winter season, while shade nets are used mainly in summer season.

Types and Designs of Greenhouses

The greenhouse design and cost range from a simple plastic walk-in tunnel costing about Rs.100/m^2 to a climate-controlled, saw-tooth greenhouse with automatic heating, ventilation and cooling, costing more than Rs. 3000/m^2. The selection of the greenhouse design should be determined by the growers' expectations, need, experience and above all its cost-effectiveness in relation to the available market for the produce. Obviously, cost of greenhouse is very important and may outweigh all other considerations. Greenhouses are classified in different shapes, which also determine their cost, climate control and use in terms of crop production. Commonly used structural designs are briefly described below.

Gable: This is the most basic structure similar to a hut-like construction and was perhaps the first version of a greenhouse with glass as the covering material. The roof-frame can be inclined at any angle to present an almost perpendicular face to the sun to minimize losses due to external reflection. The structure also allows large openings in the side-walls and at the ridge for high rates of natural ventilation. Modern gable-shaped greenhouses are multi-span units with bay widths of 6-12 meters.

Gable Type Greenhouse

Gambrel: These structures are similar to the gable but have high strength to withstand high wind loads during storms. This design is more suitable where wood or bamboo are to be used for the greenhouse construction.

Gambrel Type Greenhouse

Skillion: In this kind of structure, the roof consists of a single sloping surface. This is because the greenhouse is built as the southward extension of a building with a solid wall on the northern side. Such greenhouses have the advantage of low structural requirements.

Curved-roof –Raised High Arch / Raised Arch: The semi-circular tunnel greenhouse structures appeared with the introduction of polyethylene film as the covering material. These structures, besides being most simple and easy to construct, have the advantage of high strength with a relatively light frame due to inherent strength of the curved arch. But these structures have the disadvantage of poor ventilation efficiency since the curved roof is not amenable to incorporation of ridge ventilators. In an attempt to improve ventilation efficiency of curved roof greenhouses, raised arch type of structures have been adopted. This design has vertical side-walls, which permit high head room and improved ventilation due to wind velocity.

Saw-tooth: In these structures, the roof consists of a series of vertical surfaces separated by a series of sloping surfaces, all of which are pitched at the same angle and facing in the same direction. The vertical surfaces consist entirely of ventilating area. These types of greenhouses are the most efficient from ventilation point of view. Such greenhouses are also suitable for multi-span structures. Orientation of saw-tooth greenhouses can also be used as a means of maximizing natural ventilation. By facing the open vertical ventilation areas away from the wind, airflow over the greenhouse roof creates a negative pressure, which facilitates in sucking out warm greenhouse air. However, this air dynamics relies on the premise that there are large ventilation areas in the greenhouse walls on windward side.

Plastic Low Tunnels/Row Covers: Thesc structures are laid in open fields to cover rows of plants with transparent plastic film stretched over steel hoops of about 50 cm height spaced suitably along the rows. Polyethylene film of 30-40 micron thickness, without UV stabilization, is used which is perforated *in situ* as the season gets hotter. Row covers used in vegetable production have different purposes in temperate and tropical regions. In cold conditions, they are used to conserve warmth, stimulate germination and early growth, protect plants from frost injury and improve the quality of crops. Other beneficial effects, such as maintaining soil structure and protecting crops from the attacks of birds and pests, can also be expected. The main advantage of these covers in northern India is to grow vegetables, especially cucurbitaceous crops, ahead of normal season in winters. Experiments on muskmelon have proved it a highly profitable proposition. The muskmelon seedlings could be transplanted under such covers in the last week of January. The crop growth was sustained during the cold period. Temperature profile inside the cover indicated a difference of about 7℃ averaged over 24 hour cycle. This rise in temperature provided necessary warmth to sustain growth of plants. In hot season, however, materials used as row covers need to have adequate permeability to air and moisture, to prevent the accumulation of excessive heat inside the covers. The covering materials used in summer are woven polyester wind-break nets, cheese-cloth, and insect-proof screens. These types of covers are generally laid over the planted rows without the support of steel hoops, and are also called as floating covers. To provide adequate space, the seedlings are planted in furrows and the covers are laid over the ground. But such planting should only be adopted in light textured soils having high infiltration rates.

Plastic Low Tunnel

Naturally Ventilated Greenhouse

Naturally ventilated greenhouse is the most common and popular greenhouse type for Indian farmers. It is a zero energy model greenhouse with natural ventilation from sides and top. Saw tooth type greenhouse design has the maximum ventilation and is the most effective and suitable for crop production. This type of greenhouse can be used for crop production ranging from 9-12 months depending upon location and climatic factors. CPCT has standardized the design of naturally ventilated greenhouse and the production technologies of growing various crops inside it. The design and production technologies are tested in various parts of the country in the farmers fields.

Naturally Ventilated Greenhouse

Mulching

Mulching is done to cover the soil around plants with a protective material, which may be organic or synthetic. Organic mulches, like leaves, sawdust etc. add nutrients and humus to the soil as they decompose, improving its tilth and moisture holding capacity. Synthetic or plastic mulches have various beneficial effects on crop production. Transparent polyethylene mulch raises the soil temperature. This effect derives mostly from suppression of latent heat loss through evaporation. The mean difference in soil temperature between transparent film mulched and bare soil in early February in Delhi was observed to be 5°C in the top 7 cm of soil where most of the root zone of young muskmelon existed. It increased the plant growth by about 15% during the same period. Mulching conserves soil moisture and fertility. The former is higher with black plastic than under the transparent plastic. Mulching prevents leaching of fertilizer, because it acts as a physical barrier to rainfall, thereby conserving the fertility. Black polyethylene film also gives effective weed control by cutting down solar radiation by more than 90%, resulting in etiolated growth and eventual death of weeds under the film. Transparent film, on the other hand, has little effect on weed control unless the film is coated with weedicides. Plastic mulch is also effective in the control of pests and diseases. Silver-coloured film is used as mulch to suppress the increase in soil temperature and to control pests and diseases. The silver colour acts as a repellent to aphids, which transmit viruses. On the other hand, yellow coloured mulches attract insects, which could be killed easily. Muskmelons, tomatoes, peppers, cucumbers, squash, eggplant, watermelons and okra are vegetable crops that have shown significant increases in earliness, yield, and fruit quality when grown on plastic mulch.

Mulching with Drip Fertigation

Shade Houses

Shade nets are perforated plastic materials used to cut down the solar radiation and prevent scorching or wilting of leaves caused by marked temperature increases within the leaf tissue from strong sunlight. These nets are available in different shading intensities ranging from 25 to 75%.

Leafy vegetables and ornamental greens are recommended to be grown under shade nets whose growth rates are significantly enhanced compared to unshaded plants when sunlight is strong.

Shade Net House of Black, Green and White Color

Net houses

Shade house and net house are often synonymously used but more correctly a net house is enclosed with perforated screen primarily to act as a barrier for entry of insects and pests. Insect-proof nylon nets are available in different intensities of perforations, ranging from 25 mesh to 60 mesh. Nets of 40 or higher mesh are effective means to control entry of most flying insects and save crop from diseases. These structures permit early planting of tomatoes and capsicum without the risk of vector. UV- stabilized nets are now available which have a longer life.

The above mentioned protected structures are indigenously designed with strict guidelines for all fixed and consumable parts. The technology is available at Centre for Protected Cultivation Technology, Indian Agricultural Research Institute, Pusa, New Delhi. Protected Cultivation Technology is now linked with Drip Fertigation.

Net House for Safe Vegetable Cultivation

Water Harvesting Technology

The principle is to harvest all the rain water and subsequently used it with drip fertigation for different protected structures and horticultural crops. It is estimated that there is saving of water and nutrients in the range of 40-80% and 20-30% respectively in protected cultivation. The yield potential can be increased from 50-200 % on average in protected cultivation in comparison to open field cultivation. The present market cost of various protected structures varied from Rs 150 per square meter to Rs 5000 per square meter of very primitive and fully climate controlled greenhouse respectively. The initial cost of greenhouse and suitable design and production technology are major bottlenecks for widespread dissemination of protected cultivation technology in India. The Planning Commission, Govt of India has taken major steps to promote this technology in Twelth five year plan. Massive subsidy ranging 50 to 90 percent is now being given to farmers throughout the country to establish various protected structures with drip fertigation system.

Important Specifications for Design of Naturally Ventilated Greenhouse (1000m^2)

- Galvanized steel tubular structure and saw tooth type design
- Gutter height 4-5 meter high and gutter slope of 2%
- Aprox 50 cm high polyethylene on all the four sides with proper fixing arrangements
- Roll-up plastic curtains, openable up to 3 m on all the four sides with crank mechanism; 40 mesh nylon insect-screen (UV stabilized) to be fixed inside the curtain on all sides
- Ridge height 6 meter and Ridge Vent up to 2 meters
- GI pipe Column of 76 mm OD, 2-3 mm thickness or equivalent sections. Offer of other sizes should be supported by suitable design criteria
- Bottom and top cord 60 mm OD x 2-3 mm thickness; Truss members 48 mm OD x 2-3 mm thickness. Braces-25 NB of class 'A''GI pipe
- Purlin modulus 42 mm OD x 2 mm thickness or equivalent size/section, having about the same moment of inertia and section
- 180-200 micron thickness and UV stabilized Plastic.
- Insect Proof Net of 40 mesh size and UV stabilized.

Semi- Climate and Climate Controlled Greenhouse Design

Apart from the basic specifications required for the naturally ventilated greenhouse as mentioned above, the following design specifications are required for the semi-climate and climate controlled greenhouse.

- Exhaust fans for removal of hot air inside greenhouse
- Cellulose pads with pumps for evaporative cooling
- Sprinklers for cooling
- Provision for heating

Design Issues for Greenhouses

Greenhouse technology is a relatively new technology for India. There are very few standard indigenous companies and firms involved in the design and installation of greenhouse. Most of the firms are adopting the design guidelines of greenhouses located outside India mostly in Israel and European countries. This is the major issue in the design of greenhouse. The need of the hour is to design the greenhouse according to location specific conditions suited for the particular region. The cost and energy availability are the other major issues related to design of greenhouse.

References

Brahma Singh, Balraj Singh, Naved Sabir and M. Hasan (2013). Advances in Protected Cultivation. New India Publishing Agency, New Delhi, pp 248.

Cengel Yunus A. (1998). Heat Transfer: A practical approach, WCB/McGraw Hill, New York, p 514.

Singh D.K. and K.V. Peter (2014) Protected Cultivation of Horticultural Crops, New India Publishing Agency, New Delhi

Tiwari G.N. (2003). Greenhouse Technology for Controlled Environment. Narosa Publishing House.

Ornamental Horticulture, pp 419-462
© Editor, S.S. Sindhu, 2016
New India Publishing Agency, New Delhi - 110034, India

19

Protected Cultivation of Flower Crops

S.S. Sindhu

The floriculture industry, at present, comprises of the cultivation and trade of cut and loose flowers, potted plants, foliage and bedding plants and dried flowers under open field as well as under protected environment conditions. Globally, more than 145 countries are involved in cultivation of floriculture crops. Though the art of growing flowers is not new to India, large scale commercial cultivation especially protected cultivation in poly houses is relatively new in India. Enormous genetic diversity, varied agroclimatic conditions and versatile human resources offers India a unique scope for diversification in to new avenues which were not explored to a greater extent. India also has a strong dry flower industry which contributes a major share to the overall trade. Other segments like fillers, potted plants, seeds and planting material turf grass industry and value added products also contribute a share in the overall growth of floriculture sector.

Cut flower production is a growing business in most Asian countries as the result of higher per capita income of people. Technology for production is well known and has been utilized to the maximum in many countries. As a consequence, cut flowers are commercially produced in larger quantities in these countries resulting in surpluses during normal periods of blooming, end result of which is a lower price for cut flowers. On the other hand, such prices are much higher during periods of needs such as certain special occasions, or off-season.

Division of Floriculture and Landscaping, IARI, New Delhi, India - 110 012

Specific-Date Production

The demand for cut flowers is quite high during certain occasions, like Christmas Celebration, New Years' Day, Mothers' Day, Memorial Day, Valentines' Day, Graduation Exercise Day, Buddhist Lents' Day, etc. Thus, it will be profitable to farmers if they could produce cut flowers on these occasions.

What is Greenhouse Technology?

Today about 92% of plants, raised by man, are grown in open field. Since the beginning of agriculture, farmers had to cope with growing conditions given to them by Mother Nature. In some of the temperate regions where the climatic conditions are extremely adverse and no crops can be grown, man developed technological methods of growing some high value crops by providing protection from excessive cold and excessive heat. This is called Greenhouse Technology. Greenhouse Technology is the science of providing favourable environment conditions to plants. It also protects plants from the adverse climatic conditions such as wind, cold, precipitation, excessive radiation, extreme temperature, insects and diseases. An ideal micro climate can be created around the plants. Now the area under protected cultivation has reached from 4000 to 20,000 ha which include production of high value flowers. Greenhouses are framed or inflated structures covered with transparent or translucent material large enough to grow crops under partial or fully controlled environmental conditions to get optimum growth and productivity.

Advantages

- Crops can be grown throughout the year
- Quality of the produce is superior
- Higher input use efficiency
- Higher income per unit
- Easy to manage insect-pests and diseases
- Water requirement is minimum
- Less labour requirement

Limitations

- High cost of initial installation
- Non-availability of various components

- Needs a special postharvest facility
- All the operations are very intensive

The market prices of floricultural products are highly volatile. The production strategy should be such that growers are not at loss. The price determination in floriculture products are based on following parameters.

- Season and demand
- Variety and quality
- Free from pest, disease and mechanical damage
- Uniformity in opening
- Bright colour
- Packaging
- Regularity in export

Institutional Inventions

During the last two decades, ICAR and SAUs' made tremendous and significant growth in research and development. The comprehensive research programmes are on crop improvement using conventional and biotech tools, production in open and protected environments, technology dissemination for the benefit of farming fraternity. The IARI made a distinction of evolving the largest number of rose varieties in the country possessing desirable traits. 75 varieties comprising of Hybrid teas, Floribundas, Grandifloras, Miniatures and Climbers. Induced mutants *viz*., Abhisarika, Striped Christian Dior and Madhosh: natural mutants Chitra, Pusa Abhishek, Climbing Sadabahar, Climbing Dr. Homi Bhabha, Delhi Pink Pearl, Pusa Gaurav and Nav Sadabahar exhibit a unique spectrum of patterns and are well received by garden enthusiasts.

Many new cultivars were developed in chrysanthemum *viz*., Baggi, Birbal Sahni (at NBRI) 'Arka Ravi'(at Hessaraghatta, Solan, Udaipur and UAS Bangalore), 'Punjab Gold'(at Hessaraghatta and Pune), 'Mother Teresa'(at Hessaraghatta), 'CO.2' and 'Indira'(at Coimbatore), Yellow Gold(at Solan), 'Ratlam Selection' and 'Basanti Local'(at Hyderabad), 'Sonali Tara,' 'Sunil' 'Baggi' and 'IIHR Hybrid-11 (at Pune).

Crop Specific Production Technology

1) Rose

Rosa indica var. *odorata* rootstock was the best for Coimbatore, Delhi, Lucknow, Ludhiana and Pune conditions; *R. multiflora* and IIHR - Thornless for Bangalore, Hyderabad and Kalyani conditions; and *R. chinensis* (Titri) for Udaipur conditions. Depending upon the varieties used, 30 x 20 cm spacing was the best for Super Star and Gladiator under Pune and Udaipur conditions; 30 x 30 cm spacing for Montezuma at Kalyani conditions; and 30 x 40 cm for Super Star under Delhi, and for Sonia Meilland under Udaipur conditions, as quality flower production was maximum there.

For commercial cultivation of rose, NPK (400:200:200 ppm/plant/week) at ťun off'stage is recommended. Coimbatore centre recommended NPK [$(NH_4)_2$ NO_3, H_3PO_4 and KNO_3] 150:40:150 for cv. First Redùnder protected conditions. At Ludhiana, 60:40:40 g/m^2 NPK with cv. Super Star is recommended. Delhi centre found 10 cm pruning from Dept. 23 to Oct. 1, in rose cvs. Queen Elizabeth and Happiness, and 45 cm leaving 4 shoots, from Oct. 13 to Nov. 17 in cv. Super Star as the optimum. Against weeds, Oxyflurofen (0.5 kg a.i./ha) drenching at Kalyani, and 1 kg a.i./ha at Ludhiana; Oxyflurofen (1 kg a.i./ha) or Diuron (2 kg a.i./ha) at Delhi in cvs. Super Star'and Raktagandha, and Diuron (2.5 kg a.i/ha) or Simazine (3 kg a.i./ha) or Round Up(1 kg a.i./ha) at Yercaud and Bangalore conditions are quite effective.

Rose is one of the natures beautiful creations and is universally acclaimed as the 'Queen of Flowers'.' It is a top ranking cut flower in flower trade on the basis of average production and consumption. Great diversity in plant growth, flower colour, flower shape, fragrance, slow opening of buds and good keeping quality made rose so popular that it is grown commercially to meet the demand of cut blooms. In India, roses are grown for cut flowers, making essential oil, rose water and gulkand. It is dried petals used for making incense sticks. People are not aware of different propagation techniques of rose,

therefore they have to purchase rose planting materials from nurseries at very high cost. The main propagation methods, through which we can produce plants at high level, are as below:

Propagation

Roses can be propagated both by seeds and by various vegetative methods. But now-a-days tissue culture techniques are being used for production of disease free plants in large number.

Seed propagation

This method of propagation is generally adopted by breeders for developing new cultivars with desirable characters. Seedlings of rose species are also used as stock for grafting or budding. The rose fruits (hips) are harvested when fully ripe and thoroughly dried before extraction of seeds. In rose seed germination is very poor because most of the seeds when mature are in resting conditions requiring an after ripening period before germination. Stratification of seeds at 1.6-4.4^{0}C for 6 weeks improves germination. Germination can also be improved with scarification with different chemicals like sulphuric acid for 1-2 hours.

Care of seedlings

For raising seedlings clean seeds are sown about 5 cm apart in pairs containing pot compost with adequate organic matter or in small pots. The flower buds appeared on new seedlings should be pinched off until plants become reasonably large and sturdy. A continuous moisture and nutrient supply and control of weeds, diseases and pests contribute to successful seedling growth.

Vegetative Propagation

Roses are propagated by various vegetative methods like cutting, layering, budding and grafting.

Cuttings

Propagation by cutting is normally done to raise stocks for grafting or budding and also adopted for multiplying vigorous types of cultivar, climbers, rambler, polyantha and miniatures. Cuttings are always prepared from shoot. Root cuttings have also been found to form shoots and ultimately a new plant.

i) Stem cutting

It is one of the least expensive and easiest methods of rose multiplication. Cuttings may be single, double or triple eyed, the later being preferred as there is greater chance for more shoot and better root formation. Cutting should normally be taken during monsoon or spring. The cuttings are made from pencil size thickness (24-27 cm) shoots just below the node and each cutting should have 2-3 buds. Sand is the most ideal medium for rooting of cuttings

ii). Root cutting

Some rose species like *Rosa blanda, R. nitida, R. virginiana* can be propagated by root cuttings. The cuttings should be taken from young and healthy plants and it is important to maintain polarity when planting. The cuttings should be inserted vertically with the upper most ends at level of the rooting media.

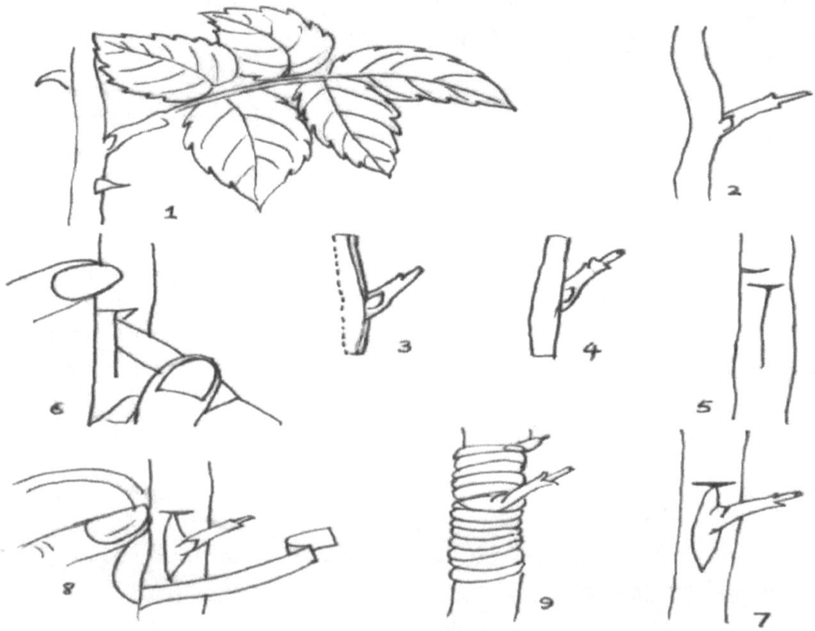

Budding Methods in Rose

Layering

This type of vegetative propagation is usually limited to climbing and rambling roses and usually practiced during monsoon or early spring and leads to own rooted plants. Mainly 2 types, ground layering and air layering are used

i) Ground layering

It is performed by bending shoot to ground and covers it with soil leaving terminal end exposed. Root promoting substance can also be used for improving root formation in layers. Rooting usually takes place in a month or so and the layered shoot is detached 15-20 days after root formation.

ii) Air layering

It consists of removing a ring of bark (about 2.5 cm long) from stem, around the shoot and wrap sphagnum moss around the ringed portion and covering with polythene film. For the improvement of root formation in air-layer some root promoting substances can be used.

Budding

Budding is the most popular and successful method for multiplying roses. It provides larger number of plants than cuttings, layering or grafting as a single shoot of the desired scion furnishing a number of buds for budding. Shield or T-bud is the method ordinarily used. On the selected rootstock, the buds are inserted into a T-shaped incision and then tied with suitable wrapping material. Several other budding techniques like forkert, patch, chip, skin *etc.* were have also been tried in rose with varying degree of success.

The time of budding varies from place to place and the right stage of budding is when plants have adequate sap flow and the cambium tissue is highly active. The best time to bud roses in eastern India is from January to March, while in northern India it is from November to February. In places with mild climate, like Bangalore, Pune budding can be done throughout the year.

Grafting: Top grafting is the common practice and time consuming in all exotic varieties. The 'Natal Barrier"is used as root stock.

Some common rootstocks in India are as follows

a) Rosa multiflora

This rootstock is tolerant to nematode and frost. It performs well in Bihar, Bengal and in the hills.

b) *Rosa bourboniana* (Edouard Rose)

Susceptible to powdery mildew and die back diseases. Plants provide straight stems and mostly used in northern plains of India for budding standard roses.

Budded Planted Roses

c) *Rosa indica* var. odorata

Well adapted to both excessively dry and wet soil. Plant raised on this rootstock is vigorous and produce flower with longer stems. It is resistant to powdery mildew and insect pests and recommended for northern plains of India.

d) *Rosa canina* or *Rosa manetti*

R.canina or *R.manetti* is the most ideal for greenhouse cultivation.

Production

Planting time

Planting time depends mainly on the climatic conditions of the region. In general, roses can be planted any time except hot sunny days and during heavy rains when the soil is wet. But winter months are the best and the safest. The suitable time for planting roses in the plains of India is during September-October and in the hills during October- November or February-March under green house.

Planting depth and Distance

The position of the bud union should be looked into. In temperate region the bud union should be slightly below the soil. In India better results are obtained if the planting is done with bud union 2-3 cm above the soil level.

Planting Method of Rose in Greenhouse

Under green house roses are planted at distance of 30×25 cm (Row×Plant) the common spacing and total 7 to 10 plants per sq. m. are accommodated in the beds. Two rows are preferred on one meter wide bed and a passage is kept of 85 cm between two beds.

Pruning/Bending

It is an important operation for maintenance of floriferousness and flower quality along with vigour of rose plants. The main objectives of pruning are to remove weak and unproductive shoots which allow air and light to reach plant centre. It keeps the plant in proper shape and size and also forces the plant buds to break and grow to produce quality flowers. It also helps in maintenance of proper balance between vegetative growth and flowering.

Pruning should be done when plants are less active or dormant. In northern plains of India, it is mainly done during October-November. The basic rule of pruning is to cut half centimeter above the vigorous, outward directed bud by giving a slanting cut because it does not retain moisture and prevents fungal growth.

In H.T. roses all dead, weak, damaged, diseased and crossing shoots are removed from the base retaining only 4-5 healthy basal shoots. Moderately severe pruning appears to be ideal. Under the green house conditions, a bending practice is common to regulate production as well to have better quality flowers.

Manuring and Fertilization

Rose is a nutrient loving plant and about 16 elements are known to play important role for its growth and development. Rose should be fed with both organic and inorganic sources. One hundred gram of mixture containing Groundnut cake (5 kg), bonemeal (5 kg), ammorphos (11:48) (2 kg), Ammonium sulphate are used after pruning.

Single super phosphate (2 kg) and potassium sulphate (1 kg) should be applied per bush for better result. It is advisable to apply the fertilizers in three split doses and application should be made when there is sufficient moisture in the soil. The first application of the fertilizers should be given 15 days after pruning when the new growth starts, i.e. in October then December and third dose in February. Under green house, 300: 250 and 200 ppm of NPK is recommended through fertigation system for quality production. Normally, water soluble fertilizers are used to feed the crop.

Disbudding

It is also an important operation by removing buds and retaining only central resulting in large flowers with better quality.

Greenhouse Roses

De-shooting

Removal of lateral shoots to allow only terminal shoots is advised. It is necessary, otherwise plants get heavily branched with large number of small flowers.

Cultivars

Golden Gates,'Grand Galla,'First Red,'Kiss,'Konfetti,' Mercedez,' Ravel,'Noblesse,'Vivaldi,'Golden Gate,' Hollywood and Starlite.'

Planting

Optimal planting time is October-December. The planting density should be between 7-13 plants/m^2 (50-60 thousand bushes/ha area). Planting should be done preferably in 2-row system.

Temperature requirement

The greenhouse temperature is generally maintained from 20^{0}C or 21^{0}C on cloudy days and 24^{0}C-28 ^{0}C on sunny days. The temperature range of 15-27^{0}C is optimum.

Diseases

Die-back

It is one of the major diseases of rose, particularly in India. The older plants are more prone to the attack of this disease as compared to younger ones. The stem and roots of the dead plants show browning of internal tissues. *Diplodia rosarum* is the causal organism. It may occur through improper use of fertilizers, improper watering, borer damage, poor drainage and lack of light. Proper cultural practice is the prevention for die back in roses. Some systemic fungicide like Carbendazim (1g/l) may also be used as soil drench.

Powdery-mildew

This major disease of rose causes much damage to roses all over the world. It is caused by the fungus *Sphaerotheca pannosa*. It can be controlled by applying sulphur but most of the roses are susceptible to sulphur injury.

Black-Spot

This is a fungal disease caused by *Diplocarpon rosae.* Dark brown and circular spots with fringe borders present on both sides of leaflets, Benlate can be used as control measure.

Harvesting and yield

Red and pink cultivars are harvested when first 2 petals are beginning to unfold and calyx is reflexed below the horizontal lane. Yellow cultivars are harvested slightly earlier and white cultivars slightly later than red and pink types.

Grading of Roses

After cutting, the stem ends are dipped in 200-500 ppm citric acid or aluminium sulphate and placed in cold storage at 1^0C till grading. Stem ends are re-cut, bunched in 20s and placed in preservative solution with sucrose up to 2%.

Packing of Roses

Flower yield

Flower yield of 250-350 stems/m^2 is the ideal. Flower yield can be increased by spraying BAP (50-100 ppm) before flowering flush. Burning of saw-dust during winter months in the morning hours (7-10 am) increases flower yield and quality significantly.

2) Carnation

Carnation (*Dianthus caryophyllus* L) is a herbaceous perennial flowering plant, belonging to family Caryophyllaceae. It has been under cultivation by man for over 2000 years. The common name Carnation (Latin, *Carnation*, from caro; *carnis*, flesh) is from a Greek word Coronation'because these flowers were used in decorating the crowns of Greek athletes.

Carnations are indigenous to the Mediterranean areas. They are grown on a large scale in Italy, Spain, Columbia, Kenya, Sri Lanka, Canary Islands, France, Holland, Germany, Israel and USA. Due to high cost of production in greenhouses in Europe and USA, the cultivation of carnations is shifting to more naturally growing areas where they are being produced at lesser cost. Owing to its ever increasing demand in the world market, high quality carnations are also being cultivated in India. The major production centers are around Pune and Bangalore. Carnations are now being commercially grown in Solan, Shimla, Mandi, Kullu, Chamba and Bilaspur districts in Himachal Pradesh. It has ample scope of its production in Uttarakhand as well.

Under polyhouse conditions in Solan, when cvs Impala, Super Star, Veleta and Fantasia were supplied with 4 hrs extended lighting, induced early flowering with maximum floral stalk. They also recommended April planting resulting in earlier flowering (160 days) while October planting delayed it (213 days) but gave better stalk length and floral diameter. Also they recommended pinch and a half method for earlier and regulated flowering. Planting at 15 x 15 cm spacing (4,44,444 plants/ha), double pinching (40 and 60 days after transplanting) and application of NPK @ 30-40:20:10 g/m^2 gave the highest yield with nitrogen in 3 splits (basal, one month after and two months after transplanting).

There are about 250-300 species of *Dianthus*, comprising annuals and evergreen perennials but only a few are cultivated. The most common cultivated species are *D. caryophyllus, D. barbatus* and *D. chinensis.*

The present day Perpetual Carnation was developed from cross between *D. caryophyllus* (female parent) and *D. chinensis* (male parent). They are grouped into standard and spray types on the basis of flower size. The standard type produces large single flowers on a longer sturdy stem. They suffer from diseases when grown under warm humid climates. The spray or miniature types produce many flowers of a smaller size on a short stem and are suitable to warm climates. The most common varieties of carnation are: Arthur Sim, Sunrise, Lemon drop, Pink Dona, Master, Liberty, Tempo, Lavender *etc.*

Climate

Carnations perform the best in relatively cool climate. It is a long day plant and requires approximately 21.5 K lux sunshine for at least 8 hours a day. Temperature is a major factor that influences growth and flowering, colour, grade and quality of cut flowers. The location having a day temperature of 10°C is good for high quality Carnations. The optimum temperature for production of standard Carnation is 18°-23°C. High temperature and high humidity produce inferior quality flowers. The optimum humidity for growing carnation is between 50-65%.When humidity is higher; plants become susceptible to fungal diseases like *Alternaria* and *Botrytis*. High light intensity and very low humidity are detrimental to a young crop and may cause leaf burning and weak foliage. It leads to infection of parasites likes red spider mites.

Soil

Selection of site is very important for successful carnation cultivation. Since roots of Carnation are highly susceptible to poor drainage conditions, it needs well drained and aerated soil. The growth of roots in heavy soil is very much restricted and adversely affects growth of plants and quality of flowers.

A rich sandy loam or loam soil is ideal for its cultivation. Soils having higher silt or clay content should be amended by incorporating organic matter or compost. The pH 6-7 and EC 0.8-1.6 ms/cm at 25°C is ideal. Both higher and lower pH to show adverse effects on flower production. The site selected for plantation should be free from weeds, nematodes or soil borne pathogens. Pre-plant sterilization with stem, sun or 5% formalin is beneficial. Reused soils may contain many pathogens especially *Fusarium*, which makes some beds unsuitable for Carnation cultivation even after disinfection.

Perpetual flowering Carnations are multiplied by vegetative means, soft terminal cuttings while seed propagation is normally practiced for the purpose of hybridization. For successful cut flower production, choose Carnations with well developed roots having minimum four leaves and not developed any flower bud. Cuttings should be of about 10-15 cm long for standard cultivars and 8-10 cm long for spray types, disease free and of best commercial variety. Leaves should be free from any fungal spot.

Planting

Soft terminal cuttings from healthy mother plant having vigorous root should be planted at shallow depth. Deep planting of cuttings may cause root and stem rot. During planting keep, the base of the cutting above the soil level as they were in the rooting medium to ensure rapid development. To obtain the maximum no. of flowers, planting time and plant density are very important. When planted for commercial scale, raised beds of 30 cm height, 1.05 m width and convenient length are preferred. A planting density of around 32 plants / m^2 (15x20 cm spacing) is generally recommended for 2-year production. When it is planted in pots, a single plant can be planted in 7.5 cm pots containing a potting mixture of equal parts of soil & well rotten F.Y.M. and half part of sand. Normally planting distance is as follows.

Planting Carnation

Row to Row distance (cm)	15
Plant to Plant distance (cm)	15
Plant population (Gross / m2)	20
Plant population (Net / m2)	36

Season and Cycle

Planting schedule is important to produce maximum number of excellent quality flowers. Carnation produces heavy flush of flowers during a short period of time, which is not desirable for commercial production. Therefore, it is regulated through planting time, planting density and method of pinching of plants. In Northern plains of India, planting is done during mid September-November, which produces flower during February April. In hilly areas where there is no snowfall, October-February planting is suitable to obtain flowers during mid-April to mid July and in hilly areas having seasonal snowfall, planting is done during February- April to get flowers from August- October. In protected conditions, Carnations can be planted round the year.

Carnation Growing Period

Carnation flowers take approx.150-180 days from planting to flower under open conditions and 120-150 days under protected conditions depending upon the season. The main propagating season of Carnation extends from November to March. Perpetual Carnations are mostly planted on a 2 year cycle after which their yield decreases.

Irrigation and Fertilization

Rooted cuttings should be watered immediately after planting. Spraying of water through misting is more effective than the surface watering for early establishment of the cuttings. After three weeks of planting, drip irrigation has to be adopted. Frequency of watering depends on the weather conditions and the soil type. Stagnation of water should be avoided to minimize incidence of diseases.

Carnation plants need regular supply of nutrients for continuous production of flowers. Therefore, application of nutrients in smaller doses but at regular interval favours better growth and yield. Low level of nitrogen and boron may cause calyx splitting. In the first three weeks after planting, no chemical fertilizers are applied. A total nutrition of 5-10 kg of FYM/m^2, 30 g N/m^2, 20 g P_2O_5 and 10 g K/m^2 is applied for Standard Carnation while same dose is applied for spray types except the Nitrogen @ 40 g/ m^2. Half dose of nitrogen is applied before planting and remaining half N one month after planting. Spray of 0.1% urea and potassium nitrate (KNO3) and 1 ppm boron at 10 days interval till flowering produces good quality flowers.

Some micronutrients like Fe, Zn, Mn, Mo and B may also be required depending upon the soil type, mineral content of irrigation water and leaf analysis of plant sample.

Support nets

In order to obtain straight stems and to prevent lodging of plants, Carnation needs supporting nets of 4-5 layers. These plants are planted within the net of mesh size 7.5x7.5 cm. Nets are gradually raised with plant growth. For every 2.5 to 3.0 m, the metal wires or nylon rope should be supported with poles. The first net should be fixed at 12.0 cm above the bed. The remaining nets, whose squares should be 12.5x12.5 cm or 15x15 cm, are placed over the first net. These nets are placed at 15 cm apart.

Pinching

For successful production of top quality Carnations, pinching is an important operation. During pinching, the tip of thc stem is removed to encourage growth of Standard or Sim Carnations. Only 4-6 well grown lateral shoots will be allowed to grow. Pinching is donc after 30-35 days of planting leaving only 6 pairs of basal leaves. If too many leaf pairs remain, stem and flower quality may be reduced considerably. Similarly, production will decrease if a few leaf pairs will be left. There are three types of pinching methods generally followed.

Single pinching: In single pinch, apical portion measuring 5-7 cm is pinched off to give about 4-5 lateral shoots, which produces flower. This is done for early crop.

Pinch and-a- half: This method involves single pinching of the main stem to give 4-5 side shoots. Only half of total number of lateral shoots are pinched when they are about 5 cm long. This method reduces the load of first crop and provides a steady supply of flowers throughout the year.

Double pinch: In this method, all the lateral shoots arising from the first pinch are again pinched when they are 8-10 cm in length at 2-3 nodes. It is done approximately 5-6 weeks after first pinch. This is generally done for late harvesting or delaying the flowering period. Approximately 8-10 shoots are retained. This method produces large number of flower bearing shoots and poor quality flowers hence not preferred.

Disbudding

Disbudding is the removal of unwanted buds. It is practiced in Carnation to obtain good quality flowers. Buds are taken away before they get too big, as they will considerably reduce flower size. In case of standard Carnation, the lateral flower buds are removed leaving only the terminal or main flower bud. In Spray or Miniature Carnation, the terminal or main flower buds are removed, to encourage lateral flower buds. Disbudding helps to produce quality bloom in standard carnation.

Post harvest handling and Plant protection

Harvesting of Carnation flowers depends on the size of bud and growth of petals. Standard Carnations are harvested when they are half open and the first petal lays horizontally or at paint brush stage. Spray types are harvested when at least two flowers have opened and remaining buds are showing colour.

Flowers are the best harvested during morning hours when they are turgid. Keep the flowers in fresh, clean water or preservative solution for about 4-6 hours and then place in cold store at 4-6°C. Do not cut the flowers too deeply as this may adversely affect production during next season.

Grading of flowers is based on stem length; flower diameter and physical conditions like stem sturdiness, freedom from pests, diseases and physiological abnormality. They are grouped in bundles of 20 and kept at 0-2°C for 12-14 hours and then packed in corrugated cardboard boxes.

Vase life and storability of flowers can be increased by placing flowers in 10 % sucrose + 1 mm STS for 8-10 hours. Carnation flowers are very sensitive to ethylene gas. Now days' anti-ethylene compound 1-MCP (1- methyl cyclopropene) is being used as floral preservative.

Insect, Pest, Disease and Physiological disorders

Insect Control: The most common insects attacking the carnation are aphids, thrips and red spider mite.

Aphids *(Myzus persicae)* suck the sap from new shoots and leaves. They leave sticky substances on leaves and flower buds. Foliar spray of Rogor (2ml/ litre) or 0.25% Demeton-methyl effective.

Thrips *(Thrips tabaci)* also suck the sap of leaves and turn them yellow and patchy. Under severe attack they also cause streaks and spots on flowers and stems. Pink and red cultivars are more susceptible. Malathion or Rogor (dimethoate) or dichlorovos, all at 0.1-0.2 % gave the best control of Carnation thrips.

Red spider mite *(Tetranychus urticae)* is the most serious pest of Carnation. They are very small red coloured insects mostly feeding on the lower sides of leaf and suck the sap. Leaves become yellow with dusty coating and fine webs on their surface.

The most effective chemical against mite is Vertimac, which is effective against adults, eggs and larvae.

(ii) Diseases: Carnations are attacked by a number of diseases like *Fusarium, Alternaria*, Bacterial wilt, Rust, *Rhizoctonia*, and Stem rot.

Fusarium wilt caused by *Fusarium oxysporum* f. sp.*dianthi*, also known as vascular wilt, is one of the most serious diseases of carnation. It is soil borne, invades the plant through roots and damages stem tissues resulting in wilting and death of plant. Soil fumigation and treating the beds with Bavistin (0.1%) and Dithan M-45 (0.1%) at regular interval is effective.

Leaf spot caused by *Alternaria dianthi* is another serious air-borne disease. It is characterized by grayish- brown spots on leaf and stem. High temperature and humidity favour the disease spread. Removal of infected leaves and foliar spray of Mancozeb (Dithan M-45) @ 0.2% or Bavistin 0.1% controls this disease.

Bacterial wilt *(Pseudomonas caryophylli)* is prevalent at high temperature. It is a soil born disease attacking older plants. Production of grayish green foliage, rotting of roots and deep longitudinal cracks on internodes are the major symptoms. Soil sterilization, treating the plant material with 0.01% Streptocycline and using disease free plants help in reducing the incidence.

(iii) Physiological disorders: Splitting of calyx is a serious disorder affecting quality of carnations. In this disorder sepals beneath the flower are unable to form a cylindrical calyx tube which supports base of petals. As the bud opens and size of petals increases, calyx may split down either half or completely because it is unable to contain the load of extra petals. The primary cause of calyx splitting is unknown but may be due to genetical, environmental like low temperature and fluctuation in day night temperature and nutritional level like low nitrogen and boron. The problem of calyx splitting can be reduced by placing of rubber band around the calyx of the flower bud, when they just started opening. Increased level of potassium and boron, choosing the cultivars which are less prone to splitting and uniformity in day night temperature can reduce the problem.

3) Chrysanthemum

Chrysanthemums, often called 'mums,' are a genus *Chrysanthemum* of about 30 species of perennial flowering plants in the family Asteraceae, native to Asia and northeastern Europe. Botanically, it belongs to *Dendranthema grandiflora*.

Uses

Modern chrysanthemums are much more showy than their wild relatives. The flowers occur in various forms, and can be daisy-like, decorative, pompons or buttons. This genus contains many hybrids and thousands of cultivars developed for horticultural purposes. In addition to the traditional yellow, other colors are available, such as white, purple and red. The most important hybrid is *Chrysanthemum × morifolium* (syn. *C. × grandiflorum*), derived primarily from *Chrysanthemum indicum* but also involving other species.

Solan Centre (H.P.) developed the technology for year round flower production through staggered planting, pinching, growth retardants, photoperiod regulation and cultural practices, as following:

PRC	Pinching	Retardant	Lighting	Blackout (days)	Blooming period
June 2	June 24	July 28	NLD upto Aug. 4	48-70	Oct.2-Nov.4
Aug. 5	Aug. 27	Sept. 30	NLD upto Aug. 26 + ALD from Aug. 27-Oct.6	48-73	Dec. 11-Jan. 1
Oct. 11	Nov. 2	Dec. 5	ALD upto Dec. 12	50-74	Feb.11-Mar.4
Dec. 1	Dec. 20	Jan. 23	ALD upto Jan. 30	94-115	May 24-June 6
Feb.2	Feb.24	April 1	ALD upto April 8	48-73	June 8-29
April 3	April 22	May 28	ALD upto April 22 + NLD from April 23-June 4	52-72	Aug.7-29

- Blackout means natural/artificial short days after termination of light to the stage till 60-70% flower buds showed colour.
- NLD = natural long days, and ALD = artificial long days.
- The range under blackout column denotes variation with the treatment.
- PRC = planting of rooted cuttings.

ALD treatments at later stages of growth (1 and 2 months after planting) extend mores vegetative growth than LD treatments immediately after planting. GA (50 ppm) advances the flowering while NAA (50 ppm) delays it. At Ludhiana, by providing natural short day conditions from Dec. to Feb. and artificial short days from March, flowering was induced in May in cv. Punjab Anuradha. Black polythene covering of the plants of cvs Birbal Sahni,' Jubilee,' Flirt,' Sunil,' Jayanti and Vijay for 14 h daily at night till 60-70% buds show their colour and induced earlier flowering.

Rooting was more in suckers than the terminal cuttings though reverse was true with respect to flowering and yield. Spacing of 30x30 cm with single pinching at 4th week or 2 pinchings at 4th and 7th week of planting provide higher yield. *Azospirillum*, VAM and phosphobacterium along with 200 ppm N, 40 ppm P and 200 ppm K provide better growth, yield and post harvest life in cvs. Bronze Spray and Red Spray.'

Chrysanthemums are divided into two basic groups, Garden Hardy and Exhibition.' Garden hardy mums are perennials capable of being wintered over in the ground in most northern latitudes. Exhibition varieties are not usually as sturdy. Garden hardies are defined by their ability to produce an abundance of small blooms with little if any mechanical assistance (*i.e.* staking) and withstanding wind and rain. Exhibition varieties on the other hand require staking and over-wintering in a relatively dry cool environment, sometimes with addition of night lights.

The exhibition varieties can be used to create many amazing plant forms; large disbudded blooms, spray forms, as well as many artistically trained forms, such as: thousand bloom, standard (trees), fans, hanging baskets, topiary, bonsai and cascades. Chrysanthemum blooms are divided into 13 different bloom forms by US National Chrysanthemum Society, Inc., which is in keeping with the international classification system. The bloom forms are defined by the way in which the ray and disk florets are arranged.

Chrysanthemum blooms are composed of many individual flowers (florets), each one capable of producing a seed. The disk florets are in the center of the bloom head, and the ray florets are on the perimeter. The ray florets are considered imperfect flowers, as they only possess the female productive organs, while the disk florets are perfect flowers as they possess both male and female reproductive organs.

Greenhouse Cultivation of Chrysanthemum

Cultivars and Pot Culture

i) **Large Flower:** Beauty, Snow Ball, William Turner, Innocence, White, Green Godess, Chandrama, Super Giant, Sonar Bangla, Pink Cloud, Shirley, Perfection, Classic Beauty, Pink Turner, Brave, Distinction, Alfred, Autumn Blaze *etc*.

ii) **Small Flower:** Mercury, Jyotsna, Birbal Sahni, Himani, Topaz, Archana, Basanti, Nanako, Sujata, Kundan, Nilima, Alison, Jaya, Gem, Jean *etc*.

Chrysanthemum culture in pot requires very strict time bound monthly intercultural practices with precise operation. Selection of disease free and high quality planting material is ideal for commercial growers. Pot culture includes soil preparation and selection of proper size pot for it.

Soil preparation for pots

Ideal soil composition for growing healthy chrysanthemum plants in pot includes one part of clay soil, one part of leaf mold and 2 parts of rotten cow dung manure. Fertilizer is also mixed at this stage which includes bone meal one part, super phosphate one part, di -ammonium phosphate one part and neem cake one part i.e. in the ratio of (1:1:1:1). The mixture should be kept as a heap and spread for solar treatment and for proper sun drying prior to filling of pots. For filling 100 pots, 5 kg bone meal, neem cake and fertilizer mixture is required. Soil is prepared well in advance i.e. a month prior to planting. A small amount of Malathion powder is added to soil to get rid from insect.

Preparation of cuttings

Cuttings are prepared from mother plants during June to August. In case of early rains, cuttings can be prepared at the end of June otherwise July to mid August is the ideal time for making cuttings. Stem cuttings from terminal portion are selected and its length restricted to 10-12 cm. Bavistin or Indofil M-45 (2% solution) is prepared for the treatment of cuttings. After quick dipping for 2-4 seconds in fungicide solution, hormone like Rootex No.1 or alcoholic solution of IBA (quick dip method) is applied at base of the cuttings. Cuttings are planted in 25-30 cm pot filled by sterilized coarse sand. For large flowered chrysanthemum especially for spider type cultivar, sand filled pots are ideal as the roots developed from stem cuttings are fine and its development is the best in sand. The rooting takes place within 15-20 days of planting. Fifty cuttings of small flowered chrysanthemum can be accommodated in 30 cm pots while forty cuttings of large flowered chrysanthemum can be accommodated in 25 cm pots.

Planting and month wise operation

Planting and month wise operation include watering, fertigation, pinching, dis- budding, removal of old and diseased leaf, protection from hot winds and excessive rains and staking. Intercultural practices from January to December for small and large flowered chrysanthemum are briefly explained.

Suckers are planted in earthen big size pots or beds in January and February and large flowered chrysanthemum a transplanted to 10 cm pots. Small flowered chrysanthemum can be transplanted in well manured beds. Similarly few plant of large flowered chrysanthemum should be planted in beds to have cuttings from them in July. In large flowered chrysanthemum beds regular watering is required in April and potted plants watering is restricted to check growth. Small flowered chrysanthemum in beds requires more branching by pinching. This is the ideal time to pinch the plants back to 10cm from ground level. In May large flowered chrysanthemum should be protected from hot wind and repotting in 15cm pot gives better results. Small flowered chrysanthemum requires second pinching in May. Large flowered chrysanthemum requires protection from hot wind in June. Small flowered chrysanthemum requires protection from hot winds during June and this is the proper time for third pinching. The pinching should be done in first week of June. In July small flowered chrysanthemum should be lifted from beds and planted in 25cm pots. Large flowered chrysanthemum requires repotting in 20cm pots in July. This is the high time for taking cuttings for raising dwarf plant with superior blooms. In August chrysanthemum plants need protection from excessive rains. Diseased foliage should be removed and fourth pinching is required in small flowered chrysanthemum.

Nutrition and Propagation

Dry Fertigation: For dry fertigation di- ammonium phosphate, super phosphate, bone meal and neem cake at ratio of (1:1:1:1) is mixed thoroughly. One teaspoon (10 g) of this mixture is applied at fifteen days interval in chrysanthemum pots for better growth and flowering.

Liquid fertigation: For better growth and quality flowering in chrysanthemum, liquid fertigation is very essential. For making 100 lt. of liquid fertigation following components are mixed in a big tank.

- Unripe cow dung… … … … … … … … … 5kg
- DAP… … … … … … … … … … … … … … … … 5kg
- Super phosphate… … … … … … … … … … … … 5kg
- Neem Cake… … … … … … … … … … … … … … 5kg
- Calatropis leaf (Madar)… … … … … … … … … 10kg

The solution should be stirred well prior to use and kept for a 15-20 days before use. Liquid fertigation should start from October and continued up to November at fortnight interval. This liquid should be diluted ten times before use.

Propagation: Chrysanthemum is propagated both by vegetative and sexual methods. The latter method is adopted by the breeder for development of new varieties. The vegetative method includes multiplication through suckers, cuttings and micro-propagation.

Suckers: Profuse suckering occurs about a month after blooms fade. After separating the suckers from mother plants, they are planted directly in the field or for root formation in pots.

Cuttings: Propagation through stem cuttings is the common method followed by the commercial growers. An ideal cutting is supposed to be 4-6 cm long with a diameter of 3.2 to 4.8 mm. The terminal cuttings are late June and July. Seradix B-1 or IBA (2000 ppm) to 0.5 cm basal portion. The cuttings root between 3 and 4 weeks.

Micro-Propagations: Chrysanthemum can be multiplied faster by culturing shoots apex, somatic callus, from peduncle and other floral parts and from leaves *in vitro.* The terminal cuttings 2-3 cm long with at least one node are taken.

Growth and flowering: Environment has great influence on growth and flowering of chrysanthemum. The major factors are light, temperature, relative humidity and air composition.

Light: It is well known that chrysanthemum plants grow vegetative till the nights are shorter than critical level ($6^{1/2}$ hrs) in most of the traditional varieties. Flourescent light of very high intensity given only for one minute every night completely inhibits flowering. A minimum of 3-5 days are necessary for bud initiation. It is observed that day-length of 14-$14^{1/2}$ hrs could initiate flower buds, but their development occurred only when day-length was of 13-$13^{1/2}$ hrs.

Temperature: The growth rate in chrysanthemum remains at low level at or below 10^{0}C. When the temperature is raised to 15^{0}C, the growth rate increases rapidly. A constant temperature of 16^{0}C is the best for rooting of cuttings. On the basis of their response to temperature regarding flowering, the varieties are grouped into 3 classes by Cathey.

Thermozoro: Flowering at any temperature between 10^{0}C and 27^{0}C temperature.

Thermopositive: At higher temperature of 27^{0}C, bud initiation is rapid but flowering is delayed.

Thermonegative: Bud initiation occurs at low to high temperature between 10^{0}C and 27^{0}C, but continuous high temperature delays bud development.

Relative humidity: A relative humidity between 70 and 90% is optimum for growth of chrysanthemum.

Air composition: In temperate regions, chrysanthemum is grown in glasshouses. During the day time, the required amount of 0_2 is available as a result of photosynthesis. During night, proper ventilation is necessary for ensuring sufficient oxygen supply. Production is found to increase significantly by artificially enriching the greenhouse atmosphere with Co_2. Supplementary Co_2 (700-900 ppm) produces longer stem length and early flowering.

Growth substances: Growth substances and various other chemicals are found to regulate growth and flowering in chrysanthemum.

Growth promotion: GA_3 shows marked effect on growth and development depending mainly on the time of application and stage of growth. Treatment with GA_3 during 3rd week of short days caused elongation of stem and peduncle when applied in 4th week of SD. Spraying chrysanthemums at pinching time with BA (Benzyeadenine at 40 ppm) increases branching.

Growth retardation: For developing well-shaped pot plants, growth retardants have proved very effective. SADH and phospon are being used commercially. The application of SADH (2000-4000 ppm) after disbudding reduces stem length below the flower head. The liquid form is used as soil drench. CCC (Chlormequat) (1.25%) suppresses plant height.

4) Gerbera

Gerbera originated in South Africa, belongs to family Asteraceae. It is an important commercial cut flower crop grown throughout the world under a wide range of climatic conditions. During 2001, at Aalsmeer Flower Auction Centre (Holland) it remained at fourth place among the top ten cut flowers of the world floriculture trade in sales volume. In India its commercial cultivation under low cost polyhouse environment is on rise and has made its cut flowers availability round the year.

Gerbera (*Gerbera jamesonii*) is commonly known as Transvaal Daisy, Berbrton Daisy or African daisy. It produces very attractive long stalk cut flowers. Its flowers are of various colours and forms, suitable for different flower arrangements. Its cut blooms have 7-12 days vase life. Apart from upcoming huge domestic consumption, it has got great potential in export also.

Growing environment

Gerbera is normally grown under open field condition, but with the introduction of new exotic varieties, which require more care in their management, it has become difficult to get production of superior quality blooms under open field conditions or even agro shade nets. Therefore, gerbera has to be grown under protected environment. Different growing environments, *viz.*, open field, 50 % shade net, insect proof net, naturally ventilated polyhouse (low cost) and fan and pad evaporative cooled polyhouse (medium cost) were compared for growing exotic varieties of gerbera under Bangalore condition. The yield of superior quality flowers was reported higher under naturally ventilated polyhouse. The ideal temperature for gerbera flower initiation is 23°C and for unfolding are 25-27°C. Flowering is harmed considerably at temperature below 12°C and above 35°C. The optimum relative humidity inside the greenhouse should be 80-85 %.

Commercial varieties

Several exotic varieties of gerbera in single, semi-double and double types are available with nurserymen. Gerbera varieties are also classified in two groups i.e., large flowered and mini gerbera with desirable characters like– large flower diameter, long and stout flower stalk, double type of flower arrangement, black center , increased vase life and ability to withstand long distance transportation and which have demand in the market is significant. Some of the suitable varieties under Indian conditions are: Cabana, Ellymay, Essandre, Evening Bells, Feugo, Gold Spot, Golden Gate, Ibiza, Jankfrau, Lindessa, Lyonella, Optima, Ornella, Paganini Piton, Polar, Red Monarch, Rosabella, Ruby Red, Sangria, Sunway, Tiramisu, Twiggy etc.

Growing medium

Well drained loamy soil, rich in organic matter and with adequate moisture holding capacity, is ideal for successful gerbera cultivation. A soil pH of 5.5 – 6.5 and E.C. 0.5-2.0 ds/cm^2 are ideal. A lower pH may result in the lack of manganese or iron and a high pH affects the soil structure. The land should be ploughed / dug deep enough 3-4 times and brought to a fine tilth. Gerbera is grown on 1 meter wide and 30 cm high raised beds of convenient length, leaving 30-40 cm space between two beds. The beds should be prepared by mixing farm yard manure (8-10 kg/m^2) and sand (3-4 kg/m^2) or beds may be prepared by mixing farm yard manure, sand and coconut coir in 2:1:1 ratio.

Disinfection of soil

Before taking up transplanting, soil sterilization or disinfection of the prepared beds is absolutely necessary. It would minimize the infestation of the soil borne pathogens, which would otherwise destroy the crop partially or completely. The raised beds should be drenched by 4% solution of formaldehyde (1-1.2 litre/m^2) and immediately covered with air tight polyethylene film for 4-5 days. Soon after, the treated beds should be watered thoroughly to drain off the chemical after disinfestation and subsequent washing out of the soil. It is advised to wait for two weeks before taking up transplanting. Soil disinfection may also be done with methyl bromide (@ 30g/ m^2 area). Growing medium may also be sterilized by application of Basamid @40g/m^2 area. The treated beds have to be covered gas proof by polyethylene film for 6-7 days. A few days before transplanting neem cake @ 1kg/m^2 can be incorporated into the prepared beds. Alongwith this, Furadan @ 10g/m^2 can be applied as a preventive measure against nematodes.

Transplanting

Gerbera is commercially propagated either through suckers or tissue culture plantlets. Its propagation through tissue culture plantlets has become the predominant method to provide uniform flowering, free branching and pathogen free plants. Faster the delivery of the young plants, it is necessary to transplant them as soon as possible. It is good thing to keep them in a room without direct sun, arrange the boxes one by one without placing above each other.

Hardened Plants of Gerbera

In this way, it is possible to guarantee a good aeration. Proper planting is important. If the plants are planted too deep the crown rot is encouraged and too shallow the exposed roots are easily disturbed and plants may wilt and die. The young plants should be planted so that the crown is slightly 2-3 cm above the surrounding soil level. The crown must be dried out between two waterings; otherwise condition will be ideal for infection of *Phytophthora cryptogea*. Planting can be adopted at a spacing of 30 x 30 cm, in three rows per bed of one-meter width. Plantlets are transplanted without disturbing the root ball. After planting, the relative humidity inside the greenhouse should be kept 70-80 % or shade nets should remain closed for 4-6 weeks to avoid desiccation of plants.

Planting	Gerbera
Row to Row distance (cm)	37.5
Plant to Plant distance (cm)	30
Plant population (Gross/m^2)	06
Plant population (Net/m^2)	09

Watering

Gerbera requires a constant supply of water but in a smaller quantity. The soil surface should dry out between waterings. The deep roots draw water from the base of beds. Excessive dampness around the leaves and crowns encourage the incidence of Botrytis, powdery mildew and crown rot diseases. Immediately after transplanting, watering should be done using overhead sprinklers, watering cane, watering hose fitted with a 'rain' nozzle up to one month. Thereafter, gradually change to drip irrigation. Generally, one dripper per plant is required. The average water requirement is 700 ml / plant /day (4.5 to 6.0 litre/day/m^2), depending on season, soil texture, light intensity and crop stage. It is necessary to avoid over watering, even though plants tend to dry out rapidly due to their large leaf area. Watering can be done in 2-3 shifts a day. A watering shift should not be too short to ensure vertical flushing of excessive salts. On soils with insufficient drainage, irrigation should be limited to only one large watering before 10 a.m. to let the soil dry out during the day.

Nutritional Requirement

Gerbera requires plenty of macro and micronutrients for proper growth and development. The first dose of fertilizers should be applied 10-15 days prior to transplanting. Application of 10:15: 20 g NPK/ m^2, respectively during first three months of transplanting will be helpful in better establishment and growth of plants. From fourth month onwards application of 15: 10: 30 g NPK/ m^2 at two months intervals is desirable. Fertilization with 80 % of the recommended dose (1: 10: 30 g / m^2) two months through straight fertilizers and remaining 20 % through the water soluble fertilizers reported cost effective to obtain higher returns and good quality of cut flowers. Improved flower yield, stalk length and flower diameter are obtained by application of 15 tone / ha vermicompost plus 75% recommended dose of NPK (15:10: 30 g / m^2 /two months). Spraying of micronutrients like boron, calcium, iron, copper and magnesium @ 0.15 % each, at monthly intervals is also desirable. Concentrated complex of micro-nutrient mixtures may also be sprayed (0.2 %) at monthly intervals for obtaining increased flower production.

Application of fertilizers through drip fertigation preferably in morning hours around 8.00 a.m., is advisable. Fertilizers should be given 10 days after transplanting. Initially 19:19:19 NPK should be given at 0.5 to 1.0 g /l of water on daily bases upto flowering stage. After appearing of flowering, the fertigation should be given as per the following ratio: *viz.*, 1:1:3:1: 0.2 NPK, Ca and Mg. Application of micro-nutrients through drip irrigation is also beneficial. It is suggested to analyze the soil and water samples once in two months regularly to fine tune the fertigation programme according to the soil and water analysis report.

Harvesting of flowers

The first flowering occurs in 7-12 weeks after transplanting. Harvesting of flowers should be done when outer 3-4 rows of the disc florets have fully developed. A few growers remove the first flower as soon as feasible. They believe this will result in more rapid and uniform elongation of the remaining flower escapes. In a young plantation, the plants may be pulled loose while picking. Therefore, the first flowers have to be picked very carefully. The cut flowers should be pulled rather than cut, because cutting will have a stem stub on the growing plant that encourages development of diseases. The base of flower stem should be pushed slightly down and to one side and then separated from the crown with gentle upward tug. There is absolutely no need to apply knife, scissors, *etc.* while harvesting of flowers. Water movement through the basal portion of stem is slower than through the segment 5-6 cm higher. Therefore, the basal 5-6 cm portion should be cut using very sharp knife removed before placing in the harvesting bucket. Gerbera has a fairly soft stem and walls of the vascular bundles are also soft so that they are prone to damage if cut with pruning shears.

Yield and Plant Protection

Yield: The flower yield of gerbera crop depends upon various factors like growing environment, varieties planted, spacing adopted, *etc.* The average yield under low cost naturally ventilated polyhouse is 30-55 flowers / plant / year

(200-400 flowers/m^2/year). If properly maintained, healthy plants give economical yield for a period of $2^{1/2}$-3 years, after that plantation will have to be replaced.

Insect-Pests

White fly (*Bemisia tabaci*) lays eggs on lower surface of leaf. Nymphs and adults suck cell sap from leaves resulting in discolouration, reduced plant vigour and development of shooty mould on leaves. Removal and Burrying of heavily infested leaves check white flies build up. Chemical control includes sprays of Confidor/Admior or Dicofol @ 1.5-2 ml/l of water at regular intervals.

Leaf miner (*Limomyza trifolii*) causes considerable damage to foliages. Adults lay eggs in the leaves by punching. Maggots feed inside the leaves by leaving characteristic mines which turn brown, brittle and dry in heavy infestation. Removal and destruction of the heavily infested leaves reduce further damage. Spray during morning hours of Pyrazophos, Metasystox, or Dimethoate @ 1.5-2 ml/l of water are quite effective measures.

Aphids (*Myzus persicae*) attack growing flowers. Both nymphs and adults colonize on lower sides of petals and spread to entire flower in heavy infestation. Aphids suck the sap from petals leading to disclouration, wilting and distortion of flowers. Spray of Monocrotophos or Dimethoate @ 1.5-2 ml /l of water at regular intervals is quite effective to control aphids.

Diseases

Crown rot or root rot is a major disease and severe infection can destroy entire gerbera crop. The crown of the plant turns dark and decays. The leaves of the infected plants wilt gradually, become dry and turn to radish brown colour. The infected roots develop a water brown colour. The disease can be effectively controlled by drenching of the root zone of infected plants with 0.2 % solution of Mataxyl, Captan or Carbendazim or Copper Oxychloride at fortnightly intervals.

Powdery mildew (*Erysiphae cichoracearum*) infection causes white powdery coating on the leaves. Spraying with thiophanale methyl @ 1.5%, Carbendazim @ 0.2%, Hexaconazole @ 1ml/l, triadimefon @ 1g/l or Dipotassium Hydrogen Orthophosphate (K2 HPO4) @ 4.3 g/l, are effective against this disease.

Wilt (*Fusarium oxysporum*) infected plants show yellowing of lower leaves. The collar portion shows vascular discoloration. Rouging and soil application of Benzamidazoles (0.2%) help in reducing the disease spread.

Root Rot is caused by *Pythium.* Root skin is easily removed and finally wilting of the plants. Avoid excess watering. Drenching with Benlate @ 2 gm/l or with Bavistin @ 2 gm/l is helpful to control root rot.

5) Alstroemeria

Also known as the Peruvian lily, lily-of-Incas or Inca lily, is a genus in the Alstroemeriace family. It consists of approximately 60 species of rhizomatous, herbaceous plants, most of which are native of Chile and Brazil. It is one of the quickest growing flowers with varying flower colours like purple, lavender, salmon, red, pink, white, orange, yellow and bicolor. Alstroemeria is generally cultivated as a cut flower crop, a garden flower and in the recent years a potted plant. Cut flowers of alstroemeria are very popular in countries likes UK, the Netherlands and North America. The cut flower excellent post harvest keeping quality, availability in all attractive colours and high productivity makes alstroemeria known and popular worldwide. Alstroemeria is yet to make its mark in the domestic floral market of India. There is considerable scope for growing this crop in the Himalayas, including higher altitudes of Northeast India. Alstroemeria flowers are third in terms of popularity and foreign exchange earnings in Kenya after roses and statice.

Importance and uses

Alstroemeria is the best known for its ability to produce long-lasting cut flowers with post harvest life of 2-3 weeks. It can be used with elegant flowers or mixed with pompons or spray carnations. Alstroemeria is grown in the garden and treated as annuals or tender perennials. The plants have ever-blooming habit in the garden and continue flowering through out the year except in the frost.

Cultivars

A few important cultivars are:

Amor	:	Yellow
Tiara	:	Red
Rosita	:	Dark pink (year round flowering)
Cinderella	:	Dark pink
Capri	:	Pink
Serena	:	Maroon
Aladdin	:	Yellow
Pluto	:	Orange
White Wings	:	White

Propagation

Alstroemeria is vegetatively propagated by rhizome divisions or through micropropagtion. Vegetative propagation allows plants to grow into true to type. Divide beds, growing its plants at least every first and second year, depending on variety and growth characters. About two to three weeks prior to division, plants should be severely pruned, leaving only the youngest 15-20 cm shoots. It is also propagated by seed for cut flower production to develop new hybrid varieties.

Climate

Temperature

Climate is very important for its cultivation and flower production. It is recommended to keep the soil temperature below 20 ^{0}C, as high temperature hinders formation of buds for second crop under long day conditions and moreover, they delay growth after first crop. In summer, try to keep the temperature below 20^{0}C to prevent plant dormancy. During darkest months, December and January, it is recommended to keep the temperature at about 10^{0}C. Growth is slow then but the low temperature has a favorable influence on flower production in April, May and June. In areas with favorable day light conditions during these months, a temperature of about 13^{0}C can be maintained. A soil temperature between 14 and 17^{0}C is a good range for many varieties to stimulate flower bud establishment.

Light

Light is an important determining factor for flower stem quality of alstroemeria. Sufficient light (5,000 foot candles) is important to prevent bud abortion which results in strong and heavy quality flowering stems. The effect of day length differs from variety to variety. A long day (14 hours or longer) combined with higher temperatures (over 18^{0}C) causes a decrease in growth, consequently number of shoots. Lengthening of light period in winter stimulates the formation of buds in shoots; buds are formed after a fewer internodes, stems become shorter and flowering starts earlier. The optimum photoperiod for all varieties is 12-14 to be the best in winter or spring.

Relative humidity

Alstroemeria is the most suited to air humidity between 65-85%. Higher humidity can create a weak crop and susceptibility to diseases.

CO_2

CO_2 produces a distinct increase in quality of flower stem. It induces better flower quality, early flowering and increased flower production. CO_2 at 900 ppm is the best for quality flower production and early blooming.

Soil

Alstroemeria can be grown on nearly all types of soil. In general the soil with a pH of 5.5 to 7.0 and electrical conductivity less than 1.0 is the best suited for its cultivation. In peat and sandy loam soil, Alstroemeria shows much quicker and luxuriant growth than in clay soil. In clay soil, it is advised to mix sand and well decomposed FYM to make it porous. Sterilization of soil is important to prevent soil borne diseases, which cause damage to the crop. The soil should be thoroughly worked to at least 40 cm deep. This will be necessary as the plants will stay for a number of years on the place.

Agro –techniques

Planting

Alstroemeria can be planted in one meter wide raised beds in two rows parallel to each other with 40 × 50 cm distance. The planting density of 4 plants /m^2 is ideal. Rhizomes are planted 7-10 cm deep in the soil.

Planting time

The planting time mainly depends upon type of cultivars and climatic conditions. Alstroemeria cut flower production; the beds should be 1-1.2 m in width with footpaths 50-60 cm. The beds should be 30-40 cm deep to allow the roots to grow nicely during two years of flower production cycle. Before planting, mix well decomposed dry FYM @ 5-8 kg/m^2 area. In cool climatic regions, alstroemeria can be planted year round, but planting time differs because of required flowering time and high temperature. The usual time of planting is in autumn (September-October).

Flowering season

In alstroemeria, two flowering seasons:

- Peak flowering in spring, with a peak production in May
- Less prolific crop in the autumn (August, September, October, November). After the first crop, plants should be kept cool to encourage the second crop to develop quickly.

Irrigation

Alstroemeria plants should be kept moist and cool, but should not be allowed in poorly drained soils. It is sensitive to salt so good quality water must be used for better growth. Rain water is highly suitable for its cultivation. The salt content in irrigation water should not be higher than 10 micromol per liter. The irrigation system must be adjusted to the soil. Normally water is given from underneath by a sprinkler system. Drip lines can also be used.

Fertilization

Alstroemeria requires heavy feeding with high nutrients once the plants established. Regular fertilization with 450 mg N, 130mg P and 300-450 mg K/ week/plant is required for good growth and development. Nitrogen should be provided in the form of nitrates under cool growing conditions. Regular feeding with Fe, Mn and Mg should also be given to overcome nutrient deficiencies.

Shading

Shading is important to protect the crop and soil from high sunlight. About 25% shading is ideal to extend cropping season. It is important to find a good balance between shading for cooling and to keep optimum light intensity in polyhouse.

Support and training

Alstroemeria grows very tall. Thus, it is important to use 2-3 layers of wide plastic supporting net (20 x 17 cm) at planting time or after initiation of new shoots from rhizome. These layers should be raised as the plants grow. The lowermost layer should be about 30 cm from the ground level.

Pinching and thinning of the thin and blind shoots brings air and light into the crop. This is also beneficial to sprout formation. In the younger stage, the thin sprouts are pinched. It is mainly preferred in spring and summer.

Harvesting and post harvest management

Harvesting

Alstroemeria can be harvested when the buds are about to open and are starting to colour. For direct sale, the first flowers should have just opened. Harvesting should be done by pulling the stems rather than cutting the stems. The yield of flowers is around 180 stems per square meter from two flushes. Alstroemeria sometime exudes a sap which can cause skin irritation. To avoid this, use of gloves is recommended

Grading and bunching

Alstroemeria flowers are graded mainly into Class A and B, Class A flowers are 80 cm long with rigid stems, show three or more stems per stalk, are free from chemical residues and parasites. Class B should have a minimum length of 60-70 cm. Alstroemeria flowers are usually sold in ten stem bunches, leaves must be stripped from the bottom of 10 cm of stem. The bunches are then placed in plastic sleeves to prevent damage to the leaves. After grading and bunching, the flowers should be kept in water and placed in the light. The flowers can be stored in a refrigerated room at a temperature of 10°C for a period of up to 6 days.

Post harvest management

Alstroemeria flowers are sensitive to ethylene and quickly become limp after harvest and as such should be placed immediately placed in water, preferably with some preservative or germicide added to it. To prevent the leaf yellowing, flowers may be placed in water treated with a growth regulator (GA_3). The vase life of 14 days is considered normal.

Insect pests and diseases

Alstroemerias are not very susceptible to diseases. However, aphids, thrips, spider mites, caterpillars and nematodes can cause problems. Systematic insecticides should be sprayed at regular intervals as a prophylactic measure. Pythium is a problem only when soil is too wet. Rhizoctonia sometimes develops under warm humid conditions. These diseases can be managed by drenching with Carbendazim @0.1% or Thiram@ 0.2%.

Botrytis is another disease that can also affect the crop. It prevails during high humidity. It can be managed by following cultural practices such as restricted watering, improved ventilation and prevention of condensation of water drops on flowers.

Disorders

Flower abortion or blasting occurs due to low light or when roots are damaged by excessive salts or over watering. Therefore, precautionary measure should be taken to control the disorder.

Summary

Alstroemeria has emerged as a high value bulbous cut flower. Its cultivation has taken up as commercial venture in open as well as protected cultivation in Himachal Pradesh and North Eastern regions, since it gives high returns to farmers.

6. Lilium

Lilium is one of the most fascinating bulbous crops used as cut flower and potted plants all over the world. It ranks fourth in the International flower auction market. It belongs to family Liliaceae and native of the Northern Hemisphere up to South Canada and Siberia. It has beautiful, attractive, bright and scented flowers of different array of colours ranging from orange, yellow, red, pink to white. Botanically, lilies are classified into nine major groups, but the three most important groups for commercial cut flower production are the Asiatic hybrids, Oriental hybrids and LA hybrids .The cut flowers remain fresh upto 7-12 days at room temperature. It has a capacity to rehydrate after continuous transportation.

Cultivars

The varieties suitable for growing well under sub-temperate and mid hill conditions of Himachal Pradesh are: Elite, Navona, Alaska, London, Pollyanna, Prato, Stargater, Sumplon, Brunello, Dreamland, Connecticut King, Shiraj, Gran Paradiso, Vivaldi, Sterling Star in Asiatic hybrids, whereas Stargazer, Sorbonne, Tiber, Rialto and Siberia are common in Oriental hybrids.

Cultivars and their characters

Asiatic Hybrid Cultivars

Cultivars	Average Bud Count	Average Height (inches)	Average Days to Bud Color	Primary Flower Color
Connecticut King	5	21.0	59	Yellow with brown spots
Enchantment	3	20	60	Orange, separated petals
Gran Paradiso	4	27	67	Burnt orange
Jollanda	7	32	64	Orange
Monte Rosa	4	23	62	Pink, yellow
Pollyanna	8	30	65	Dark and light yellow
Salmon Beauty	8	32	62	Salmon, brown spots
Snow Star	7	27	60	White, yellow
Red Tiger	11	33	62	Dark orange, black spots
Roma	5	23	54	Cream
London	8	35	58	Yellow
Alaska	6	30	54	White
Prato	6	40	60	Orange Red

(Contd.)

Cultivars	Average Bud Count	Average Height (inches)	Average Days to Bud Color	Primary Flower Color
Oriental Hybrid Cultivars				
Casa Blanca	3	26	93	White
Everest	5	28	98	White
Mona Lisa	3	14	73	Pink, white
Stargazer	6	25	88	Pink, white
Sans Souci	6	24	94	Pink, white
Primeur	8	17	64	White
Lareve	4	18	65	Pink, yellow
Journeys' End	6	23	93	Dark pink
Acapulco	6	30	95	Pink
Tiber	7	25	92	White

Climate

Pre-cooling Requirements

Asiatic and Oriental lilies require a cold treatment of a minimum of 6 and 8 weeks at 2 to 4°C before planting for rapid shoot emergence and flowering. Additional cooling is not harmful. Bulbs which will be used for late forcing (usually after February) are usually frozen in peat and maintained at -2°C. Keeping the bulbs at this temperature prevents sprouting, reduces loss of the bulbs' energy reserves and minimizes disease problems. When frozen bulbs are received from a supplier, defrost the bulbs slowly at 7 to 13°C for one to three days.

Temperature

Temperature is very important for its cultivation and flower induction. Optimum day and night temperature for quality bloom are 15-25^0 C and 8-10 ^{0}C, respectively. Temperature beyond the limit affects the length and quality of spike and resulting into flower bud drop and poor development of bulbs.

Light

The crop is the most comfortable at a relative humidity of 70-85%. Sufficient light (2500-3000 foot candles) is important to prevent bud drop which results in strong and good quality flowering stems. The effects of day length differ from variety to variety. Long day speeds up development of flower. The optimum photoperiod is 12-14 hours.

Humidity

The crop is the most comfortable at a relative humidity of 70-85%.

CO_2

CO_2 at (1000 ppm) is effective for improved plant and flower quality and reduced bud abscission.

Soil

The soil with a pH of 6-7 with electrical conductivity less than 0.5 mS/cm is the best suited for its cultivation. In peat and sandy loam soil, lilium shows much quicker and luxuriant growth than in clay soil. In clay soil, it is advised to mix sand and well decomposed farmyard manure to make it porous.

Propagation

Lilium is vegetatively propagated by bulblets, axil bulbils, and separation of scales and through micro- propagation. Vegetative propagation allows plants to grow into true-to-types. Propagation through scales is a rapid means of multiplication. The bulbs should be lifted carefully in early autumn and scales removed from the outside of bulbs. They are treated with fungicides (Bavistein 0.1%) and growth regulators (IBA 200-500 ppm) and planted in sand or in coca peat and kept at constant temperature of 25^0C for 8 weeks. Small bulblets appear at base of scales and these should be kept for further growth and lifted only next year to their flowering location.

Planting and planting time

Lilium cut flower production, planting may be done on raised bed of 15 cm above ground level. The beds should be 30-40 cm deep to allow roots to grow nicely. Lilium must be planted in 1 m wide beds at 8-10 cm deep at a closer distance of 10 x 10 cm. Before planting, mix well-decomposed dry manure @ 5.0 –8.0 kg/m ^{2} area. Lilium can be planted year round, but planting may be planned depending upon the required flowering time and prevailing temperature. October - November is the best time for plains and March –May in hills to get normal crop.

Bulb size

Lilium bulbs are sold in different sizes at a circumference of 10-12 cm, 12-14 cm, 14-16 cm and larger than 16 cm in Asiatic and 16-18 cm, 20-22 cm and above 22cm in oriental hybrids. For cut flower production, smaller sized bulbs are preferred.

Management practices

Watering

Lilium is a salt sensitive crop. Good quality water is used for its better cultivation. Rain water is highly suitable for its cultivation. The salt content in irrigation water should not be higher than 1.5 mS (milli Siemens) per liter. It is advisable not to mix fresh animal manure through the soil because it contains a lot of salt and increases incidence of insect/pest in the soil. Uniform moisture level is important for proper growth and development of plants.

Fertilization

Nitrogen should be applied @ 1 kg in the form of CAN per 100 m^2, three weeks after planting. For better growth and development the optimum fertilizer doze should be 1.4 kg N, 2.8 kg P and 2.0 kg K per 100 m^2 area.

Shading

Shading is necessary to protect lilium crop and soil from high sunlight. About 50% shading net is ideal. It is important to find a good balance between shading for cooling and to keep optimum light intensity in polyhouse.

Support

Lilium grows more than 80 cm and requires support. To get quality flowering stem, it is recommended to use 2-3 layers of wide plastic supporting net at planting time or after initiation of shoots from the bulbs. These layers should be raised as plants grow.

Insects and diseases

The bulbs are eaten by maggots, nematodes or mites. To get rid of them, the bulbs should be surrounded with sand but fresh manure should not be used.

Caterpillars, aphids and thrips attack the leaves and flower bulbs which can effectively be controlled by an application of Endosulfan at the rate of 2 ml in one liter of water.

The plants are affected by *Fusarium* bulb rot, *Botrytis* rot and *Rhizoctonia*. As soon as the disease is noticed, the affected plants should be removed and they should be treated with fungicides like Bavistein (1gm) or Captan (2gm) per 100 ml of water.

Physiological disorders

Bud drop

Bud drop is characterized by a withering and bleaching of the flower bud followed by necrosis. It generally occurs when the bud is 2-3 cm in length. It is associated with low light intensity and short photoperiod and due to high temperature.

Leaf scorch

It occurs at the critical visible bud stage. First of all, the young leaves curl slightly inward and then a few days later greenish yellow to whitish spots appear on scorched leaves. Leaf scorch occurs when there is a disturbance in the balance between absorption and transpiration of water causing calcium deficiency in young leaves. A foliar application of 1.0% calcium chloride just before visible buds reduces the leaf scorch.

Harvesting

Lilium flower should be harvested when the buds are just opening and showing colour in morning hours. The spikes should be harvested 15 cm above the ground level and are graded as per number of flower buds per stem, length and firmness of stem. Flowers should be kept immediately in the fresh water. The lilies are bunched in a bundle of six stems. The foliage must be removed from 10 cm from bottom of stem. To store cut flowers, place them in good quality water at 2-3°C temperature for 2-3 days.

Storage

After bunching the spikes of lilies are placed in a suitable preservative and optimum storage temperature of cut flower is 2.0°C to 3.0°C for a shorter period, otherwise brown spots on the outside of the petals may be developed if harvested at higher temperature. In these conditions, it is advisable to store at a temperature of minimum 4.0°C. The bulbs should be properly packed in perforated trays in peat moss or sawdust and placed at 2°C-3°C with 70% R.H. for minimum 2-3 months during summer in cold storage. After harvesting the cut flowers, the bulbs should be dug out carefully without damaging, clean them and treated with fungicides (Baviste in @0.1%). The bulbs should be properly packed in perforated trays in moist peat moss or saw dust and store at 2-4^{0}C with 70% relative humidity for minimum of 6-8 weeks.

Prospects of Floriculture Industry Under Protected Environment

Worldwide consumption of floriculture products is estimated 70 billion US dollars. Cut flowers contribute about 60 percent of the world trade and the remaining are live plants, cut foliage, dry flowers, dried plants etc. India is situated comparatively closer to major flower consuming countries that its Asian counterparts. It has good scope and potential in flower trade. The severe winter in major flower producing European Countries is also an advantageous factor to India, especially areas like Bangalore, Pune, Hyderabad, Nasik, North East (for Orchids and Anthurium) which enjoy moderate climate all throughout the year. The prospects of an auction market for floricultural products coming up shortly at Singapore give an added advantage to India. (Sindhu, 2010).

In India nearly 98.5% of flowers are grown under open cultivation and hardly 1.5% flowers are grown under greenhouse. Beside the export growth, the domestic demand for cut flowers is expanding at an impressive rate of 7-8%, which opens up new avenues for greater investments in this sector. Until this time, the floriculture industry in India is over dependent on cultivation of roses and it is time for us to diversify into other crops like orchids, anthurium, lilium, alstroemeria, carnation, etc. (Sindhu and Misra 1997).

The Aalsmeer in Holland is the largest market for cut flowers. To ensure quality flower production of cut flower, greenhouse technology is required. Before taking cut flowers under greenhouse, certain considerations should be kept in mind for flourishing business.

- Greenhouse is to be constructed according to the location, climate and crop to be grown
- Availability of labour, quality water and planting material
- Development of Post harvest handling techniques and efficient marketing channel

Ruling Exotic Varieties

Grand Gala

First Red

Ravel

Hollywood

Frisco

Prophyta

Jazz

Bianca

Existing production centres in India

Flowers	Existing Location	Future Location
Rose	Bangalore, Pune	Kodaikanal, North -East
Carnation	Sangali, Satara, Bangalore Otty, Kullu, Simla	Uttaranchal, North -East, H.P
Gerbera	Sangali, Satara, Pune Bangalore	New Delhi, Kolkata, North -East
Orchids	Tamilnadu. Kerla, Maharashtra, Mangalore, Guwhati	Goa, North -East
Anthurium	Croog, Kerla, Bangalore, Ali bagh	North East, Bangalore
Lilium	Ooty, Bhimtal, Kullu	North-East, Utarakhand
Bird-of Paradise	Bangalore	North-East, Pune, Utarakhand

Allied Activities other than Cutflowers

Cut Green –An Evergreen Tool for Diversification

Flowers and foliages fit perfectly in the changing social sector of the world over. The billion-dollar floriculture industry has projected itself as worldwide phenomena in the recent decade. In the sphere of agro-export, cut foliage trade is now identified as an extreme focus segment for rejuvenating the stagnant economy. Cut foliage is used as a filler, lining and background material in various flower arrangements. They are also used for bringing life to the bouquets, wreaths and garlands. Cut foliage has the following advantages:

- Lesser cost of production because most of them can be grown in open conditions
- Year round production of a particular crop
- No strict time limit for storage
- Lesser risk of damage to quality while transport
- Longer shelf life

Flower seed production

Seed production of seasonal flower crops is a lucrative business and is practiced in considerable area in Punjab and Haryana. This offers higher returns from unit area. Of late, demand is increasing in the domestic market also. Research work is required on the development of high yielding varieties including F_1 hybrids, agro-techniques for producing uniform seed with higher certification standards.

Crops Suitable for Cut Flower Production Under Greenhouse

Crops	pH	EC (mMhos/cm)	Temperature (D/N)	Photoperiod (Hr)	Co_2 (ppm)	Yield/m^2/ year	Varieties
Rose	6.2-6.8	0.7	28/18	14	1000-1500	150-300	First Red, Grand Grand Gala, Konfetti, RavalVivaldi, Golden Gate
Carnation (Standard)	6.0-7.0	1.2	18/13	16	500-1500	200-300	Scania, Garbo, Tanga, Raggio-de-Sole, Calypso
Chrysanthemum	6.2-6.7	1.0	16/10	10	700-900	250-450	Snowball, Snowdon,Thiachen Queen
Lilium (Asiatic &Oriental)	6.5-7.0	1.0	18/12	50% shade	1000-2000	100-200	ConnecticutKing, London, Pollayana, Stargazer, Nepal
Gerbera Rosula,	6-7	1.4	16/12	14	800-1200	200-300	Diana, Nettle, Sangria, Gloria, Sun Way
Alstroemeria	6-7	1.0	18/15	16	1000-1500	-	Pluto, No.-14 Alladdin, Tiara, Cardinal

Pot Plant Production

There is growing demand of potted plants for various purposes in India and for export to other countries. At present India exports potted plants nearly 50% and it is 10% higher than the cut flowers. Many nurserymen are exporting foliage plants to West Asian countries. Flower/foliage leaves have to take advantage of the changed socio-economical situation in various parts of the country. The per capita consumption is on rise worldwide including India, but proportion of spreading is not so high in India. The trends indicate that production is steadily increasing which is essential for growth of pot plant industry in India. The pot plants are grouped into four broad categories.

References

Anonymous (2006a). Indian Horticulture Database 2009. *NHB,* Ministry of Agriculture: 18-20.

Barreto, M.S. and Jagtap, K.B. (2007). Assessment of substrates for economical production of gerbera flowers under protected cultivation. *J. Orn. Hort.,* 9(2): 136-138.

Bhattacharjee, S.K. and De, L.C. (2003). Advanced Commercial Floriculture. Aaviskar Publishers, Jaipur, Vol -1, pp 4-5.

Bose, T.K. and Yadav, L.P. (Ed.) (1989). Commercial Floriculture. Naya Prokash, Calcutta.

Chadha, K.L. and Choudhury, B. (1986). Ornamental Horticulture in India, ICAR, New Delhi.

Chaudhary M.L. (2000). A Flower City Banglore special issue on high-tech horticulture *Phal Fool* -Hindi (17 -21).

Das, P., Samanta, P.K.S. and Parthasarathy, V.A. (2002). Gerbera. In. *Commercial Floriculture* Vol. I. (Eds). T.K. Bose, and others. Naya Prokash. pp 163.

Dutt, M., Patil, M.T. and Sonawane, P.C. (2002). Effect of various substrates on growth and flowering of chrysanthemum. *Ind. J. Hort.* 59(2): 191-195.

Kumar S. (2000). Floriculture industry. Ready for takeoff. *Agriculture Today*, (June 2000) PP 15-17.

Mishra R.L. and Naveen Kumar, P. (2007). Biennial progress report of AICRP, ICAR New Delhi.

Ramesh Kumar (2010). The Genesis of Directorate of Floriculture Research, *A Bulletin*, ICAR, New Delhi.

Sindhu, S. S. (2009). National Conference on Recent Trends and Future Prospect in Floriculture 5-8 March 2011 held at Meerut. *A Souvenir*. 173.

Sindhu, S.S. and Misra, R.L. (1997). Problems and Prospects of flowers trade in India. *Indian Horticulture* Vol. (2) : *Golden Jubilee issue* (50 Years of India's Independence).

Sindhu, S.S. and Choudhary, M.L. (2001). Commercial Flower Production, Division of Floriculture & Landscaping, IARI, New Delhi-12.

Singh H.P. (2000). Current Status of Floriculture in India, Floriculture Today, Sept., 6-14.

Ornamental Horticulture, pp 463-485
New India Publishing Agency, New Delhi - 110034, India

20

Future Ornamentals

V. L. Sheela

Ornamentals are beautiful plants used in home and urban landscaping to enhance quality of our lives. Ornamental crops are widely cultivated worldwide and appreciated for their characteristics and unique appreciable qualities such as attractive color, impressive shape and size of flower and for fragrance of flowers and foliage. Earlier ornamental plants were largely associated with our social life. Of late trends has changed towards commercialization and now they are used for decoration, personal adornment and value added products *viz.* dry flowers, extraction of essential oils, perfumes and dyes and in landscaping. Further, it helps in creating more job opportunities for youth and women.

The growing demands for quality flowers, introduction of new ornamental plants and diversification needs more attention in ornamental floriculture industry. Novelty is the key to success in ornamental horticulture. Introduction of exotic ornamentals and the domestication and release of new genotypes of wild ornamentals is always welcome in this connection. Some exotic ornamentals hitherto not widely grown in Indian gardens as well as a few native ornamentals with capability to bring in a color revolution and having the potential for exploitation are discussed below.

Exacum bicolor

Family: Gentianaceae

The *Exacum bicolor*, a member of family Gentianaceae, is an extremely beautiful flowering herb of the Western Ghats. It has remarkable potential for

College of Agriculture, Kerala Agricultural University, Thiruvananthapuram, Kerala

its uses in annual flower- beds in home gardens. Of the 20 species distributed in Africa, China, Malaysia and India, 16 are found in India. Majority of them are found in grassy western slopes and plateaus in Western Ghats from Konkan south wards to Malabar and Travancore in Kerala. The genus has beautiful showy flowers. This species is worth trying as flowering annual in garden in humid tropics. Its easy adaptability to cultivation and lack of such flowering annuals in tropics duly support for its further research and development as an annual flowering herb.

Thespesia lampas

Family: Malvaceae

A natural mutant, an under shrub of about 3 m height with yellow flowers, is found almost throughout in India and frequently in inland evergreen forests of Western Ghats. The variant with promising garden type plants having ornamental value can be introduced in home gardens and parks.

Osbeckia stellata

Family: Melastomataceae

Osbeckia stellata, a brilliant ornamental shrub, bears bowl- shaped large flowers in shades of pink, purple, mauve, and rosy lilac. The shrub has a great potential to be used in pots, borders and beds. It consists of about 60 species of small shrubs native to India, Nepal, Malaysia, China, Africa and Australia. In India it is found in Shimla, Sirmour and Mandi districts of Himachal Pradesh, and earlier it was reported from Dehradun and Mussorie.

It can be propagated by softwood and semi-ripe cuttings. In nature, propagation is through seeds. The seeds are very minute. It can be grown as a specimen plant in small campuses. If properly exploited, it can come up as new ornamental crop. Its leaves are used in the treatment of tooth ache.

Viburnum opulus

Family: Adoxaceae

Viburnum is a genus having 175 species of shrubs or small trees, native through the temperate Northern Hemisphere, with a few species extending into tropical monlaneous regions in South America and South East Asia. About 17 species are found in India.

Viburnums are valued for their attractive flowers, decorative foliage and coloured fruits. Viburnum have both kind of flowers; sterile and fertile ones. Viburnums are well suited as borders of shrubberies, informal hedges, for screening and planting along roads. In prehistory, long straight shoots of Viburnum were used for arrow-shaft. Most of the flowers emits musk -like fragrance and are used in perfumery. Fruits are rich source of vitamins and minerals. Delicious Jam and jellies can be prepared from the berries.

Rhododendron arboretum

Family: Ericaceae

Burans (*Rhododendron arboretum*) are extensively grown in entire Himalayan tract. It is an evergreen tree with about 0.5-1.0 m girth and 6-8 m height. These trees are liked by all for their attractive red and pink coloured flowers and medicinal properties. Flower juice has medicinal properties, and has the ability to overcome many problems related to heat, digestion and respiration. So, it could be used for making squash. The recipcs and process of squash making have been standardized at VCSG College of Horticulture, Bharsar (Singh and Bist, 2008).

Gymnostachyum febrifugum

Family: Acanthaceae

It is an elegant, rare, native and ornamentally important plant in Kerala. It is a small, stem less attractive herb. It is propagated by seeds and suckers. Seeds obtained from the previous season may be sown before the onset of monsoon. River sand is ideal for growing. They germinate within 23- 30 days. After initial hardening seedlings are transferred to field. It spreads rapidly through suckers during the rainy season. The flowering commencing from October and lasts for 2 -4 months depending on the habitat. Subsequently fruiting is completed and ripe seeds are produced usually in 3-5 months. It is a shade loving plant and can be used in indoors or can be used in public garden.

Syzygium zeylanicum

Family: Myrtaceae

Syzygium zeylanicum is a small evergreen tree which can be widely used in garden design. It is commonly found in the west coast of Indian Peninsula, from Konkan southword and Andaman Island. The plant is stimulant and anti-rheumatic and the bark yields a black dye. It has enormous horticultural value in terms of being an ornamental plant. If these aspects are blended with conservation programme its protection would be easier as it comes under unutilized category of fruits of the tropics. This plant species can be used as effectively as *Phyllanthus myrtifolius,* an exotic garden plant, which is being widely used as hedge plant. This aromatic plant is not seriously attacked by any pest and disease. The naturally stunted nature of this tree also makes it as an ideal material for beginners in bonsai making.

Ficus krishnae

Family: Moraceae

Sacred Fig tree probably a form of *Ficus indica*, is native to India. It is a small tree with grey bark, leaves cup- shaped, deep green and the limb of the cup having raised ivory mid rib with 4-5 pairs of lateral nerves, fruits axillary, sessile, solitary or sometimes in pairs, yellow about 2.25 cm in diameter, plant usually propagated by air layering during rainy season.

On looking carefully, we will find a pocket-like pouch at the base above every leaf stalk. In other words, every leaf is bent into a little cup or scoop. It is believed that Lord Krishna used the leaves to scoop out butter. Because of the butter-cup legend, the tree has been immortalized as *Ficus krishnae*. These leaf pouches could be nature's way of catching little bits of water and releasing them slowly. This drip irrigation would certainly have helped the plant in its struggle to survive.

Hoya retusa

Family: Asclepiadaceae

Hoya retusa found in Himalayas as well as in Western Ghats. Asclepiadaceae belongs to family. It is a shade loving climber. It can tolerate direct sunlight of early morning and late evening. It is mostly grown as epiphytes. Hoya can't tolerate water stagnation condition, but can withstand dryness of 15- 20 days. Hoya can be propagated by stem cutting. The most interesting feature of Hoya is that flower emits aroma at night only, in day time flowers do not emit aroma.

Cochlospermum religiosum

Family: Cochlospermaceae

Yellow silk cotton (*Cochlospermum religiosum*) belongs to family Cochlospermaceae and is found in India along with twelve species of *Cochlospermum*. It is widely distributed in India and Myanmar. It is planted and cultivated in Sri Lanka near temples, since it is considered sacred and hence the specific name *religiosum*. During winter season it looks beautiful with its profuse obvious bright- yellow flowers born on terminal loose clusters at the end of leafless branches. As an ornamental tree, it is best suited for planting in public gardens and as avenue planting. The wood contains an inflammable, amber coloured gum known as Karaya'used in book binding, shoe- making and substitute to gum arabica.

Fouquieria splendens

Family: Fouquieriaceae

Fouquieria splendens commonly known as ococtillo, coach wood or vine cactus is also referred to as candlewood, slimwood, desert coral, Jacob's staff, Jacob cactus. It is a plant indigenous to the south western United States and northern Mexico. It is used to raise an impenetrable ornamental hedge and for landscaping. The cut flowers last for two to three days and can be used for making bouquets. The old stem yields valuable ococtillo wax. The waxy wands can be ignited like candles. These plants can be effectively used in landscaping as accent plant or as silhouette. They also attract humming birds and planting ocotillo can be done the year around with care. Ideal ocotillo plants have been grown from stem cuttings or from seed. Ocotillo plants prefer well drained sandy or gravely loam soils with light to moderate amounts of organic content. To help prevent a newly transplanted ocotillo from falling over or blowing down in a storm, large stones may be placed over the root area instead of staking, which often scars the stems. Transplanted ocotillo plants require irrigation to become established, but once established, they can survive on 240 mm of rainfall per year. Watering should be stopped during winter season. Plant has spines or sharp edges; use extreme caution when handling. The plant requires minimal maintenance. Individual stems may reach a diameter of 5 cm at the base, and the plant may grow to a height of 10 m. The plant branches very heavily at its base, but above that the branches are pole-like and only infrequently divide further, and specimens in cultivation may not exhibit any secondary branches forming an open vase shaped clump. Shearing permanently destroys form of the plant. The leaf stalks harden into blunt spines, and new leaves sprout from base of spine. The bright crimson tubular flowers appear especially after rainfall in spring, summer, and occasionally fall. Flowers are clustered indeterminately at the tips of each mature stem. Individual flowers are mildly zygomorphic and are pollinated by humming birds.

Ornithogalum thyrsoides

Family: Asparagaceae

Also known as ‘wonder-flower’ and ‘star-of-Bethlehem,’ Chincherinchee’s genus name comes from the Greek phrase ‘birds’ milk’–ornithos (bird) and gala (milk) –which was used to describe something outstanding. South Africans know this flower as ‘tjenkenrientjee’ due to the rustling sound made by the leaves which is where the strange-sounding English name, ‘chincherinchee,’ originates from.

Ornithogalum thyrsoides is native to South Africa. It is also endemic to the Cape Province in South Africa. Chincherinchee is a perennial herb that grows to between 20 cm and 50 cm tall, with round bulbs that bloom in the late summer and become dormant in winter months. The plant has around five to seven fleshy leaves that die back after flowering and range from between 15 cm and 30 cm in length. The leaves are lance-shaped and smooth in texture. Their compactly arranged star-shaped, bell-like flowers are very decorative and tend to be extremely fragrant, blooming individually in circles carried at the top of leafless stems. The florets have six petals and alternately long yellow stamens dilated above the base. Under field conditions they close at night and open in the morning with the rising sun. Floret opening is 100 percent even in the base. Senescence of individual florets is marked by permanent closure of petals, which do not drop off even months after drying.

Scapes to be used as cut flower should be harvested when the lower most flower is showing colour and is about to open. Do not damage the growing bulbs while harvesting. The flowers have such a long vase life, and can very easily last a whole month after being cut. Due to this the flowers are sometimes called a 'florists' nightmare".

Chincherinchee is basically a tropical plant. It is ever green in the tropic and never becomes dormant. It can be grown in full sun under field conditions and as a potted plant in semi shade. Best time for planting is mid November. Neutral loam and sandy loam soils are ideal. Soil should have good drainage and aeration for successful growing,

Chincherinchee blooms between June and September. The most common is *Ornithogalum thyrsoides. Ornithogalum umbellatum*, on the other hand, has larger star-shaped white flowers and narrower leaves, as well as a beautiful and very distinctive green stripe on each flower. *Ornithogalum arabicum* has a long stem with a distinctive umbrella of waxy white flowers carrying a green-black ovary in the centre. *Ornithogalum dubium* is a lovely clear orange shade, and has cup-shaped flowers.

Water seldom, about once every 1-2 weeks with 1-2 glasses of water, allowing the soil to dry out for several days before watering again. Chincherinchee should be grown in direct sunlight and cannot abide temperatures below zero.

Astrantia major

Family: Apiaceae

With a name derived from the Latin word for 'star'(aster), Astrantia is well-known and loved for the star-like flower heads.

Astrantia major, maxima and *carnicolica* are also known affectionately as the 'Melancholy gentleman,' 'masterwort' and 'Hatties' pincushion.' It belongs to family Apiaceae. This herbaceous perennial is a native to Central and Eastern Europe.

Astrantia historically flourished through widespread growth in Central, Eastern and Southern Europe and the Caucasus. Growing rapidly to be about 30 or 60 cm tall in just one season, Astrantia comes in a variety of colours, ranging from green through to cream, pink and deep reddish-purple. Astrantias' tiny, delicate, star-shaped flower heads are held on fine stems radiating from a central point with a ruff of bracts protecting them and their seeds. Each flowerhead (6-9 cm) features dense, compact, domed umbels of tiny greenish-white flowers which form a delicate pincushion-like center subtended by a showy collar of papery, petal-like, creamy white involucral bracts. This shade-loving perennial will bloom about three months after spring planting, or six to nine months after autumn planting, if grown in partial shade and with adequate moisture. The flowering period extends from June through September. The flowers are, entomophilous, best grown in organically rich, medium to wet, well-drained soils in part shade.

Astrantia is relatively pest-free, so once it's basic shade and moisture needs are taken care of, it can practically be left to its own devices. Astrantia is at its happiest when in damp, heavy, regularly fertilized soil and dappled shade, jostling for space with other plants. When planting in sunnier sites it's important to incorporate more humus into the soil. Ensure plentiful water supply in dry conditions, especially if the plants are young.

Bouvardia

Family: Rubiaceae

Bouvardia is a genus of flowering plants in the Rubiaceae family. It contains about 30 species of evergreen herbs and shrubs native to tropical America. Several species of *Bouvardia* are grown as ornamental plants, both in the tropics and indoors as houseplants in temperate regions. Several cultivars and hybrids have been selected. When grown as houseplants, a minimum winter temperature of 7°C is required, with a minimum of 12°C while in flower. Propagation is by cuttings taken in late spring or summer, which need to be kept at a temperature of 20°C by night and 25°C during the day, and shaded. Grown largely for its scented flowers, this small shrub can be untidy when not in flower but spectacularly aromatic and pretty when blooming. It's stems and branches are brittle so it does need shelter and care. The flowers are good for cutting and are finely elongated as buds that burst open into a four petalled star shape creamy white flower. It is quite frost tender. Bouvardias like fertile, well drained soil, and a sheltered sunny position. Hard prune flowered stems in late winter to encourage new growth. The beautiful white flowers have a marvelous perfume. It is said to represent enthusiasm and is given as a gift for those who have a zest for life. The Bouvardia is native to Mexico and tropical America, namely Arizona, New Mexico and Texas. They are chiefly grown in sub-tropical zones and under glass in Holland.

A relatively uncommon bloom, Bouvardias' delicately scented loose clusters of tubular starlet flowers are fringed with leaves and carried on tall stalks, interspersed with large, deep-green pointed leaves. Each stem resembles a small bouquet in bright, cheerful colours, in various shades of pink, yellow, orange, red and white. Bouvardia will typically bloom between May and October and can live for up to 10 years. This flowering shrub is also known as 'Firecracker bush,' 'Trumpetellia' and 'Hummingbird flower.' Their closest relatives are the fabulously fragrant Gardenias.

The soil should be fertile and contain roughly equal proportions of clay, silt and soil, allowing for easy drainage and the ideal site is preferably in full or at least partial sunlight. These plants require watering daily in the summer and fertilising monthly with a water-soluble plant food. This plant was virtually forgotten from the 1800s until about twenty years ago when a Dutch breeder began developing a dwarfed version of it, which was produced for use as a houseplant in 1997.

Eremurus

Family: Xanthorrhoeaceae

A genus of between 35-40 flowers, *Eremurus* has a long, almost tuberous shape. They are also known as Desert Candles or Foxtail Lilies, these graceful spires of dense flowers are one of the most spectacular, early-summer flowering blooms. These tall willowy spires are comprised of densely-packed flowers that open progressively from the bottom of each spike up, with a sparse, ground-level rosette of long strappy leaves and the preferred types of environments, sunny and dry and belong to the lily family. Colors range from snow white and bright yellow to pastel pink, orange and copper. Some plants within the genus have two or more of those colours. The tip of most species resembles a long spike. Eremurus originated in central Asia, especially in Iran, Turkey and Afghanistan. However, some varieties are known to have thrived in the Himalayan regions including *Eremurus himalaia*.

Eremurus prefers to remain undisturbed and therefore is not good candidates for containers. So recommend planting them in beds and borders rather than in containers.

Preferring rich, well-draining soil with bright sunlight and protection from wind, these woody, tuberous rootstocks should be planted upon receipt 118 cm apart (never overcrowd) and covered with just 6 to 9 cm of soil. Eremurus dry out a bit after harvest and rehydrate once planted.

To grow them correctly, Eremurus plants require a lot of sunlight, a cold snap to help with flowering and good drainage. For drainage, lay the roots above a 3-6 cm in thick layer of sharp sand.

These plants are usually available during spring, summer and autumn, although you can start growing them during winter, especially in the event of a cold snap. The most well-known species is the *Eremurus robustus* simply because of its size, although others look just as striking. *Eremurus bungei*, native to Iran, have yellow 68-90 cm long inflorescences, while *Eremurus olgae* is a dwarf variety which flowers as late as June or July.

Drainage is hugely important with Eremurus plants. By placing the roots on top of a layer of sand, you help to prevent any damage that could be done to the roots, as they often thrive in sandy conditions. Planting them during winter will help with flowering and by doing just that, they could last for a long time. The *Eremurus robustus* species can grow to over 3 m in height, making it one of the tallest flowers known to man.

Gentiana ornata

Family: Gentianaceae

The *Gentiana* genus, also known as Gentian, is a large genus with around 400 species. It was thought to be named after Gentius, King of Illyria (present-day Albania), who was the first to use this plant medicinally. Some varieties are harvested commercially for use in aperitif wines, liqueurs and tonics.

The trumpet-shaped flowers usually come in shades of violet and purple, which grow from short stalks. They grow primarily on grassland and can also be grown between shrubs, but care should be taken that they do not become overwhelmed.

Most species of *Gentiana* can be found growing in temperate alpine climates, particularly in Europe and Asia. They are also found in parts of the Americas and Australasia.

An autumn flower, *Gentiana* can be found at any time between September and November, but they can sometimes thrive during mild winters or late summer. The two most notable species are the Sino-ornata and the *Gentiana septemfida*. Both thrive in the autumn months, but they have notable differences. While both grow on grassland, the former has short stalks while the latter has longer ones. Both species can grow as tall as 20 cm. *Gentiana septemfida* is a woodland plant which complements other more vivid plants.

Ideally, they need to be planted between March and May for them to bloom in full by autumn. The Sino-ornata species is, unlike its relatives, quite easy to grow, as it does not need as much sunlight. Lime free soil is seen as a must for growing. The Sino-ornata species was given the Award of Garden Merit by the Royal Horticultural Society.

Helenium autumnale

Family: Asteraceae

Helenium commonly called Helens flower is a genus of about 40 species of annuals and deciduous herbaceous perennials in the daisy family Asteraceae native to the Americas. They bear yellow or orange daisy-like composite flowers. A number of these species (particularly *Helenium autumnale*) have the common name sneezeweed, based on the former use of their dried leaves in making snuff, inhaled to cause sneezing that would supposedly rid the body of evil spirits. Larger species may grow up to 2 metres tall. They are hardy, colourful. Easy to grow, tolerate difficult growing conditions, provide months of colour, are attractive to pollinating insects and unaffected by most pests and diseases. are often displayed with other plants such as solidago, rudbeckias and heliopsis to introduce colour to an arrangement. They are seen as being part of the daisy family, which is due to their small flowers and multiple petals. Most species in the genus resemble daisies, in that they have small petals and a long, thin stalk. However, some species have a flower where the centre sticks out while the petals tend to form a circle below. They come in a number of shades, including crimson, pink and orange. Most species flower during the summer between June and August, although some varieties flower later on in the year. Heleniums tend to like full sunlight, so need plenty of exposure during the summer months to thrive.

Helenium originated in North America, although it was brought to Europe in the 18th century, where many species can still be found today. They were most commonly found in the eastern regions of the United States and parts of Southern Canada with more temperate climates. The most common of the 40 or so species of Helenium is the *H. autumnale*, which is said to have healing qualities. Other common species include the *H. bigelovii,* which adapts very well to wet, autumnal conditions, and Moerheim Beauty,' which usually blooms during hot, dry summers.

Numerous cultivars have been developed for garden use - mainly from *H. autumnale* and *H. bigelovii*. They are useful for late summer and fall bloom, usually in less formal compositions. They are appropriate for native gardens in areas where they are indigenous, and they look wonderful in bouquets. Annual species are easily grown from seed, and perennials should be divided every year in order to retain their vigor. It is more commonly referred to as the Christmas Rose. It was given this name according to a legend that it sprang out of the ground from the tears of a young girl who had no gift to give to Jesus Christ. Hybridising (deliberate and accidental) between *H. orientalis* and several other closely related species and subspecies has vastly improved the colour-range of the flowers, which now extends from slate grey, near-black, deep purple and plum, through rich red and pinks to yellow, white and green. The outer surface of the sepals is often green-tinged, and as the flower ages it usually becomes greener inside and out; individual flowers often remain on the plant for a month or more. The inner surface of each sepal may be marked with veins, or dotted or blotched with pink, red or purple. 'Picotee''flowers, whose pale-coloured sepals have narrow margins of a darker colour, are much sought-after, as are those with dark nectaries which contrast with the outer sepals.

Recent breeding programmes have also created double-flowered and anemone-centred plants.

The soil should be fertile with a generous amount of organic manner in the form of compost, manure or other decayed organic matter in addition to, perhaps, an application of a complete fertilizer in spring. Heleniums should be grown in full sun average to moist soil with good drainage. They are drought tolerant, but should be watered on planting and regularly until established. Most varieties of Helenium require great amounts of sunlight to help them grow. They need alluvial loam and soil which can soak up plenty of moisture as well. Should your soil be light, using organic matter could help to improve growing conditions. It's best to plant most species around March, but if taken care of properly, some species can last well into November.

Heliniums are much under-rated hardy perennial garden plants. They are a mainstay of the summer and autumn border but are so often overlooked by gardeners, garden journalists and growers.

Helleborus niger

Family: Ranunculaceae

The Helleborus has large flowers which are shaped like a bowl. The petals have a translucent appearance and dotted stamens which have pointed delicate leaves. The stems are olive-green in colour and only reach a height of approximately 20 cm, resulting in a somewhat dainty overall appearance. Hellebores are widely grown in gardens for decorative purposes. They are particularly valued by gardeners for their winter and early spring flowering period.

Helleborus is now cultivated all over the world, with specialists producing the flowers in Japan, Australia, New Zealand and Europe. The flowers' origins are hard to specify, but its ancestors have been found growing naturally from South Asia to the west coast of the Mediterranean Sea.

There are two ways of propagating the Helleborus –by seed or by division. You can plant seeds in a pot when they are ripe and then place in a cold frame. Alternatively, divide the plant after it has flowered, in the spring or summer. The Helleborus flowers between November and February.The *Helleborus* belongs to the Ranunculaceae family. There are three varieties of cut flowers. The *Helleborus niger* has flowers in either a deep, warm purple or a bright pure white. The Helleborus corsicus has cream or green bell-shaped cups and the Helleborus orientalis has streaked or speckled flowers in striking shades of green, cream, purple and pink. Once cut, hellebores do not last for long. It is best to provide them with a good drink of water before arranging. They will also live longer if kept cool. The black roots of Helleborus niger are highly poisonous so it is important to wash your hands after handling them.

If you are planting them out, choose a spot which is sheltered and partly shaded. They perform best in moist and fertile soil which is well drained and alkaline to neutral in composition. An ideal combination is heavy clay soil set in an area which has partial shade. They benefit from annual mulch, which should be done in the autumn. The botanical name for the Hellebore is Helleborus but Commonly known as hellebores, members of the genus *Helleborus* comprise approximately 20 species of herbaceous or evergreen perennial flowering plants in the family Ranunculaceae, within which it gave its name to the tribe of Helleboreae. Many species are poisonous. Despite names such as "Christmas rose" and "Lenten rose", hellebores are not closely related to the rose family (Rosaceae).

Helleborus odorus (at NYBG), this genus is native to much of Europe, from western Great Britain, Spain and Portugal, eastward across the Mediterranean region and central Europe into Romania and Ukraine, and along the north coast of Turkey into the Caucasus. The greatest concentration of species occurs in the Balkans. One typical species (*H. thibetanus*) comes from western China; another atypical species (*H. vesicarius*) inhabits a small area on the border between Turkey and Syria.

The flowers have five "petals" (actually sepals) surrounding a ring of small, cup-like nectaries(petals modified to hold nectar). The sepals do not fall as petals would, but remain on the plant, sometimes for many months. Recent research in Spain suggests that the persistence of the sepals contributes to the development of the seeds (Herrera 2005).

The plants are surprisingly frost-resistant and many are evergreen. Many species of hellebore have green or greenish-purple flowers and are of limited garden value, although *Corsican hellebore* (*H. argutifolius*), a robust plant with pale green, cup-shaped flowers and attractive leathery foliage, is widely grown. So is stinking hellebore or setterwort (*H. foetidus*), which has drooping clusters of small, pale green, bell-shaped flowers, often edged with maroon, which contrasts with its dark evergreen foliage. *H. foetidus* 'Wester Flisk', with red-flushed flowers and flower stalks, is becoming popular, as are more recent selections with golden-yellow foliage.

The so-called Christmas rose (*H. niger*), a traditional cottage garden favourite, bears its pure white flowers (which often age to pink) in the depths of winter; large-flowered cultivars are available, as are pink-flowered and double-flowered selections.

Hypericum

Family: Hypericaceae

St. Johns̓ Wort can be found in a variety of forms. Many are small herbaceous herbs while other species can be larger shrubs or even trees that can grow up to 12 metres in height. The simple oval leaves can be deciduous or evergreen and the flowers –generally consisting of five petals and similar to buttercups ranging in colour from a pale to a dark yellow. The fruit tend to be fleshy and orange or red in colour and are rather similar to berries, although in some species they can be dry pods.

Hypericum origins can be traced back to the Himalayas and China. Unfortunately, many of the species are considered as weeds that, while only a minor problem to farmlands and gardens can be a major problem on pasture land. , hypericum has grown in popularity over recent years. While just a few years ago only one variety was widely available, there are now seven different varieties which have all been individually named.

Also known as Magical Passion'Hypericum is a genus with over four hundred different species

Hypericum takes its name from the father of the sun, according to Greek mythology. He was named Hyperion, so the brightly colored yellow flowers of Hypericum were in honor of him.

Leucospermum cordifolium

Family: Proteaceae

Leucospermum is a genus of about 50 species of flowering plants in the family Proteaceae, native to Zimbabwe and South Africa, where they occupy a variety of habitats, including, forest and mountain slopes.

Commonly known as the pincushion due to its cluster of needle-like flower heads, Leucospermum is a shrub that produces flowers in orange and red. Leucospermum takes its name from the Greek words leukos and sperma, which mean white and seed respectively, a reference to the white seeds of many

Leucospermum species. As its common name suggests, the flowers of the *Leucospermum* resemble a pincushion with a rounded head of springy hooks that curl upright. It is an evergreen shrub that grows to a height of between 0.5 m and 5 m. The leaves of the plant are tough and leathery and grow in spirals around the stem.

The *Leucospermum* is native to South Africa and Zimbabwe, where they grow in a variety of different habitats including mountain slopes, forest and scrubland. When they flower their nectar attracts a variety of insects, which in turn attract a large number of insectivore birds, making it popular attraction wherever it is grown. These birds help the pollen transfer between the plants and are vital for them to reproduce. Excellent ornamental spreading shrub, 4x6' with pincushion flowers and grey-green leaves. Various hybrids available with colours ranging from yellow-pale orange- deep orange-red and orange. Good long lasting cut flower. Prefers well-drained light to heavy soils. Flowers best in open sunny position. Tolerant to light frost. Some horticultural selections are: *Leucospermum* Vlam,' *Leucospermum cordifolium* - red. The flowers are produced in dense inflorescences, which have large numbers of prominent styles, which inspires the name

Leucospermum is propagated by semi hardwood cuttings from current seasons growth that are taken from November to March and then placed in a growing house until they are ready to be planted out. The flower heads appear from August to October. They can be grown from seed but it is a fairly complicated process that involves a lot of preparation because the seeds need very specific conditions for germination, namely a smoke seed primer that contains substances that stimulates seed germination.

Leucospermum comes from the Proteaceae family and is a genus that contains around 50 different species. The common feature of members of this family is the inflorescences of small flowers that are densely packed into a compact head. The flowers attract insects because of the presence of nectar and in turn draw towards it numerous carnivorous birds.

Leucospermum is nan excellent garden plant, as a focal point or planted in groups and provides good long stemmed cut flowers. The plant is rather short lived becoming leggy after about eight years. *Leucospermum* prefers milder climates and while it is a popular garden plant in South Africa, if it is grown somewhere colder it should be moved indoors over the winter months. It needs well-drained soil that has low levels of phosphates and nitrates and should be watered moderately to keep it looking at its best.

Kniphofia

Family: Xanthorrhoeaceae

Kniphofia is commonly known as torch lily or red hot poker, although the species comes in a variety of colours such as yellow and cream. This plant often features in garden borders and in bouquets because of its striking flowers.

Kniphofia is an evergreen plant with thin, grass-like leaves and an erect stem. It has a number of tubular flowers that make up a spike resembling a torch. The colouring becomes more intense towards the tip of the spike. The more common varieties are vibrant red and yellow, although subtler creams and greens are also available. The flowers themselves produce an abundance of nectar when in bloom, making them attractive to wildlife. Kniphofia are native to South Africa, although a number of varieties such as Bees'Sunset and Little Maid are grown commercially. Kniphofia are resilient plants and should be grown in a sandy soil that is well drained. In their native South Africa, Kniphofia would receive plenty of rain to stimulate growth, but good drainage is essential, especially in winter as the flower enjoys very dry winters in South Africa. Kniphofia flowers in late summer and autumn when planted in gardens. Cut flowers are usually available between May and October.

There are over 70 species of kniphofia and they are an easy-to-grow and hardy garden plant. They are part of the Xanthorrhoeaceae family in the subfamily of Asphodeloideae.

It is best to plant them in full sun, although they can tolerate partial shade. Some thought should be given to their position in borders, as they can grow up to 1.2 metres tall and 0.6 metres wide. After the flowers have died, their spikes should be removed and mulch applied. The foliage should be cut back to the ground in spring in order to keep it fresh looking. With cut flowers it is important to protect the stems, as these can be damaged in transit. Lower flowers should be removed as they die off.

Convallaria majalis

Family: Asparagaceae

It is a sweet scented flowering plant and a native of the cool temperate areas in Northern hemisphere in Asia Euope and Southern United States. It belongs to family Asparagaceae and subfamily Nolinoideae.

Lily of the valley is found across Europe, Asia and in North America, in dry woodland, usually on alkaline soils and sometimes in the crevices or grikes of limestone pavement. In common with many plants that are impossible to eradicate once theyve made themselves at home –Reginald Farrer calls it 'the worst of all delicious weeds"

Flower colour and form can vary too. With a heady, strong fragrance, the Lily of the Valley is popular not only for its scent but also for its ground-covering, spreading properties in shady conditions. It also produces small berries, which tend to be red or orange.

Suitable for light (sandy), medium (loamy) and heavy (clay) soils and can grow in heavy clay and nutritionally poor soils. Suitable pH: acid, neutral and basic (alkaline) soils. It can grow in full shade (deep woodland) semi-shade (light woodland) or no shade. It prefers dry moist or wet soil.

It can be planted from January to December and usually flowers between June and September. The seeds can be sown in containers or trays in a cold frame or greenhouse. Lily of the Valley can also be propagated by dividing mature plants in the autumn.

Replant the divided sections in their new location at their original depth and with their shoots above ground. It will happily tolerate acid, neutral or alkaline soil and is ideal for use as ground cover in informal planting arrangements. It will willingly grow beneath other plants, such as roses or shrubs, adding interest to otherwise barren areas of the garden.

The Lily of the Valley has been the national flower of Finland since the late 1960s and used to be the floral emblem of the old Yugoslavia, which was dissolved in 1992. The Lily of the Valley flower supposedly symbolises the return of happiness, sweetness and a feeling of contentment.

Chrysanthemum frutescens

Family: Asteraceae

Marguerites have green, woody stems and typically produce dainty white flowers, though pink and yellow varieties also exist. The plants are usually around 60cm tall, although they can grow up to a metre in height if planted in suitable conditions. Originating in the warm Canary Islands, Marguerites will thrive if planted in full sunshine. The flowers grow best in sandy, clay or loam type soils and adapt well to different levels of acidity.

Potted Marguerites flower freely throughout the summer months and, although they are a perennial shrub, they are likely to require replanting after several years. They are readily available from garden centres and nurseries during the spring and can be planted outside once you are certain the last frosts are over. Alternatively, they can be propagated in a greenhouse.

Marguerites are of the frutescens species. Similar to Leucanthemum, or ox-eye daisies, they were formerly considered to be part of the Chrysanthemum genus but have been reclassified as Argyranthemum, making them part of the Aster family.

Marguerites do not tolerate cold climates and will lose their green leaves if temperatures drop too low. Regular dead-heading will encourage the growth of new flowers. Adding mulch, manure or other forms of organic fertiliser to the soil before planting can also help to promote strong, healthy flowers. Marguerites can be susceptible to pests such as leaf miners and occasionally suffer from disease. They may benefit from careful treatment with insecticides or fungicides.

Moluccella laevis

Family: Lamiaceae

A member of the mint family, Molucella is a genus that contains only four species. The most popular species, the Moluccella laevis, is commonly referred to as Bell of Ireland on account of its green colour and bell-shaped leaf bracts, not because it is native to Ireland.

Bells of Ireland or shellflower, *Molucella laevis*, is a half-hardy annual that produces unusual pale green to emerald green, funnel-shaped "bells" along green stems in summer. The persistent bells are the showy calyx (cup-shaped leaves around the base of the flowers) which surrounds tiny fragrant white flowers. The papery 2-3 cm bells are densely packed along most of the length of the square stems that reach 60-100 cm tall. Clusters of 7.5 cm" long leaves alternate between the bells, with pairs of small thorns or spines below each calyx.

Despite the common name, this plant in the mint family (Lamiaceae) is not from Ireland, but is native to western Asia, around Turkey, Syria and the Caucasus. Even Linnaeus, who named the plant, was a bit confused about its origin, naming the genus after the Molucca Islands in Indonesia where it was mistakenly thought to be from. It's supposed association with Ireland probably has to do with the color (and associated marketing potential). It is cultivated since 1570; the flowers are a symbol of good luck. Both the flowers and rounded, pale green leaves with slightly scalloped edges have a distinctive, difficult-to-describe scent.

Bells of Ireland are nice when combined with other colorful annual flowers in the border or cutting bed. They also do well in containers, offering a vertical component. The light green color complements purple-leaved plants, or can be used in monochromatic schemes with other green-flowered plants such as Zinnia Envy Double'and *Nicotiana langsdorffii*, or along with bright green coleus (such as Super Chartreuse). They even look interesting late in the season when the old spikes become dry and bleached (if they haven't been pruned out to encourage new growth).

Blooming stems can be cut to be used fresh or dried in flower arrangements. They are especially effective when used in all-foliage and contemporary arrangements. When dried, eventually the color fades to pale beige. To dry, pick the stems when the bells are fully open and hang upside down in small bunches in a dry, airy place until fully dry.

This bedding plant is easily started from seed. Sow seeds indoors 8-10 weeks before last frost. Barely cover the seed as they need light to germinate. Germination may be slow (up to a month). To speed up germination, chill the seeds for two weeks (seal the sown seeds in a plastic bag and place in a refrigerator) first. The small plants can be transplanted outdoors once night time temperatures are above 40°F. Place in sun to partial shade in ordinary garden soil and space about a foot apart. Bells of Ireland often self-seeds in the garden if flower stalks are left on the plants until dried. The dark colored seeds can be collected to save for sowing the following year or allowed to scatter on the ground. Small seedlings can be transplanted to other locations, if desired (although they may wilt temporarily until re-established).

Bells of Ireland prefer full sun and regular water. Water during drought periods and fertilize monthly for best results. They do not do well in summer in hot, humid climates. Taller spikes may need staking in windy sites. Bells of Ireland has few pest problems and is not favored by deer or rabbits.

The plant has a tall, woody stem around 60 cm long. *Moluccella laevis* produces tiny white flowers encased in small, pale-green calyxes –cup-shaped leaves which resemble bells.

Moluccella laevis is native to the Mediterranean, particularly in the east, and Syria. While the plant can adapt to different soil types, it grows best in a compost-rich soil with good drainage.

Moluccella laevis is an annual summer-flowering plant that can be propagated indoors or planted outside in the springtime, once the frosty season is over. The plant has an average bloom duration of around ten weeks. The species is *laevis* and the genus *Moluccella.* A smaller breed of moluccella laevis is the Pixie Bell, which typically grows to around 30cm long. Growing it requires a certain degree of skill, as the plant can be difficult to germinate. However, once the shoots have appeared, Bell of Ireland is a hardy plant which can be grown in a variety of different soil types. *Molucella laevis* seeds must be sown under cover in a propagator or greenhouse. Once the seeds have germinated, the seedlings must be allowed to acclimatise gradually to outdoor conditions. *Moluccella laevis* are excellent as dried flowers –simply hang them upside down after you've cut them in bloom.

The white seeds of the plant can be found at base of the petals, which is why it is sometimes referred to as a 'shell flower.'

Muscari

Family: Asparagaceae

Muscari is a genus of perennial plants that produces distinctive bell-shaped flowers that hang downwards and resemble upturned grape bunches during springtime, and has a grapey fragrance. This is why it is also known as the Grape Hyacinth. The name Muscari derives from the Latin word 'muschos' or 'musk,' which refers to the fact that the flower has a strong scent.

Muscari plants usually produce a blue or purple flower that, in warmer climates, can be present all year. They grow up to 30cm tall and some plants have a double flower. Most of the varieties of Muscari have a grape-shaped flower and some have flowers that are two tones in color.

The *Muscari* flower is native to the Mediterranean, the eastern Aegean region, Turkey and Morocco. They prefer to be in a partially shady to slightly sunny locations with a neutral soil and good drainage.

It blooms from March onwards and flowers can easily last for a month or longer. The *Muscari* bulbs are easily cultivated as they naturalise, or multiply, with ease. The longer you leave your bulbs in the ground, the more flowers and bulbs you will have the next year. It should be planted in autumn.

There are believed to be around 40 species in the *Muscari* genus, which belongs to the Asparagaceae family and the Scilloideae subfamily. The Muscari genus was formally established by Scottish botanist Philip Miller in 1754, although Muscari has been cultivated in England since 1576.

Muscari are incredibly easy to care for. The plants like well-drained and neutral soil. The bulbs should be planted at a depth that is twice the height of the bulb itself.

Scabious colombaria

Family: Dipsacaceae

Scabious is better known as scabiosa, the butterfly blue or the pincushion flower. It produces large numbers of blue flowers that resemble pins stuck in a pin cushion. It is popular as a garden flower as it produces a lot of nectar, which effectively attracts butterflies and bees.This is a long-stemmed flower with an abundance of pin-like florets within a one-and-a-half-inch diameter bloom on top of a wiry stem that may grow between one and two feet high. It has thin, knife-like leaves of a grey-green color and its roots can grow to around eight inches.

The *scabious* is found in Europe and Asia. They have been cultivated since 1800, when they were brought from Hungary and Mediterranean Russia. In its native habitat, scabious grows on calcareous soils, chalk pits and sheep-grazed pastures. In a garden, it thrives in soil that is well drained and warm, preferably more dry than damp. It prefers a neutral to alkaline soil, mostly sand or clay loam.Attractive to bees and butterflies. Good for cut flowers. Grow in full sun but protect from winter wetness

The plant flowers between July and September and grows well in a rock garden or in a container. New scabious plants can be created by dividing the existing plant every few years in early spring or summer. If you need more plants, take cuttings in the summer.

The species *scabiosa* is a genus of the teasel or Dipsacaceae family of flowers and shrubs. Teasels are recognised by their prickly leaves and stems with lavender, purple and blue flowers. These are native to temperate climates and are found throughout Europe, North Africa and Asia.

Scabious should be planted where it will receive full sunlight. The best time to sow from seed is just after the last frost of the winter. Perennial *scabious* may be sown in spring or late autumn. The flowers will grow continuously if dead-headed. The wiry stems need staking if grown in a windy area. *Scabiosa* can be annuals, biennials or herbaceous or evergreen perennials, with simple or pinnately lobed leaves and characteristic flower-heads with enlarged outer florets. *S.columbaria* is a branched perennial to 75 cm, with simple or pinnately lobed leaves, the upper ones more divided, and light bluish-purple flowerheads 4cm across in summer and early autumn.

References

Jacob Thomas, Santhosh Kumar E.S. and Abdul Jabbar, M. (1996). *Gymnostachyum febrifugum* var *bracteatum*- A potential wild ornamental plant from the Western Ghats. *Indian Horticulture* 41(1): 22.

Jacob Thomas, Santhosh Kumar E.S., Abdul Jabbar, M., Jose, P.A. and Raj Vikraman, R. (1998). Wild ornamental plants of Western Ghats (Kerala). *Journal of Economic and Taxonomic Botany* 22(2):477-489.

Kumaraswamy M.V, and Satish S. (2008). Anti-oxidant and anti-lipoxygenase activity of *Thespesia lampas* Dalz & Gibs. *Advances in Biological Research.* 2: 56-59.

Mohanan N. and Benoy, P.C. (1996). Yellow Silk Cotton, a tree of elegance. *Indian Horticulture* 41(2):25.

Sreelatha, U, Baburaj, T.S., Narayanan Kutty, C., Nazeem, P.A. and Jyothi Bhaskar (2007). Cultivation prospects of *Exacum bicolor* Roxb. An endangered, ornamental and anti diabetic herb. *Natural Product Radiance* 6(5):402-404.

www.gardeningknowhow.com

www.missouribotanicalgarden.org

www.classicviburnum.org

www.plantzafrica.org

www.plants.usda.gov

www.suuculentplant.com

www.desert-tropicals.com

www.theplantlist.org

FSC
www.fsc.org
MIX
Papier | Fördert gute Waldnutzung
FSC® C083411

Kolibri 360 GmbH
Ferdinand-Jühlke-Straße 7
99095 Erfurt, Deutschland
produktsicherheit@kolibri360.de